SYSTEMATICITY

OXFORD STUDIES IN PHILOSOPHY OF SCIENCE

General Editor
  Paul Humphreys, University of Virginia

Advisory Board
  Anouk Barberousse (European Editor)
  Jeremy Butterfield
  Peter Galison
  Philip Kitcher
  James Woodward

The Book of Evidence
Peter Achinstein

Science, Truth, and Democracy
Philip Kitcher

Inconsistency, Asymmetry, and Non-Locality: A Philosophical Investigation of Classical Electrodynamics
Mathias Frisch

The Devil in the Details: Asymptotic Reasoning in Explanation, Reduction, and Emergence
Robert W. Batterman

Science and Partial Truth: A Unitary Approach to Models and Scientific Reasoning
Newton C.A. da Costa and Steven French

Inventing Temperature: Measurement and Scientific Progress
Hasok Chang

The Reign of Relativity: Philosophy in Physics 1915–1925
Thomas Ryckman

Making Things Happen
James Woodward

Mathematics and Scientific Representation
Christopher Pincock

Systematicity: The Nature of Science
Paul Hoyningen-Huene

Causation and Its Basis in Fundamental Physics
Douglas Kutach

Reconstructing Reality: Models, Mathematics, and Simulations
Margaret Morrison

The Ant Trap: Rebuilding the Foundations of the Social Sciences
Brian Epstein

Systematicity

THE NATURE OF SCIENCE

Paul Hoyningen-Huene

UNIVERSITY PRESS

OXFORD
UNIVERSITY PRESS

Oxford University Press is a department of the University of Oxford.
It furthers the University's objective of excellence in research, scholarship,
and education by publishing worldwide.

Oxford   New York
Auckland   Cape Town   Dar es Salaam   Hong Kong   Karachi
Kuala Lumpur   Madrid   Melbourne   Mexico City   Nairobi
New Delhi   Shanghai   Taipei   Toronto

With offices in
Argentina   Austria   Brazil   Chile   Czech Republic   France   Greece
Guatemala   Hungary   Italy   Japan   Poland   Portugal   Singapore
South   Korea   Switzerland   Thailand   Turkey   Ukraine   Vietnam

Oxford is a registered trade mark of Oxford University Press
in the UK and certain other countries.

Published in the United States of America by
Oxford University Press
198 Madison Avenue, New York, NY 10016

© Oxford University Press 2013

First issued as an Oxford University Press paperback, 2016.

All rights reserved. No part of this publication may be reproduced, stored in a
retrieval system, or transmitted, in any form or by any means, without the prior
permission in writing of Oxford University Press, or as expressly permitted by law,
by license, or under terms agreed with the appropriate reproduction rights organization.
Inquiries concerning reproduction outside the scope of the above should be sent to the Rights
Department, Oxford University Press, at the address above.

You must not circulate this work in any other form,
and you must impose this same condition on any acquirer.

Library of Congress Cataloging-in-Publication Data
Hoyningen-Huene, Paul, 1946–
Systematicity : the nature of science / Paul Hoyningen-Huene.
p. cm.—(Oxford studies in philosophy of science)
Includes bibliographical references (p.).
ISBN 978–0–19–998505–0 (hardback : alk. paper); 978–0–19–029833-3 (paperback : alk. paper)
ISBN 978–0–19–998506–7 (updf)
1. Science—Philosophy.   2. Philosophy.   I. Title.
Q175.H8825 2013
501—dc23
2012034682

Contents

Preface ix

1. *Introduction* 1
    1.1 Historical Remarks 1
    1.2 The Question "What Is Science?" in Focus 8

2. *The Main Thesis* 14
    2.1 Science and Systematicity 14
        2.1.1 A Little History 14
        2.1.2 Preliminary Remarks 21
    2.2 The Concept of Systematicity 25
    2.3 The Structure of the Argument 30

3. *The Systematicity of Science Unfolded* 35
    3.1 Descriptions 37
        3.1.1 Some Preliminaries 37
        3.1.2 Axiomatization 41
        3.1.3 Classification, Taxonomy, and Nomenclature 42
        3.1.4 Periodization 43
        3.1.5 Quantification 45
        3.1.6 Empirical Generalizations 47
        3.1.7 Historical Descriptions 49
    3.2 Explanations 53
        3.2.1 Some Preliminaries 53
        3.2.2 Explanations Using Empirical Generalizations 56
        3.2.3 Explanations Using Theories 59

3.2.4 Explanations of Human Actions 62
3.2.5 Reductive Explanations 63
3.2.6 Historical Explanations 68
3.2.7 Explanation and Understanding in the Humanities in General 71
3.2.8 Explanations in the Study of Literature 75
3.3 Predictions 78
3.3.1 Some Preliminaries 78
3.3.2 Predictions Based on Empirical Regularities of the Data in Question 80
3.3.3 Predictions Based on Correlations with Other Data Sets 82
3.3.4 Predictions Based on (Fundamental) Theories or Laws 83
3.3.5 Predictions Based on Models 85
3.3.6 Predictions Based on Delphi Methods 87
3.4 The Defense of Knowledge Claims 88
3.4.1 Some Preliminaries 88
3.4.2 Nonevidential Considerations 92
3.4.3 Empirical Generalizations, Models, and Theories 94
3.4.4 Causal Influence 98
3.4.5 The Verum Factum Principle 102
3.4.6 The Role of Mathematics in the Sciences 103
3.4.7 Historical Sciences 107
3.5 Critical Discourse 108
3.5.1 Some Preliminaries 108
3.5.2 Norms and Institutions 110
3.5.3 Practices in Science Fostering Critical Discourse 111
3.6 Epistemic Connectedness 113
3.6.1 Preliminaries: The Problem 113
3.6.2 Failing Answers 116
3.6.3 The Concept of Epistemic Connectedness 118
3.6.4 Revisiting the Examples 121
3.7 The Ideal of Completeness 124
3.7.1 Some Preliminaries 124
3.7.2 Examples 126
3.8 The Generation of New Knowledge 132
3.8.1 Some Preliminaries 132
3.8.2 Data Collection 134
3.8.3 The Exploitation of Knowledge from Other Domains 139
3.8.4 The Generation of New Knowledge as an Autocatalytic Process 141

3.9 The Representation of Knowledge   141
   3.9.1 Some Preliminaries   141
   3.9.2 Examples   142

4. Comparison with Other Positions   148
   4.1 Aristotle   150
      4.1.1 The Position   150
      4.1.2 Comparison with Systematicity Theory   151
   4.2 René Descartes   152
      4.2.1 The Position   152
      4.2.2 Comparison with Systematicity Theory   154
   4.3 Immanuel Kant   155
      4.3.1 The Position   155
      4.3.2 Comparison with Systematicity Theory   158
   4.4 Logical Empiricism   159
      4.4.1 The Position   159
      4.4.2 Comparison with Systematicity Theory   160
   4.5 Karl R. Popper   161
      4.5.1 The Position   161
      4.5.2 Comparison with Systematicity Theory   162
   4.6 Thomas S. Kuhn   163
      4.6.1 The Position   163
      4.6.2 Comparison with Systematicity Theory   164
   4.7 Paul K. Feyerabend   165
      4.7.1 The Position   166
      4.7.2 Comparison with Systematicity Theory   168
   4.8 Nicholas Rescher   170
      4.8.1 The Position   170
      4.8.2 Comparison with Systematicity Theory   173

5. *Consequences for Scientific Knowledge*   176
   5.1 The Genesis and Dynamics of Science   176
      5.1.1 Conceptual Clarifications   177
      5.1.2 The Genesis of a Science   180
      5.1.3 The Dynamics of Science   183
   5.2 Science and Common Sense   187
      5.2.1 The Preservation of Common Sense   187
      5.2.2 The Deviations from Common Sense   190
      5.2.3 Additional Remarks   193

5.3 Normative Consequences 196
5.4 Demarcation from Pseudoscience 199
5.4.1 A Little History 200
5.4.2 Systematicity Theory's Demarcation Criterion 203

6. Conclusion 208

Notes 213
Literature Cited 259
Index 279

Preface

IN THIS BOOK, I will explicate and argue for the following thesis: key to understanding what makes scientific knowledge special is the notion of systematicity. More to the point, I will argue that the essential difference between scientific knowledge and other forms of knowledge consists of the higher degree of systematicity of the former. This particularly applies to the relationship between scientific knowledge and everyday knowledge. At first, it is necessary, however, to explain what is meant by stating that scientific knowledge is more systematic than other forms of knowledge because the notion of systematicity is far from clear; in fact, it is vague and ambiguous. Arguing for the thesis will be systematic in itself, as I will attempt to provide backing for every aspect that is involved. Part of the overall argument for the thesis will also consist of a comparison with other attempts to describe the characteristics of scientific knowledge. Eventually, these alternative attempts should be contained in my thesis. In other words, they are not false, but rather single-sided. As opposed to many attempts in the twentieth century that dealt with questions about the specificity of science, I will not start with the presumed opposition between science proper and pseudoscience or metaphysics. By contrast, I will compare everyday knowledge with scientific knowledge. Only after full development of this comparison will I be able to meaningfully comment on the so-called demarcation criterion. This criterion aims at articulating and explicating the difference between science on the one hand and pseudoscience and metaphysics on the other.

My work on this subject dates back to four years into my professional beginnings in philosophy. In 1979, I was asked by members of the Institute of Geography at the University of Zurich, Switzerland, to give a talk in a seminar entitled "Theory in Geography" that was planned for the winter term 1980–1981. At that time, there were heated controversies about the nature of geography as a science, and I, a physicist-turned-philosopher of science, found this topic quite interesting. Perhaps I could act as a neutral instance and provide an external view. The view would certainly have been from an outsider, because my knowledge of geography and its character as a scientific discipline was, politely understated, rather limited. In the literature on fundamental questions in geography to which the professionals had directed me, I had found a specific attempt that laid out the scientific foundations of geography in a way that seemed plausible to me. However, this approach was perceived to be outdated and untenable by most professional geographers. As I had the suspicion that their rejection was not well founded, I set out to defend this attempt. My strategy in my talk was to describe what science was in general, and to demonstrate subsequently that the given geographical attempt satisfied this explication. I therefore asked the question, "What is science?"; and my preliminary answer was, "Science is systematic knowledge." Then, I proceeded to develop the answer over a couple of pages, before applying it to the geographical approach in question.

For almost twenty years, I did not explicitly return to the question, "What is science?" However, while engaging in Paul Feyerabend's philosophy, I was always dissatisfied with his answer that there is nothing specific about science, that it is just one form of knowledge among others, just one part of culture, without any special characteristics that make it unique in one sense or another. I always had the feeling that science was more successful than any other kind of knowledge, if only perhaps in a limited and instrumental way. It took me quite a while, in fact more than twenty years, to realize that Feyerabend's main argument against any special cognitive status of science was fallacious. In essence, Feyerabend argues that traditionally, the special character of science has been founded on science's use of "the scientific method" (or "scientific methods") in the sense of binding rules of procedure. However, according to Feyerabend, the attempt to identify any binding rules in the actual history of science fails, especially in some of its highlights, like in Galileo's physics. Feyerabend's conclusion from this fact (I accept this as a fact, although it is somewhat controversial) is that there is nothing very special about science—it is just a form of knowledge with specific advantages and disadvantages, like any other kind of knowledge. This conclusion, however, is unwarranted, because the failure to identify method as a (sufficient) criterion for science does not imply that there is no such criterion whatsoever that marks a unique quality of scientific knowledge.

Feyerabend's negative result (and it is not only his) constituted and continues to be an enormous challenge to philosophy of science. If one still thinks that science is somehow special, and if it is not method that makes it special, what makes it special? Although I consider this question to be central to the discipline, philosophers of science did not really take it up in the past decades. What question if not the question, "What is science?" (or the question that I take to be synonymous with it "What is the nature of science?") should be discussed—and possibly even be answered—by a discipline legitimately called philosophy of science? In the past decades, this discipline was so busy in discovering special sciences and their disunity that the more general questions moved into the background. However, this more or less abstract challenge became very concrete for me when I was invited to give the first keynote speech at Forum I of the World Conference on Science that took place in June 1999 in Budapest. The title suggested to me was "The Nature of Science," and I accepted it. In my speech, I proceeded to develop what I had started two decades earlier. I had planned to expand this topic into a book directly after the conference. However, too many things intervened and continued to do so. I therefore applied to the VolkswagenStiftung in Hannover for a research professorship during the academic year 2003–2004 in order to write this book. The year was granted and liberated me from (most) administrative and teaching duties. However, it took some additional time to finish this book.

In stark contrast to an earlier book of mine, this book has neither footnotes nor numbered endnotes. This does not indicate that I haven't read anything before or during the preparation of this book and that therefore there is nothing I would like to or am obliged to refer to. Readers interested in references and further notes that do not centrally belong to the main argument will find them in a separate section at the end of the book. I do hope that this format contributes to an easier reading without compromising scholarly standards or substance. However, I cannot begin to apologize for the selected references. As the ground covered is so vast, my choices are bound to be quite accidental, as indeed they are. Many specialists will not only complain that I have not cited their books, but also that I did not pick the most important pieces of literature in the respective fields. Many of these complaints will be quite justified. Nevertheless, I had to find a compromise between writing a more or less perfect book and ending it within my lifetime. This is the price of an enterprise that gave me more than once the impression of being almost suicidal intellectually. Another weakness of this book concerns the choice of examples from varying research areas. As I shall explain in greater detail in section 1.2, my aim in this book is to cover all research fields, not just the natural sciences. This does not only lead to significant methodological difficulties (which I will discuss in detail in section 2.3), but also to problems with the appropriate choice of illustrative examples. They should, of course, have come from every area of learning. Unfortunately, my knowledge is extremely limited. I must admit that there are many research fields of whose existence I only learned in the course of writing this book, like lipid science or granite science. There are certainly many more, like otorhinolaryngology, whose names I might have mistaken for scholarly names of exotic practices such as anthropophagy. Thus, there is a statistically skewed distribution of examples in this book. I can only hope that this unevenness does not harm the claimed evenness and generality of my thesis.

Furthermore, as I cover so much ground that has already been treated by philosophy of science in the past decades, my presentation of some of these topics may appear superficial. Apart from those cases where I even do not realize my own ignorance, my discussion is often intentionally superficial. I am glossing over many interesting philosophical problems and ramifications that are connected with a given topic. However, this is justified by the specific focus of this book. Throughout this book, I am arguing for one and only one general thesis, namely, that scientific knowledge is more systematic than other forms of knowledge. Whenever I discuss a particular subtopic—say the characteristics of scientific explanation—I am only trying to expose that they are more systematic than, for example, comparable everyday explanations. Many features of scientific explanations that are not relevant in this particular respect, be they philosophically controversial or not, will not be discussed. Therefore, an impression of superficiality may result.

I have to thank many people who contributed in one way or another to this project. First of all, whenever I gave talks on the topic of this book, I received critical questions and comments, which invariably revealed a variety of weaknesses in my argument. In their temporal order since 1999 (and thus some will be repeated), audiences in the following cities were very helpful: Budapest, Hungary; Krakow, Poland; Bielefeld, Germany; Ål, Norway; Düsseldorf, Germany; Munich, Germany; Tartu, Estonia; Hannover, Germany; Athens, Greece; Zurich, Switzerland; Constance, Germany; Freiburg, Germany; Nancy, France; Belfast, UK; London, UK; Cambridge, UK; Helsinki, Finland; Brussels, Belgium; Tempe, USA; San Diego, USA; Reno, USA; Los Angeles, USA; Essen, Germany; Pittsburgh, USA; Ithaca, USA (with Richard Boyd as commentator); Boston, USA; Münster, Germany; Berlin, Germany; Minneapolis, USA; South Bend, USA; Bloomington, USA; Chicago, USA (with Jonathan Tsou as commentator); Lund, Sweden; Bristol, UK; Bonn, Germany; Taipei, Taiwan; Erlangen, Germany; Delft, Holland; Bremen, Germany; Osnabrück, Germany; Zurich, Switzerland; Berlin, Germany; Munich, Germany; Barcelona, Spain; Hannover, Germany; Bielefeld, Germany; Tilburg, Holland (with Fred A. Muller, Julian Reiss, and Manfred Stöckler as commentators), and Munich, Germany. Due to their interventions, many shortcomings in earlier versions of this book could be overcome, for which I am grateful.

The following individuals were especially helpful in many different ways: Marcus Beiner, Karim Bschir, Werner Eisner, Kirsten Endres, Stephan Hartmann, Helmut

Heit, Harold Hodes, Gero Kellermann, Martin Killias, Noretta Koertge, Martin Kusch, Simon Lohse, Eric Oberheim, Adrian Piper, Katie Plaisance, Ulrich Pothast, Joseph M. Ransdell, Nicholas Rescher, Thomas Reydon, Peter Richter, Markus Scholz, Joachim Schummer, Maya Shaha, Daniel Sirtes, Robert Stephanus, Thomas Sturm, Ken Waters, and Marcel Weber. Urs Freund contributed a cover design that, according to my taste, perfectly fits the book. Furthermore, I would also like to thank the VolkswagenStiftung for granting the research professorship in 2003–2004 and the Leibniz University of Hannover for granting me sabbatical terms in 2005–2006 and in 2011. I needed that time off my other academic duties. Two critical but sympathetic anonymous referees selected by Oxford University Press made very useful suggestions, and I gratefully followed their advice. Now that I know that these referees were Alexander Bird and Howard Sankey (as disclosed by their endorsements on the book's back cover—they never told me), I am even more grateful. I strongly disagree with both of them on some deep philosophical matters, and they did not hold this against me. I would also like to thank the whole team at Oxford University Press who transformed an imperfect manuscript into a beautiful book. Finally, I feel some gratitude about circumstances for which I cannot thank anybody. It is the existence of the World Wide Web, without which I could not have researched all of the resources that I needed in order to write this book.

PREFACE TO THE PAPERBACK EDITION

In this paperback edition, several misprints and infelicities of expression were eliminated. Some of them were brought to my attention through the reviews by Professor Darrell P. Rowbottom (*Notre Dame Philosophical Reviews* 2013.10.21) and Dr. Markus Seidel (*Zeitschrift für philosophische Literatur* 2 (4):33-38, 2014). In addition, Professors Brad Wray and Oliver Scholz sent me lists of errors they had discovered when reading the book for their reviews that appeared in *Metascience* 23 (1):141-144, 2014 and in *Zeitschrift für philosophische Forschung* 69 (2), 2015, respectively. I am grateful to all four colleagues.

Zurich, April 2015

SYSTEMATICITY

1

Introduction

1.1 HISTORICAL REMARKS

The central question to be answered by a general philosophy of science is: What is science? As obvious as this may seem, only few works in today's philosophy of science deal with this fundamental question in its full scope. In order to understand this peculiar situation, we need to look at the history of philosophy of science, or more to the point, at the history of the answers to this question. However, I should warn historians of science who read this. Usually, they hate this sort of historical presentation because it does not pay attention to a myriad of details. For the purpose of introducing the reader to my main question, however, I allow myself to view the historical process from a considerable distance where many details disappear and only grand contours become visible. I only need the grand contours in order to make the purpose of this book evident.

The history of answers to the question, "What is science?"—when viewed in a very schematic way—covers four phases or periods. The first phase starts around the time of Plato and Aristotle in the fourth century BC and ends with the beginning of modern times, around the early seventeenth century. The second phase begins in the early seventeenth century and stretches well into the middle of the nineteenth century. The third phase begins in the late nineteenth century and stretches into the last third of the twentieth century. And the fourth phase begins during the last third of the twentieth century and continues until today. Of course, this subdivision of some 2,400 years of history of philosophy of science is extremely crude. It has by no means sharp boundaries, and it is potentially badly misleading. The phases are meant to be characterized by a certain mainstream class of answers that comprises divergent answers. In particular, the proposed periodization should not be understood to claim that answers to "What is science" *typically* given in some period are *only* given in this period. Instead, answers from one period may also be present in earlier or later periods. So-called anticipation may be operative, as might be inertia in keeping answers alive beyond the advance of the mainstream. So we should handle the proposed periodization of philosophical conceptions of the sciences with a grain of salt.

Before discussing the four phases themselves, we should pause for a moment to turn our attention to the question of what the principal factors are that may lead to historical change in philosophical conceptions of science. There are two principal sources for the historical change in question. First, there is the evolution of the sciences themselves, which may trigger historical differences in the answers to our question, "what is science." Clearly, adequate answers to "What is X?"—questions will have to change whenever X changes significantly. Indeed, this factor is, to a large degree, responsible for the historical change in the characterizations of the sciences. However, a second factor also plays a role that is independent of the historical change of the sciences. It concerns changes originating in philosophy (or historiography) of science itself, which can occur in principle independently of any possible change of the sciences, i.e., its subject matter. The main examples are changes of philosophical conceptions of science due to corrections or refinements, i.e., due to better insight into the subject matter. Although these two sources for changes of characterizations of science are conceptually independent of each other, they may and often do go hand in hand. Historical development of the sciences may trigger deepened reflections on them, which in turn may reveal earlier inadequacies in philosophical conceptions of science. These somewhat sketchy remarks must suffice at this point, and we may now turn to a discussion of the envisaged four phases of scientific development. We will find that transitions between them are indeed shaped by the two factors mentioned above.

In the first phase, starting around the times of Plato (about 428–348 BC) and Aristotle (384–322 BC), two traits for scientific knowledge are postulated that are relevant in our context. It is, first, the epistemic ideal of the absolute certainty of knowledge and, second, the methodological idea of deductive proof as the appropriate means to realize this ideal. Scientific knowledge conceived in this manner, or with the Greek word, *episteme*, stands in sharp contrast to mere belief, or *doxa*. Only *episteme*, by being certain, qualifies as scientific. Its certainty is derived from being based on true first principles and deductive proofs. The truth of the first principles is, in some (problematic) sense, taken to be evident. Deductive proofs are used to demonstrate the truth of theorems, that is, of propositions derived from the principles.

Formal logic as the theory of truth-transferring deduction is developed at the same time (by Aristotle).

It is no accident that—at about the same time these ideas emerge—what is known today as Euclidean geometry was codified as well. Euclidean geometry precisely exemplifies the ideal of scientific knowledge articulated above. It is based on apparently self-evident axioms, and it proves its theorems by logical deduction from these axioms. Such an ideal of scientific knowledge would today be called the result of meta-theoretic reflection on a discipline like Euclidean geometry, i.e., a consideration of the theoretic nature of the science of geometry. Being based on self-evident principles and theorems that are cogently derived from them, geometry presents a kind of knowledge far superior to all of the other kinds of knowledge and beliefs known beforehand. As there is no obvious reason that this form of knowledge should be restricted to geometry only, it may serve as a model for rigorous science in general. During all of Western antiquity, from Plato and Aristotle onward and during the Middle Ages, as far as these periods are influenced by writings in the Western tradition, this ideal of scientific knowledge has been universally upheld.

The second phase in our schematic history of philosophy of science begins in the early seventeenth century and ends sometime in the second half of the nineteenth century. It continues with the first phase in equally subscribing to the epistemic ideal of the certainty of scientific knowledge. However, it is discontinuous regarding the means by which this ideal is to be achieved. Whereas in the first phase, only deductive proof is a legitimate means to attain the certainty of knowledge, the second phase liberalizes this requirement to what will eventually be known as the "scientific method." This expression either denotes one single method, or it is taken as a collective singular referring to a certain set of methods; what is meant exactly is typically left unanswered. Deductive proof is, of course, still a part of the scientific method, but the most important extension concerns inductive procedures. They somehow proceed from data to law and are, when applied properly, mostly perceived to lead also to secure knowledge. The most famous protagonists of this scientific method are, of course, Galileo Galilei (1564–1642), Francis Bacon (1561–1626), René Descartes (1596–1650), and, a little later, Isaac Newton (1642–1727). The scientific method is mainly conceived of as strict rules of procedure, and it is the strict adherence to these rules that establishes the special nature of scientific knowledge. During the second half of the nineteenth century, however, the belief in the possibility of secure scientific knowledge erodes, even if this knowledge is produced under the rigid auspices of the scientific method. This leads us to our third phase.

Timing the start of the third phase is quite an imprecise matter as it is the result of a process of slow erosion of the belief in scientific certainty. For reasons whose details still await in-depth historical research, especially with respect to their interaction, the conviction of the certainty of scientific knowledge already decays in the late nineteenth century. This is true both with respect to the mathematical, the natural, and the human sciences, although mathematics is able to restore its claim for conclusiveness by a decisive turn.

For the mathematical sciences, the discovery of non-Euclidean geometries in the course of the nineteenth century is dramatic. It demonstrates that the belief in the uniqueness of Euclidean geometry, and thus the conviction of its unconditional truth, is unfounded. However, the conclusiveness of mathematics is restored if the axioms of any mathematical theory are taken as assumptions whose truth or falsehood is not up for grabs. Mathematical claims then no longer concern the categorical truth of theorems, but only their conditional truth; they are conditional upon the hypothetical acceptance of the pertinent axioms. This constitutes the turn of mainstream mathematics in the nineteenth century, making it immune to factors that bring about changes to the fourth phase, which affect other sciences.

In the natural sciences, the process of erosion of scientific certainty is often only associated with the advent of the special theory of relativity and of quantum mechanics. This process is therefore assumed to begin later, namely in the first quarter of the twentieth century. This association is highly plausible, because the idea of scientific certainty is tied to the successes and the apparent definiteness of classical mechanics. If classical mechanics turns out to be only a model of reality with limited applicability and is in fact false, how can certainty about any product of science ever be achieved justifiably? Cannot the same also happen to any succeeding theory of classical physics? However, the process of erosion of the belief in scientific certainty appears to have started even earlier, when physics was still in its fully classical phase. At any rate, especially after the revolution in physics in the first quarter of the twentieth century, the belief that scientific knowledge is not certain and can never be, but is hypothetical and fallible, becomes dominant both in scientific and philosophical circles. For instance, both inductivist and, later, deductivist philosophies of science, though relying on strict methodological procedures for confirmation or testing of hypotheses, stress the hypothetical nature of scientific knowledge from the natural sciences.

Also in the nineteenth century, various influences contribute to the flourishing of the social sciences and the humanities. Partly, these disciplines are created in this period, like sociology; partly, they are redeveloped as the historical humanities. All these disciplines are seen as methodologically opposed to the natural sciences. However, despite their methods, it is the so-called historicism of this period that stresses that all knowledge is historically bound and thus fallible.

At present, we are in the fourth phase, which started sometime during the last third of the twentieth century. In this phase, belief in the existence of scientific methods conceived of as strict rules of procedure has eroded. Historical and philosophical

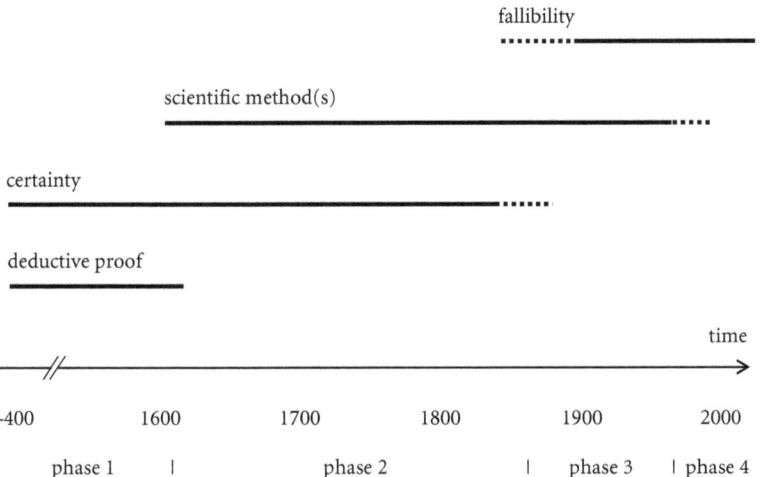

FIGURE 1.1 Schematic representation of the history of the conceptions of science

studies have made it highly plausible that scientific methods with the characteristics posited in the second or third phases simply do not exist. Research situations, i.e., specific research problems in their specific contexts, are so immensely different from each other across the whole range of the sciences and across time that it appears utterly impossible to come up with some set of universally valid methodological rules. The diversity of the sciences precludes the hope that the sciences are united, if at all, in some methodological way, unmasking as a myth the idea of the scientific method as constitutive for science. Methodological prescriptions are, at best, rules of thumb and as such cannot found a special nature of scientific knowledge. Skeptics such as Paul Feyerabend have also concluded (unwarrantedly, I believe) that due to the absence of specific scientific methods, scientific knowledge fails to have *any* demarcating properties from other kinds of knowledge and therefore is epistemologically completely on par with them. Be that as it may, the fact remains that there is no consensus among philosophers or historians or scientists about the nature of science at the beginning of the twenty-first century. If one at least entertains the hypothesis that scientific knowledge indeed has special features and that it is, in certain respects, a unique cultural product, then one needs to ask the question about the nature of science anew. At the very least, one should be aware that the question about the nature of science is *the* central question of general philosophy of science, and one should develop a stance toward it. Simply avoiding the question will not do.

Let me graphically summarize the main features of this sketchy history of the conceptions of science as it applies to the natural sciences (see figure 1.1).

It should be noted that there is no single feature that exists across all phases, thus establishing continuity. Rather, there is always one feature connecting two consecutive phases, whereas another feature is discontinuous. It is the sort of continuity of a thread that Wittgenstein discussed in the context of his exposition of family resemblance in his *Philosophical Investigations*: "[T]he strength of the thread does not reside in the fact that some one fiber runs through its whole length, but in the overlapping of many fibers." Thus, the first two phases are connected by the ideal of certainty for scientific knowledge, but deductive proof is replaced by scientific method(s) in the second phase. The second and third phases are connected by the idea of scientific method(s), but the ideal of certainty is replaced by fallibility in the third phase. The third and fourth phases are connected by the idea of the fallibility of scientific knowledge, but in the fourth phase, the belief in scientific method(s) as constitutive for science ceases. Note that only in the present fourth phase, the question about the nature of science becomes dramatic, because the only feature left for science, namely fallibility, is by no means a sign for its uniqueness. Therefore, it is no exaggeration to state that although we are familiar today with the phenomenon of science to a historically unparalleled degree, we do not really know what science is.

This state of affairs is what triggered the present book. Here is an outline of the book's content. In the following section 1.2, I shall address the question "What is science?" itself by making explicit what is asked in the question. In philosophy, it is almost always useful to reflect upon the question one is asking before attempting to answer it. In chapter 2, I will answer the question with a thesis about scientific knowledge. After some historical remarks regarding this thesis, I will start to qualify and clarify it. Section 2.2 is devoted to a discussion of the key concept of the thesis, the concept of systematicity. After having reached some basic clarity about the content of thesis, I will outline the structure of the argument for the thesis in section 2.3. Chapter 3 will unfold the thesis further by making it more concrete in nine dimensions. These dimensions are as follows: scientific descriptions, explanations, predictions, the defense of knowledge claims, critical discourse, epistemic connectedness, the ideal of completeness, the generation of new knowledge, and the representation of knowledge. In each of these dimensions, the thesis will be supported by various examples from many different sciences. Chapter 4 will connect my thesis with some older answers that have been given to the question, "What is science?" It will turn out that these older answers are not just wrong, but only one-sided. In other words, my thesis is a generalization of the older answers that adapts them, where necessary, to the universe of the sciences in the twenty-first century. This chapter is thus not an exercise in historical scholarship but part of the argumentative support of my thesis. I will discuss Aristotle, Descartes, Kant, Logical Empiricism, Popper, Kuhn, Feyerabend, and Rescher. In the final chapter, I shall draw some consequences of my thesis regarding the genesis and dynamics of science, the relationship between science and common sense, normative consequences, and the demarcation of science from pseudo-science.

In this book, I shall somehow deviate from the traditional treatment of the most general questions in the philosophy of science. Relative to the dominant tradition in philosophy of science, I shall suggest three shifts in orientation, which I shall briefly summarize here by way of anticipation; details will be discussed later at the appropriate places. First, during much of the twentieth century, the fundamental and general questions in the philosophy of science were mostly asked in what may be called an essentialist spirit. For instance, with respect to the structure of scientific explanations, the question of the goals of science, or the demarcation of science from pseudo-science and metaphysics, the expectation was to formulate *criteria* that were *general with respect to time*, i.e. ahistorical or suprahistorical, and *general with respect to disciplines*, i.e., discipline-independent. However, many of these attempts at reaching general criteria, preferably in terms of necessary and sufficient conditions, were not successful. In view of the fact that these discussions involved the best philosophers of science, that they sometimes lasted over several decades, and that the lack of success emerged independently in different areas, it may be suspected that something is fundamentally wrong with the whole approach. The viewpoint that I am suggesting in this book is to expect answers of a different kind to these general questions. A general question asking for some specific communalities of a large class of items may not have an answer in terms of necessary and sufficient criteria, applying to each and every item in the whole class. The lack of this kind of answer to the question definitely does not disqualify it as a pseudo-question. Rather, the items in that class may be connected by family resemblance only, just as exemplified by the different phases in the history of the conceptions of science set forth above. Wittgenstein's prime example, as is well known, was the concept of a game. He thought there were no criteria that all games have in common. Rather, one group of games has some criteria in common, another possibly overlapping group has other criteria in common, and a third again possibly overlapping group has yet other criteria in common. The result is several different but overlapping criteria, none of which is completely general, but which provides enough coherence to the whole class in order to distinguish it from other classes. It is this sort of answer I shall give when discussing the question about what unites all the sciences (section 2.2), what epistemic connectedness is (section 3.6), and what demarcates science from pseudoscience (section 5.4). It should be noted, however, that at this point, I have only uttered a declaration of intent. Every time I shall claim this particular character of a concept, I will have to provide arguments.

The second shift in orientation I will be suggesting in this book concerns the scope of the term "science" when discussing the most general questions in philosophy of science. As I shall suggest in the next section, it may sometimes be beneficial to include in the range of disciplines discussed not only the natural sciences but also the formal sciences, the social sciences, and the humanities. It seems that this comprehensive group of disciplines typically assembled in any full-size research university may be a more appropriate unit in the discussion of some general questions rather than the sciences in the narrow sense, i.e., the natural sciences. At least in this book, this larger group of disciplines will be our subject matter.

The third shift in orientation will concern the main contrast one has in mind when discussing the question of what science is. For almost a century, the main candidates for this contrast were nonsense, pseudoscience, and metaphysics, the latter seen by some as overlapping with nonsense. I shall, however, suggest that for the discussion of the most general questions in the philosophy of science, the main contrast to science should be other forms of knowledge, everyday knowledge in particular.

Enough of anticipations—let us now turn to our main question, "What is science?"

1.2 THE QUESTION, "WHAT IS SCIENCE?" IN FOCUS

Before proceeding directly to answering the main question, it makes sense—as it almost always does in philosophy—to dwell on the question itself and to try to clarify it. We will have to ask and answer the following preparatory questions that will put our central question into sharper focus. What disciplines do we wish to include in our discussion? What aspects of science do we focus on? What contrasts to science do we have in mind? What formal properties should an appropriate answer to the question have? Can we expect that an answer to the question will delineate the sciences as an area with thoroughly sharp boundaries? I shall deal with each of these questions in turn.

First, with respect to disciplines covered by the term "science," I want to understand the term and thus the question "What is science?" in their broadest possible sense. Therefore, not only all sciences in the (English) standard sense shall be included, namely the natural sciences, but also mathematics, the social sciences, the humanities, and the theoretical parts of the arts. Unfortunately for my project, there is no appropriate single English term denoting this broad variety of disciplines. We might collectively refer to them as "research fields" or "research disciplines." In German, there is the term "*Wissenschaft*," which covers all research fields that I intend to cover here. However, for lack of a better word, the term "science" will be subsequently used, although it does not represent well the semantic shift proposed here. Other authors pursuing studies of a similar breadth and being confronted with the same difficulty have also resorted to the very broad usage of the term "science." Therefore, one should keep in mind that throughout this book, contrary to common usage, the term "science" will denote an extremely broad range of research fields: roughly everything that is taught at a research university. I shall occasionally remind the reader of this altered usage of the term "science."

Second, with respect to aspects of science that are in our focus, I want to discuss science in a somewhat restricted sense. I will mostly deal with the epistemic aspects of science, i.e., science in the sense of scientific *knowledge*. However, I will not only address fixed bodies of scientific belief, but also so-called methodological issues, that is, the ways in which scientific knowledge is generated, processed, and applied. The sociological perspective, in which science is seen as a social system with particular social relations among its actors, will mainly be considered in the section on critical discourse (section 3.5). Social relations of science to its environment, especially its wider social, political, or economic context in which it is embedded, will not be our subject matter. Highlighting and isolating mostly the epistemic aspects of an enterprise that is fundamentally social and is strongly connected with other aspects of society certainly is a brutal abstraction of the reality of science. This abstraction is not particularly fashionable these days. Whether this abstraction is fruitful cannot be judged in advance, although many colleagues from science studies will probably oppose this statement. From their perspective, it has been demonstrated in abundance that the envisaged abstraction, typical of an old-fashioned form of philosophy of science, is principally incapable of capturing any essential and interesting properties of science. On the contrary, it can only produce a highly distorted and dangerously misleading image of science. It will not come as a consolation to these colleagues that, as already indicated, the abstraction from social factors in the present investigation is not a total one. When it comes to descriptions and explanations of certain features of scientific knowledge and its dynamics, social facts will be discussed, although they will not dominate the overall picture. Regarding the main question, whether the envisaged abstraction is legitimate and useful, I require some patience from my readers. I think that abstractions can never, or only very rarely, be justified at the beginning of an investigation. At best, we will discover at the end whether the abstraction was useful or not. At worst, the abstraction may appear useful at the end, but only an alternative research program using other points of view and associated abstractions could show the inadequacy of our result. I cannot avoid this menace, not in this book or in any other.

Third, with respect to contrasts to scientific knowledge, my question about the nature of science should not be understood in the same way that it has predominantly been understood in the second half of the twentieth century. The question,

"What is science?" has usually denoted a request for a so-called demarcation criterion. A demarcation criterion distinguishes science from other fields that are pointedly perceived as nonscientific, such as metaphysics or pseudoscience. Karl Popper's *Logic of Scientific Discovery* has probably been most influential in this regard. Popper saw the quest for a demarcation criterion as one of the most fundamental problems of epistemology, and many writers followed Popper at least in the sense that they have understood the question, "What is science?" as a question that aims at demarcating science from pseudoscience. This, however, should not be taken for granted, because the question, "What is science?" can also be understood as aiming at different contrasts. For instance, the question "What is science?" can also be understood as aiming at a contrast between scientific knowledge and other forms of knowledge, especially everyday knowledge: what is it that makes scientific knowledge a different and possibly superior form of knowledge when compared to other forms of knowledge, especially everyday knowledge? Note that in this differentiation, the contrasting pole to scientific knowledge is not denied the status of knowledge altogether, whereas pseudoscience is usually seen as simply not really being knowledge.

In this book, the question, "What is science?" is understood as aiming at the contrast between scientific knowledge and other forms of knowledge, especially everyday knowledge. The question therefore reverts to: by which features is science most characteristically distinguished from other kinds of knowledge, especially everyday knowledge? Although the question concerning the demarcation criterion will not at all be dismissed, it will not guide our investigations from the beginning. We will only be able to confront the question of a demarcation criterion late in the course of our discussion. A different approach to this problem seems to be advisable anyway, as all of the twentieth-century efforts to articulate a demarcation criterion, which immediately started with a contrast between science and pseudoscience or metaphysics, were quite unsuccessful.

Fourth, the formulation of the question, "What is science?"—and even more so the equivalent question, "What is the nature of science?"—may be seen to carry unwanted presuppositions or at least associations. Since antiquity, "What is X?" questions are mostly understood as searches for a definition of X. This definition should determine the nature, or essence, of X. An adequate determination of the essence of X would present a list of properties, each of which X necessarily possesses inherently by being X and whose conjunction is sufficient for determining X. Expectations of an answer to a "What is X?" question have evidently been based on metaphysical assumptions about the existence and properties of the essence of things. In modern times, these assumptions have become problematic, to say the least. As I want to avoid controversial philosophical presuppositions when they are not needed in the context of this book, I will not discuss the general question of whether there are essences or not, and which properties they might have. Such controversial presuppositions constrain the acceptance of what is based upon them. By contrast, I want the question, "What is science?" to be understood in a sense that is as free from these metaphysical presuppositions as possible. In particular, I am not anticipating the form of admissible answers to this question, i.e., whether the answer has to specify an essence or not.

Fifth, before answering the question, "What is science?" we should clarify whether we can expect that an answer to the question will delineate the sciences as an area with thoroughly sharp boundaries. I think there are at least two related areas in which sharp boundaries between science and nonscience (whatever is exactly subsumed under this term) do not exist. Rather, in these areas, there is a smooth transition between science and nonscience. In these areas, it would be unreasonable to expect an adequate answer to the question, "What is science?" to draw absolutely sharp dividing lines. Instead, the answer should be compatible with and even make plausible why these transition areas between science and nonscience exist.

The first transition area is constituted by the fact that scientific knowledge and scientific procedures are often applied to clearly nonscientific purposes. For instance, scientific models of the Earth's weather system are routinely applied to provide daily weather forecasts, and drugs developed on the basis of extensive pharmacological and clinical research are routinely used in medical treatments. Such application procedures are structurally identical with (parts of) the test procedures of the respective body of scientific belief. For example, the predictive accuracy of some meteorological model may be tested by applying it to a known weather situation in the past and comparing its predictions with the known development of the weather. Similarly, clinical tests of new drugs involve, among other things, their application to groups of patients and the observation of their effects. In cases like these, some actions that are parts of scientific test procedures are physically identical with some actions involving the application of the respective scientific knowledge for extra-scientific purposes. These actions differ only in their aims. In the former case, it is the scientific aim of testing some scientific hypothesis; in the latter case, it is the aim of an application for practical purposes. However, it may be difficult or impossible to decide whether such actions are a part of science or whether they belong to an extra-scientific context. This may be due to missing clarity with respect to the nature of the goals of a given activity. In addition, in the case of unexpected results, even the aims of such actions may change; what has been planned as a straightforward application may turn into an experimental investigation because of unexpected complications.

In the second transition area, the difference between research and application is even more blurred. It concerns what is known as "R & D," research and development. In this area of technology, there is often a smooth transition between applied sciences, engineering sciences, and product development. It is sometimes quite difficult—and often quite unnecessary—to decide whether a particular activity, for example an experiment, belongs to engineering science or is instead already part of product development. I will briefly present three examples: the development of a fusion reactor, earthquake engineering, and chocolate manufacturing. The development of a fusion reactor for energy production is certainly a technological goal; it has been investigated since the 1940s. However, in the course of this development, it turned out that many properties of the relevant plasmas were insufficiently understood. In this context, research into these plasmas is then located in a transition area between science, even basic or fundamental science, and technological development. The second example is earthquake engineering. As Vitelmo Bertero noted, it is concerned with "planning, designing, constructing and managing earthquake-resistant structures and facilities." There are more theoretical parts of earthquake engineering that clearly belong to the engineering sciences. An example is the experimental study of the seismic behavior of certain types of assemblages. However, there is also a transition regime toward more practical tasks because in this area, "[r]esearch alone is not enough; analytical and experimental studies must be augmented by development work" (V. Bertero). In other words, the research result must be implemented into building design and construction. The main purpose for design and construction of these buildings is, of course, an extra-scientific one. Nevertheless, the seismic response of these buildings under severe earthquake ground motions has been an important source of data for further improvement of their design and construction. Thus, there is an area where the distinction between science and product development cannot be meaningfully drawn. Finally, here is the nice and delicate example of chocolate manufacturing. No joke, there is a "science of chocolate." This science deals with the fundamental questions of chocolate manufacturing. These questions are typically explored at departments of food sciences at research universities. However, a continuous transition exists to the sort of research conducted by the research and development departments of the large chocolate manufacturing companies. Their research is partly closer to product development, and, of course, there is a fairly smooth transition to the development of new products and to innovation of production techniques.

The upshot of this diagnosis of the existence of transition areas between science and nonscience is that we should not expect an answer to the question, "What is science?" to deliver sharp boundaries in just every case. Where a thing itself is not sharply delineated, a theoretical representation aiming at an accurate description of that thing should reflect these transition areas. In such cases, the vagueness of the description is not a weakness but a tribute to accuracy. Nevertheless, we will not just be helpless when it comes to distinguishing the clear cases of scientific knowledge and those applications of science that have nonscientific goals (see section 3.6).

Now, after having worked through the necessary preliminaries, we should turn our attention to answering the question, "What is science?"

2

The Main Thesis

2.1 SCIENCE AND SYSTEMATICITY

Here is the main thesis that I shall explicate and defend in this book:

Scientific knowledge differs from other kinds of knowledge, in particular from everyday knowledge, primarily by being more systematic.

2.1.1 A Little History

Before exploring this thesis, it should be noted that some of its direct precursors may be found scattered throughout the literature of various disciplines, although very rarely in philosophy. In the last century, however, an early articulation of the idea that science is characterized by systematicity is found in the philosophical literature. It extends over twelve lines and is by John Dewey (1859–1952). In 1903, he wrote an article entitled "Logical Conditions of a Scientific Treatment of Morality." A discussion of the logical conditions of a scientific treatment of morality clearly presupposes some understanding of what the term "scientific" means. Dewey thus begins his article with a section entitled "The Use of the Term 'Scientific,'" which starts as follows:

> The familiar notion that science is a body of systematized knowledge will serve to introduce consideration of the term "scientific" as it is employed in this article. The phrase "body of systematized knowledge" may be taken in different senses. It

may designate a property which resides inherently in arranged facts.... Or, it may mean the intellectual activities of observing, describing, comparing, inferring, experimenting, and testing, which are necessary in obtaining facts and in putting them into coherent form. The term should include both of these meanings.

In this quotation, Dewey contends that the notion that science is a body of systematized knowledge is a "familiar" one. There are two aspects to the claim of systematicity: it characterizes both the product of science ("arranged facts") and the various procedures of science leading to this product. In Dewey's quotation, there is no explicit comparison of scientific knowledge with other kinds of knowledge, but it suggests that scientific knowledge exhibits these traits of systematicity, which other kinds of knowledge lack.

The topic of systematicity is also present in an influential book entitled *An Introduction to Logic and Scientific Method*, published in 1934. The book is authored by two American philosophers, Morris R. Cohen (1880–1947) and Ernest Nagel (1901–1985). In the concluding chapter on the scientific method, they write that "scientific method pursues the road of systematic doubt." Systematicity also appears in a different context, namely, regarding the whole of science: "The ideal of science is to achieve a systematic interconnection of facts." The latter quote is similar to what we already saw in Dewey. It is possible that there is indeed a direct connection between these similar statements about systematicity: Dewey and the book's senior author, Cohen, interacted at least in publications.

George Sarton (1884–1956) was not a philosopher, but he was a very important historian of science in the first half of the twentieth century who reflected on science and its history. He taught at Harvard University and founded one of the leading history of science journals, *Isis*. Among many weighty and influential books, he published a booklet entitled *The Study of the History of Science* in 1936, in which he presents his view of how the history of science should be conducted. Very early in the book, Sarton states a definition that, according to his own report, he had "published in various forms in earlier writings since 1913." It is stated as follows:

Definition. Science is systematized positive knowledge, or what has been taken as such at different ages and in different places.

In our context, three points of this definition deserve to be particularly highlighted. First, the characterization of science as systematized knowledge presents systematicity as a necessary feature of scientific knowledge. Second, Sarton leaves room for historical change and geographical variation regarding at least the positive knowledge that counts as scientific knowledge. Possibly, he even leaves room for different kinds of systematization, depending on time and geography. Third, the fact that Sarton presents his statement as a "definition" suggests that he thinks of it as not being in need of any explicit defense.

In 1960, the American philosopher Charles Morris (1901–1979) wrote a short article entitled "On the History of the International Encyclopedia of Unified Science." Morris was one of those American philosophers who had close contact with several members of the Vienna Circle already in the 1930s and was sympathetic toward their ideas. In his article, Morris outlined Neurath's general plan of the *International Encyclopedia of Unified Science*, as Neurath had described it to him in several letters in the 1930s. Otto Neurath (1882–1945) was one of the leading members of the Vienna Circle, and the *Encyclopedia* was his brainchild. About the plan for sections 2 and 3, Morris wrote:

> Section 2 was to deal with *methodological* problems involved in the special sciences and in the systematization of science.... Section 3 was to concern itself with the *actual state of systematization* within the special sciences and the connections which obtained between them, with the hope that this might help toward further systematization.

I note in passing that these sections were planned to contain sixty and eighty monographs, respectively! In 1939, Rudolf Carnap (1891–1970), Charles Morris, and Otto Neurath wrote a statement to try to obtain advance subscriptions for the *Encyclopedia*. Here, section 2 of the *Encyclopedia* is very similarly described: "This unit will stress the problems and procedures involved in the progressive systematization of science." The meaning of "systematization" in this context can easily be extracted from the task that section 2 was meant to fulfill: "to take stock, as it were, of the contemporary situation in the analysis and unification of scientific knowledge." Systematization consists of assigning specific pieces of knowledge to a place in the overall system of science, thus achieving the unification of scientific knowledge. It seems, however, that Carnap, Morris, and Neurath did not take systema*ticity* to be a defining characteristic of science. The aim of systema*tization* rather concerned the overall architecture of science as a whole, but it was controversial how far this goal could be carried out:

> Volumes III–VIII will especially stress the controversial differences in regard to special sciences (physics, psychology, etc.), in regard to the possibilities and limitations of scientific unification, and in regard to the methods involved in scientific progress and systematization.

Another philosopher who in passing referred to the systematicity of science some fifty-five years after Dewey, in 1958, is Carl Gustav Hempel (1905–1997). Hempel uses the terms "deductive systematization" and "inductive systematization" for a specific class of arguments. In these arguments, a fact is deductively or inductively inferred from statements of other facts together with statements of general laws. Hempel's interest in these arguments derives from his conviction that "scientific explanation, prediction, and postdiction all have the same logical character: they show that the fact under consideration can be inferred from certain other facts by means of specified general laws." What Hempel refers to as "postdiction" in this passage is mostly called "retrodiction" today, and it denotes an assertion of past facts on the basis of knowledge of later facts. Similarly to what Dewey had in mind, Hempel states that there are "*systematic* connections among empirical facts" that are established by general laws; these systematic connections allow for scientific explanation, prediction, and retrodiction. The systematizations therefore play a central role in science. A few years later, in 1965, Hempel stresses in the concluding section of his famous essay, "Aspects of Scientific Explanation," that "all scientific explanation ... seeks to provide a *systematic* understanding of empirical phenomena by showing that they fit into a nomic nexus." By "fit into a nomic nexus," he means the integration of empirical facts into a net of natural laws. Still later, in 1983, Hempel reports a widely held conception of science to which he probably subscribes, too: "Science is widely conceived as seeking to formulate an increasingly comprehensive, *systematically* organized, world view that is explanatory and predictive." The interrelationship with his earlier statements about explanations and predictions as systematizations suggests itself. Clearly, a world view that utilizes systematizations as its basis for explanations and predictions can itself claim to be systematically organized.

Similar to Hempel was Ernest Nagel (1901–1985), whom we already saw as (junior) coauthor with Morris Cohen in their 1934 book, *An Introduction to Logic and Scientific Method*. In 1961, Nagel published the classic *The Structure of Science*, a massive book of more than six hundred pages. In many aspects, this book reflects the state of philosophy of science just before the advent of Kuhn's *The Structure of Scientific Revolutions* in 1962. In the introductory chapter, Nagel contrasts common sense with science regarding explanations: "It is the desire for explanations which are at once *systematic* and controllable that generates science." And: "A number of further differences between common sense and scientific knowledge are almost direct consequences of the *systematic* character of the latter." A first difference concerns common sense's typical lack of awareness of its own limits, for instance the dependence of its validity on the subsistence of certain conditions, whereas science aims at removing this incompleteness. Second, common sense often contains incompatible beliefs whose elimination is one of the stimuli to the development of science. Third, the sciences mitigate the indeterminacy of ordinary language by introducing a more precise vocabulary in order to expose their statements to more thorough critical testing by experience. Finally, whereas common sense knowledge is quite concerned about its relation to human values, science's systematic explanations deliberately neglect this dimension. However, because Nagel's book mainly deals with questions of scientific explanation, his discussion of the systematicity of science is restricted to this topic. As such, the discussion of the *systematicity* of scientific explanation primarily has an introductory function as it leads the reader to the main subject of his book.

To the best of my knowledge, the only philosopher who extensively considered systematicity and its relationship with science in the last one hundred years is Nicholas Rescher (born 1928). In 1979 he published a book entitled *Cognitive Systematization: A Systems-Theoretic Approach to a Coherentist Theory of Knowledge*, and in 2005 he published *Cognitive Harmony: The Role of Systemic Harmony in the Constitution of Knowledge*. I will not discuss Rescher's position at this point. A full section will be devoted to Rescher's treatment of systematicity in chapter 4 where I compare the position developed in this book with other positions.

Finally, I want to present six examples from different areas of science (in the wide sense) in which "systematicity" is connected with "being scientific." As in many other examples, the connection is not emphasized; it is mentioned in passing and therefore implicitly presented as something that is self-evident. Let me start with two physicists. In their hotly debated book entitled *Fashionable Nonsense: Postmodern Intellectuals' Abuse of Science*, Alan Sokal and Jean Bricmont "stress the methodological continuity between scientific knowledge and everyday knowledge." Continuity does not mean identity, and the difference between scientific knowledge and everyday knowledge is stated as follows:

> Historians, detectives, and plumbers—indeed, all human beings—use the same basic methods of induction, deduction, and assessment of evidence as do physicists or biochemists. Modern science [obviously, they mean natural science, P.H.] tries to carry out these operations *in a more careful and systematic way*, by using controls and statistical tests, insisting on replication and so forth. (emphasis added, P.H)

They present no argument for this claim, because to them, it is presumably just evident.

The second example is from mathematics. In a recent article, Amir Alexander, the historian of science—and of mathematics in particular—describes the result of the "Rebirth of Mathematics" in the early nineteenth century as follows: "The main concern of nineteenth-century mathematicians was not finding useful new results but *systematizing* and developing the internal structure of mathematics itself.... [T]his has largely remained the concern of professional mathematicians to this day." And a little later: "Mathematics came to be seen as a science unto itself, whose value could be judged only by its own internal standards. Now, it seemed, mathematics could be worthy of its name only if it was rigorous, self-consistent, and *systematic*." In other words, at least since the nineteenth century, systematicity was seen as one of the constitutive elements of the science of mathematics.

The third example concerns cognitive psychology. In an article entitled "Mental Models of the Earth: A Study of Conceptual Change in Childhood," Stella Vosniadou and Bill Brewer describe two ways that the prescientific understanding of the world, the "intuitive knowledge" or "naïve physics" of children, are conceptualized:

> Some researchers believe that children's intuitive knowledge can be conceptualized as consisting of a coherent and *systematic* set of ideas which deserve to be called a theory.

Note that it is the coherence and the systematicity of a set of ideas that seem to justify the use of the term "theory" without any further comment. The belief that children have theories of this sort is not universally held, however:

> Other researchers think that naïve physics consists of a fragmented collection of ideas which do not have the *systematicity* that is typically attributed to a scientific theory.

Again, the allegedly typical attribution of systematicity to a scientific theory does not elicit any further argument, apparently as if it was self-evident and could simply be taken for granted.

The fourth example is from history. In his perceptive introduction to the methodology of history entitled *The Pursuit of History*, John Tosh writes about "what the experienced scholar [in history] does almost without thinking":

> [H]istorical method may seem to amount to little more than the obvious lessons from common sense. But it is common sense applied *very much more systematically* and skeptically than is usually the case in everyday life, supported by a secure grasp of historical context and, in many instances, a high degree of technical knowledge.

This statement is very close indeed to what I stated in the beginning of this section as the main thesis of this book.

The fifth example of a connection of "science" with "systematicity" is taken from pictorial semiotics or, more to the point, from an *Introductory Lecture to Pictorial Semiotics* by Göran Sonesson. In order to approach the question of what this possibly strange-sounding discipline is, in the first part of his lecture, the author considers the question of what sort of enterprise semiotics in general is. It is the study of signs, of course, but is it a science? So we should know what a science is:

> As a first approximation, one may want to say that science is a particularly orderly and *systematic* fashion for describing and analyzing or, more generally, interpreting a certain part of reality, using different methods and models.

Although the author is not satisfied with that explanation because of the "a certain part of reality" clause (it won't fit semiotics), he does not question the systematicity aspect—and again, without further argument.

The sixth and last example may be the most surprising one because it concerns theology. Also in theology, the aspect of systematicity is used in order to its character as science (in the wide sense of "science," of course). For instance, the entry "Problems of Philosophy of Religion" in *The Encyclopedia of Philosophy* of 1967 characterizes theology in the following way:

> Theology, in a narrow sense of that term, sets out to articulate the beliefs of a given religion and to put them into systematic order ....

It should be noted that in this (widely shared) view, theology is not a discipline the subject matter of which is God (as the word "theology" suggests), but religious beliefs. As the quoted sentence suggests, theology gains its status as an academic discipline (and thus as a science in the wide sense) by a guiding idea of systematicity.

All of the examples I have discussed share one important feature. The systematicity of science or of some parts or aspects of it is somehow taken for granted. It does not seem to incite further comment or elaboration, let alone justification. Nevertheless, I think it is worthwhile to investigate this property of science, to explicate and to argue for it, and to consider its consequences, because we can organize many existing insights about science in this way and gain new ones. This is the aim of the present book.

In the following section 2.1.2, I will start to explicate the main thesis that I put in raw form at the beginning of this chapter:

> *Scientific knowledge differs from other kinds of knowledge, in particular from everyday knowledge, primarily by being more systematic.*

I will clarify some properties of the thesis asserting the higher degree of systematicity of scientific knowledge, and I also want to qualify it in some important respects. I am only going to approach the notion of systematicity that is the central concept of my thesis in section 2.2. Given a better understanding of what the thesis asserts, we can then reflect on the structure of the argument we need in order to defend the thesis; this will be done in section 2.3. Only in chapter 3 will I be able to argue concretely for the thesis.

2.1.2 *Preliminary Remarks*

Here are five remarks intended to begin the task of clarification and explication of the thesis.

First, there are two minor terminological remarks. As an alternative to stating "is more systematic," I shall use the expression "has a higher degree of systematicity" in the course of this book. When using the latter expression, I do not mean to imply that there is a *quantitative* measure of systematicity. Whether such a measure of systematicity exists I shall leave open throughout the book (though I doubt there is). Whatever systematicity is—I will turn to this question in the next section—"more systematic" and "a higher degree of systematicity" shall be used interchangeably. The second terminological remark concerns my use of the word "knowledge." In philosophy, "knowledge" is often used in a rather strict sense, i.e., in contrast with "(mere) belief." Knowledge is then understood as a particular kind of belief, namely (roughly), as belief that is true and for whose truth a particular warrant exists. In other words, knowledge in this sense implies truth. Nevertheless, whatever the concept of truth means precisely—again a very tricky question—it would be quite dangerous to presuppose that what counts as scientific knowledge today is literally true. By using the word "knowledge" I do not mean to imply this. Rather, I use it in the sense of a "body of … belief that is well-established, widely held in the relevant community, not regarded as tentative or falsified." In other words, I shall use the term "scientific knowledge" in the same way scientists do. The reason for my preference of "scientific knowledge" over expressions that certainly do not imply truth, like "scientific belief" or "scientific knowledge claims," is that these expressions tend to have connotations in the opposite direction. They tend to suggest the definitive absence of truth, something I do not want to assume or imply here either. Whether any scientific (or any other kind of) "knowledge" is really true in the final analysis, is a question I want to leave open, because it falls outside the intended scope of this book.

Second, the thesis is meant to primarily have a *descriptive* content and not a normative (or "prescriptive") content. Therefore, it does not *pre*scribe what property or properties (good) science should or must have, but *de*scribes how science actually is.

This well-known difference becomes particularly relevant when I turn to arguments for my thesis, because arguments for descriptive claims must be different from arguments for normative claims. As descriptive claims are mostly empirical, arguments for them will also have to be empirical. By contrast, arguments for normative claims may proceed, for example, by recourse to higher (more comprehensive) norms, specific goals, or ideals of rationality. After we are clearer about what my thesis really claims, I will return to the question of how to argue for it in principle (see section 2.3). The main part of the direct argument for the thesis will be presented in chapter 3. However, I shall not completely dismiss a possible normative content of the main thesis. I shall discuss it under the rubric of "normative consequences" in section 5.3.

Third, it is important to note that the thesis is *comparative* in character. It does not state that science is systematic and that other kinds of knowledge are not, but that science is *more* systematic than other kinds of knowledge. It is therefore possible that other kinds of knowledge are also systematic, even if to a lesser extent. This point should in fact be obvious. Look at our everyday practices of determining, say, the number of people who finally turn up at our party. We count, and no mathematician would do otherwise. But counting is a perfectly systematic procedure if we take care not to omit anybody and to count every person only once. Or consider our knowledge of a city with which we are somewhat familiar. Surely our knowledge of streets, buildings, restaurants, cinemas, bus stops, and the like is not as systematic as the complete body of this knowledge when perspicuously represented in a map, but it is not simply chaos either; it does have some degree of systematicity. Or take the knowledge that is often referred to as "local" or "traditional" knowledge possessed by other cultures. It would be absurd to assume that these kinds of knowledge about animals, herbs, fishing grounds, weather changes, and so forth were entirely unsystematic given that this knowledge has been capable of securing the survival of populations over long periods of time.

An immediate consequence of my thesis in this context is that it allows for a smooth transition between prescientific (or nonscientific) knowledge and scientific knowledge. Given my thesis, the science of a particular subject could have developed out of prescientific knowledge in a gradual process by an increase of systematicity. If this process happens fast, the impression of discontinuity may arise, but there is not necessarily discontinuity. Compare this with the earlier characterizations of science regarding the existence of proof for scientific claims or the application of the scientific method or scientific methods (see section 1.1). There is not really a gradual transition between not having a proof for some statement and having one (although in actual mathematics, there are sometimes cases resembling such transitions). The same seems to be true for methods: either a knowledge claim is backed by scientific methods, or it is not. And with regard to methods themselves, in general, there does not seem to be a gradual maturing of a method until it is scientific. Hence, the earlier characterizations of science more or less strongly imply a discontinuity between scientific knowledge and other kinds of knowledge, whereas my thesis allows for the possibility of a gradual transition (without implying it). I shall come back to the question of the transition of prescientific knowledge to scientific knowledge as seen by the systematicity theory in section 5.1.2.

Fourth, for the sake of the brevity of the thesis, I have not explicitly asserted that the higher degree of systematicity of scientific knowledge refers only to other forms of knowledge *about the same subject matter*. Here is an example that illustrates what I mean. Everyone has certain—largely unconscious—ways to get to know somebody they meet for the first time. Many immediately check sex, approximate age, social status, and many more traits within seconds. Many continue this exploration by looking at the eyes, the figure, the clothing, the hands and their movements, the rest of the body language, getting an impression of the voice, and so on. Now compare this to the situation when you apply for some managerial position and the company sends you to some personality assessment center in order to find out who you are and, more to the point, whether you fit the position they want to fill. The scientific approach used in such centers to find out who you are and whether you fit the job will consist of a well-thought-out series of tests and interviews that will be evaluated following a very orderly scheme. In cases like this, our thesis asserts that the scientific procedures are more systematic than the everyday practices, and this is fairly obvious in the given case (more detailed arguments for the thesis will be provided in chapter 3). The thesis therefore maintains something about the differing degrees of systematicity of *corresponding* bodies of knowledge. By contrast, if one compares scientific knowledge about one domain with other kinds of knowledge about another domain, then scientific knowledge is not necessarily more systematic.

Take the example of the Violent Crime Linkage Analysis System (ViCLAS) that was developed in the 1990s by the Royal Canadian Mounted Police (which is the Canadian national police service). By now, ViCLAS is in operation in many countries around the globe. The main purpose of ViCLAS is to identify and track serial violent criminals across borders. Each individual case has to be described in a way that makes the comparison and linkage with other cases possible. For this task, a list of 262 questions was formulated. The questions cover details of all aspects of an incident including victimology, modus operandi, forensics, and behavioral information. The content of the questions would provide investigators with the ability to link offenses based on the offender's behavior. All cases that enter ViCLAS' database, wherever the crime had been committed, have to be described according to this list of questions. In 2004, the ViCLAS database comprised several hundred thousand cases worldwide.

It is obvious that ViCLAS is an extremely systematic endeavor. It is science-based, but the concrete knowledge gained, namely, that a certain number of crimes can be ascribed to the same offender, does not belong to science: it is not published in the scientific literature but is used in further criminal investigations and in court. Compared with loosely structured scientific fields, the information contained in ViCLAS is immensely more systematic than the information stored and processed in such scientific fields. However, this fact does not contradict our thesis in any way, as we are dealing with knowledge about different subject matters. Our thesis only asserts that given some scientific knowledge about some subject and some extra-scientific knowledge *about the same subject matter*, then scientific knowledge will exhibit a higher degree of systematicity.

Fifth, due to its comparative nature, the thesis cannot be applied directly to fields that are generally perceived to be scientific, but *lack counterparts in other kinds of knowledge*. There are countless examples for such fields in practically all disciplines. The esoteric character of those fields results from frontiers of research driven to areas of which common sense knows nothing and does not even realize its ignorance. For instance, theories about black holes, or theories about the folding of proteins into their tertiary structure, or statistical delicacies in biomedical or social science research, or complicated relations between grammars of different languages with the same origin, or intricacies of the transmission of ancient texts to the present do not have any counterparts in other kinds of knowledge because their subjects happen to be exclusively scientific. This has to be understood in a purely sociological sense at this point: only people with a scientific education, mainly working in scientific or science-related institutions, deal with these subjects. How can our thesis, nevertheless, be made applicable to fields like those whose scientific character seems to be beyond dispute but have no nonscientific counterpart?

The argument for the scientific character of knowledge in these fields must be essentially historical, and at this point, I am able to outline it only very crudely. Let me provide an example. At first, black holes were theoretical predictions of the general theory of relativity. Later, indirect observational evidence for their existence was found. The general theory of relativity supposedly is a *scientific* theory about gravitation and, if our thesis is true, it should be more systematic than, for example, everyday ideas about gravitation (which are, at least in our culture, massively influenced by earlier scientific theories about gravitation). If the application of the general theory of relativity to black holes is a scientific step when judged by the standards of our thesis—how this judgment can be passed may be clarified at a later point—then one is justified to claim that the theories about black holes are scientific, too. The structure of the argument therefore is as follows. If some new field without nonscientific counterparts originates from a field that is scientific according to my thesis, and if the steps that lead to the development of the new field that are also scientific according to my thesis, then my thesis may also be considered to be applicable to these new fields, if only in an indirect sense.

By now, we have achieved at least a little improvement of our understanding of the main thesis of this book:

Scientific knowledge differs from other kinds of knowledge, in particular from everyday knowledge, primarily by being more systematic.

According to the preceding remarks, the thesis is descriptive, not normative; it is comparative, but not quantitative in character; it compares different kinds of knowledge about the same domain only; and it applies only indirectly to scientific knowledge lacking extra-scientific counterparts. So far, so good. However, the meaning of the core expression of the thesis, i.e., "more systematic," is still totally vague. Therefore, I shall now turn to an analysis of the concept of systematicity.

2.2 THE CONCEPT OF SYSTEMATICITY

Hardly any preparatory work exists to enhance our understanding of the terms "systematic" and "systematicity" that might be useful in our context. Anglo-Saxon philosophical dictionaries and encyclopedias do not feature "system" in general, at least not without any accompaniments like "axiomatic," "formal," "interpretive," "logical," or as in the combination "systems theory." German philosophical reference works, however, always feature "system" with and without accompaniments. This is due to the importance of the idea of a philosophical system in the tradition of continental philosophy. "Systematic" is again typically missing in Anglo-Saxon philosophical reference works, at least as an individual entry, whereas German works feature it occasionally. However, after some general statements about "systematic"—for instance that it means "having the form of a system"—the discussion typically turns quickly to the rather particular contrast between "systematic" and "historical." Again, this is due to the continental context in which the tension and/or complementarity between a historical and a systematic approach in philosophy is important and often discussed. However, "systematicity" does not seem to deserve an entry in any of the German philosophical reference works. By contrast, it has been awarded the knighthood of receiving an entry in the Oxford English Dictionary of 1989. There, it is explained to be "the quality or condition of being systematic; systematicness." "Systematic" is described as "done or acting according to a fixed plan or system; methodical." The term "systematicity" occurs about twenty times in the 1998 *Routledge Encyclopedia of Philosophy*, mostly in connection with Kant or with its somewhat special use in cognitive science. The sobering upshot is this: in our analysis of the concept of systematicity, we must start from scratch.

We therefore do not know very much about the core terms of the thesis, i.e., "systematic" or "systematicity." They appear to be vague and therefore are in need of more precision and concretization. I will attempt the required clarification in two steps that seem to be demanded by the nature of the concept. First, I will discuss and try to clarify "systematicity" on an abstract level. Unfortunately, this attempt will not lead very far. Therefore, a second step is needed in which the meaning of systematicity will be made more concrete for particular contexts. As we shall see, it is in fact a plurality of meanings that we will discover at the more concrete level, covarying with contexts.

Such a particular procedure of conceptual clarification may appear somewhat unfamiliar. It may therefore make sense to make a quick detour and illustrate the procedure first with another concept. Take "refinement" as an example. What does "refinement" mean? On an abstract level, "refinement" refers to something that is, before the refinement, in some unspecified sense, coarse, rough, vulgar, or undifferentiated. To refine a thing means to modify it to such an extent that it gains some higher quality. But "to modify something in such a way that it gains some higher quality" may mean very different things in different contexts. To refine one's writing style by using a larger vocabulary, or to refine a sauce by carefully adding cream, or to refine an optical apparatus by a careful adjustment of all parts are completely different activities. These various concretizations of the verb "to refine" that are dependent on the particular contexts enhance its meaning, making it richer than its abstract meaning. Attempting to find common elements among the concrete meanings of "to refine" (or "refinement") is probably a futile exercise. All that can be expected is a family resemblance among them (I shall come back to this claim below).

So let us start at the abstract level. What are the specific qualities that something possesses that makes it systematic? Instead of answering this question directly, it is simpler to state what something that is systematic is not. Something that is systematic is not purely random or accidental, it is not chaotic, not arbitrary, not anyhow made or risen, not completely unmethodical, not completely unplanned, nor completely unordered. Rather, it embodies some kind of order. I do admit, however, that neither the last positive characterization nor the contrasting terms have enhanced our understanding of the concept of systematicity substantially, and nothing more would be gained by adding any more items to this incomplete list. Seemingly, nothing more can be achieved in principle on the abstract level where we are dwelling. The feeling of fairly thin air, of vagueness, of not really knowing what we are talking about would persist. It seems to me that this impression is both correct and unavoidable. The reason is that in the *actual* use of terms like "systematic" or "systematicity" (or "refinement"), some context is always given. The terms then receive a richer and more concrete meaning due to that context. For instance, judging an oral presentation to be very unsystematic is immediately intelligible: the presentation's elements lack a defined and expected order. The pertinent kind of order is provided by the context: a talk is expected to consist of parts (trivially: in linear succession of each other) that are somehow connected with one another so that a train of thought emerges. However, the strong association between systematicity and a connection between linearly ordered parts, pertinent to this context, may be completely missing in another context where, for instance, systematicity may be associated with an idea of completeness. For instance, this is the case when a large number of items must be classified. The systematicity of a classification has nothing to do with linear order but rather with well-defined classes and completeness. Therefore, no reference about the type of order required can be provided on the more abstract level.

In order to carry our analysis of "systematicity" further, I will have to determine the contexts in which I intend to use the term and then make its more concrete meanings in these contexts explicit. I will claim nine dimensions (or "areas" or "aspects") of science in which science is more systematic than other kinds of knowledge. These dimensions are the contexts in which we will find, in a natural way, richer, that is, more concrete meanings of "systematicity." My main thesis will decompose correspondingly into nine distinct theses, claiming higher systematicity (in a context dependent sense) for each of those dimensions. This will increase the difficulty of our argumentative task when defending my thesis (or more precisely, my the*ses*). At this point, this problem will not be addressed (I will return to it in the next section). Here, the nine dimensions of science within which science is claimed to be more systematic than other kinds of knowledge, will be introduced in a preliminary way.

The nine dimensions of science to be discussed are

- descriptions,
- explanations,
- predictions,
- the defense of knowledge claims,
- critical discourse,
- epistemic connectedness,
- an ideal of completeness,
- knowledge generation, and
- the representation of knowledge.

An in-depth discussion of these nine dimensions, both as a set and as each dimension individually, must be postponed to the next chapter. At this point, it is sufficient to repeat that in these nine dimensions, the meaning of the corresponding concepts of systematicity will be richer and more concrete.

What kind of connection exists between these richer and more concrete concepts of systematicity corresponding to the nine dimensions? Of course, they all have something in common because they are descendants of the abstract concept of systematicity. They all embody some kind of order or methodicity or any other of the general characteristics of the abstract concept of systematicity. However, they will specify the abstract characteristics in very different ways. For instance, the abstract feature of "order" can take on the concrete form of, to name just a few examples, a temporal order of events, a logical order of thoughts, a systematic classification of species, a two-dimensional spatial pattern, or a map from one set into another. Their commonalities, therefore, comprise only the abstract features of systematicity in general. The impression of their diversity will be much more dominant than the very thin band of abstract features that unites them. For example, what counts in a discipline as a systematic *description* will be fairly uninformative for what is considered a systematic *explanation* in the same discipline, or a systematic *defense of knowledge claims*, and so on. We therefore cannot expect to find features that are common to *all* of the different concepts of systematicity on the level of concrete features. In other words, we will not be able to find concrete characteristics that are candidates to figure in a definition of systematicity in terms of necessary and sufficient conditions. Rather, we will have to expect a family resemblance relation between these different concepts in Wittgenstein's sense. We will find concrete features that *some* concepts of systematicity have in common, but not all of them; other features will be shared by *other* concepts of systematicity, but not by all of them, and so forth.

There is a second source of diversity regarding concrete systematicity concepts. Not only do these concepts covary with the nine dimensions of science, but for any given dimension, they will also covary with different disciplines. Take as an example the concept of systematicity that is pertinent to descriptions. The particular systematicity of a mathematical description is different from the particular systematicity of a botanical, historical, meteorological, or art-theoretical description, as well as others. To push matters even further, for any given dimension and within any given discipline, the same concept of systematicity will not be relevant for all of this discipline's subfields. Consider, for example, descriptions in history, and consider some of its subfields. The systematicity of descriptions, say, in economic history will differ from the systematicity of descriptions in the history of mentalities, the former containing a strong quantitative element in contrast to the latter. Again, the connection between these differing concepts of systematicity regarding descriptions in different subfields is one of family resemblance. In addition, there are family resemblance relations to other disciplines and their subfields, e.g., in the case of economic history to economics, or of the history of mentalities to certain areas of psychology or of textual theory. All of these family resemblance relations will exhibit various strengths, indicating different degrees of affinity between a discipline's subfields and between disciplines.

There is even a third source for some variation in the systematicity concepts. We cannot expect that a concept of systematicity, as it may be characteristic for one of the dimensions for some scientific subfield, is entirely stable over time. Scientific disciplines change in the course of history, and with them, their dimension-specific ideals of systematicity to be realized by good science. To pick out a single example: whenever some discipline or a discipline's subfield becomes successfully quantified, the standards of the desired systematicity in many of the dimensions change, for instance its descriptions, explanations, and predictions.

I would like to close this section by initially outlining one consequence of my approach. By asserting a higher degree of systematicity as the main distinguishing feature of scientific knowledge in comparison to other kinds of knowledge, it is implicitly claimed that a certain unity of all the sciences exists—remember, in the wide sense of "science," including the humanities. This may be surprising, because in recent years, the idea of a disunity of science has been much more popular than the older idea of a unity of science that was characteristic for the positivist phase of philosophy of science. However, the kind of unity of science resulting from my approach should be carefully noted. If the concept of systematicity that is the source of the claimed unity of the sciences were a concept defined in terms of necessary and sufficient conditions, the resulting unity would be very strong. All the sciences would share certain characteristics, inherited from the defining characteristics of their common systematicity. But the pertinent concept of systematicity cannot be defined in terms of necessary and sufficient conditions. It splits up into a huge number of differing concrete variants, covarying with the nine dimensions, with disciplines, and with subfields. All of these different variants are connected by an extremely complex network of family resemblances. The resulting unity of the sciences is of a very tenuous sort. Two given sciences may have almost nothing of relevance in common on a concrete level. They can only be described as belonging to the same unity on an abstract level, namely, as enterprises that generate knowledge in accordance with certain cognitive goals that has a particular relational property, i.e., of being more systematic than other kinds of knowledge. All of the sciences are united by relations of family resemblance only. Put in metaphorical and somewhat colorful terms, the unity of science can thus be described as the unity of a rather chaotic, multidimensional fabric made of Wittgensteinian threads.

To illustrate how tenuous this kind of unity is, compare it with the unity of all things that are a refinement of something (see my discussion of "refinement" at the beginning of this section). Refined things can be vastly different from one another. Take a particularly creamy sauce and a high-resolution spectrograph that may both be elements of the set of refined things. To claim that these two things are similar to each other is, under normal circumstances, ridiculous. Although both can be considered refined things, their commonality is so abstract that we only accept it reluctantly. On a more concrete level, the notions of refinement operative in the two cases appear to have nothing to do with each other. It remains true, however, that all more concrete notions of refinement are connected by family resemblance, generating a tenuous sort of unity.

2.3 THE STRUCTURE OF THE ARGUMENT

By now it should be obvious that we cannot argue for our main thesis, the higher degree of systematicity of scientific knowledge, in one sweep. Rather, we must somehow subdivide the totality of the sciences (in the wide sense) into smaller units and show that our thesis is true for any of these subunits in any of the nine dimensions (excluding predictions where they do not apply). This task, however, is confronted with formidable difficulties.

First, it is not at all clear what those subunits within the totality of all the sciences should be. On what level should we try to identify them? In order to make our argumentative task feasible, these subunits should be such that for each of them, the same type of systematicity is operative. More precisely, in each subunit, nine different concepts of systematicity should be homogeneously operative, one in each of our nine main dimensions (descriptions, explanations, etc., excluding predictions where they do not apply). But what are those subunits? Individual sciences, disciplines, subdisciplines, scientific fields, research areas, specialties, clusters of specialties, or scientific subjects? The multitude of these expressions that can all be found in the pertinent literature, and the lack of clear contrast among most of them, indicate already that the choice of the appropriate, allegedly homogeneous subunits is extremely tricky. For example, is it appropriate to ask for a justification for the thesis of the higher degree of systematicity of science in comparison to other kinds of knowledge for physics as a whole, or for solid-state physics, or for solid-state physics of amorphous materials, or for solid-state physics of some subclass of amorphous materials?

Second, whatever the choice of the appropriate subunits of the totality of the sciences will be, their sheer number will be very difficult or even impossible to handle. Unfortunately, there is no established academic discipline nowadays that informs us about the landscape of all the sciences. However, there is some consensus that in the coarsest division of the totality of all the sciences, there are four large groups: (1) the formal sciences, (2) the (natural) sciences, (3) the social sciences, and (4) the arts and humanities. This basic division is mirrored in the organization of many universities into schools, or of funding agencies into departments, or of signatures that are used by libraries collecting scientific works. The Thomson Reuters Company that, among other things, composes indices of the citations of scientific works in selected journals (presumably the most important ones) uses the same top-level classification. At the next level of classification are the individual sciences (in the wide sense), also called disciplines or scientific subjects. Following Thomson Reuters classification of scientific subjects in the *Science Citation Index*, the group of all of the sciences in the narrow sense, i.e., the natural sciences, comprises 170 categories, starting from acoustics and ending with zoology. The *Social Science Citation Index* comprises fifty-four categories, ranging from anthropology to women's studies. What I called above the formal sciences is collected in the *CompuMath Citation Index 2002* and covers fifteen categories, from automation and control systems to statistics and probability. Finally, the *Arts & Humanities Citation Index* comprises twenty-seven categories, from archeology to theater. On this level of organization, we therefore have 266 scientific subjects altogether. Of course, this classification is far from unique: other sources end up with different numbers.

The upshot of this excursion into the art of science classification is this. Let us assume, perhaps contra-factually, that the appropriate, i.e., sufficiently homogeneous, subunits of science for which we have to verify our main thesis are scientific disciplines. We then have to run through roughly a couple hundred items. For each of them, we have to demonstrate that scientific knowledge is more systematic than corresponding knowledge from other sources. As "systematicity" in the abstract is the umbrella term that covers nine separate dimensions and each dimension has to be argued for separately, we therefore have to argue for a number of theses that is well above one thousand all together. It may even be the case that we conclude that in order to argue for the (higher) systematicity of some discipline in comparison to other kinds of knowledge, we must subdivide it into even smaller units. Take, for instance, all issues subsumed under the rubric of law by the *Social Science Citation Index*:

> Law covers resources from both general and specialized areas of national and international law, including comparative law, criminology, business law, banking, corporate and tax law, constitutional law, civil rights, copyright and intellectual property law, environmental law, family law, medicine and the law as well as psychology and the law.

Even if we could demonstrate that scientific thinking exhibits a higher degree of systematicity than corresponding everyday thinking in some of these subdomains of law, it cannot automatically be inferred for other subdomains of law as well. In other words, a seamless argument in favor of our thesis concerning law would require its subdivision into smaller units. In 1987, one source in which these smaller units are termed "fields" counts 8,530 such fields (sociology alone, for instance, consists of 58 fields!). If we multiply it by nine for the nine distinct dimensions we are considering, we end up requiring argumentation for several ten thousands of separate theses. Such an enormous number of arguments is impossible to provide. We are therefore forced to settle for less.

A third difficulty is created by disciplines that lack a more or less general consensus about their basics. Many of the social sciences and humanities clearly possess such a character as they typically consist of several competing schools (of course, there is a comparatively large grey area between "agreement" or "disagreement" regarding foundations). Whenever I want to argue for the more systematic character of some field without a scientific consensus—whatever the definition—the first problem is that the choice of examples becomes tricky. What looks to one school to be a good example of scientific practice in this field may appear very dubious to others. Even if it can be shown that the respective school-bound practice is more systematic than nonscientific kinds of knowledge, opponents of that school will dismiss the argument as irrelevant. It is therefore not really clear what the relevant evidence is that could support my thesis. Furthermore, the overall state of a field without consensus may appear quite chaotic because the intellectual interaction between the various schools may be entirely unsystematic and chaotic. In that instance, the claim of my thesis is not immediately evident. Does it contend that each school's contribution is more systematic than, say, everyday knowledge about the same domain? Or does it contend that the entire body of scientific knowledge that is produced by the various schools, although internally contradictory to a high degree, is more systematic than everyday knowledge? The latter contention, however, appears not be true in general. There are many different modes how different schools in science can coexist—some more structured than others. So I should constrain my thesis about the higher degree of systematicity in controversial research areas to each individual school and not apply it to the area as a whole. Of course, that makes our argumentative task even more difficult because in such areas, each individual school needs to be discussed separately.

A fourth difficulty consists in the fact that science is a deeply historical enterprise such that any division of science into smaller units—be it disciplines, scientific fields, research specialties, or whatever subunits of science—has historically changed and will continue to do so. Thus, a thorough argument for the higher degree of systematicity of science, applying to all times during its existence, would require an appropriate breakup of the totality of all the sciences at all of those times and not only for the present.

Hence, as we cannot systematically (!) treat the totality of all of the sciences, our argument is bound to be severely limited. It is tempting to suggest that we should only test our thesis in a number of exemplary cases. Even this more moderately sounding claim is difficult to meet, because we will be unable to demonstrate that the chosen "exemplary cases" are really exemplary. Strictly speaking, this would be equivalent to our original task of a systematic justificatory procedure because the exemplary character of some case may only be argued for with reference to all other cases. Thus, we end up in the unfortunate but unavoidable situation that I will not be able to argue for my main thesis in a conclusive way. All I can hope to attain is some plausibility for the thesis, based on a somewhat uncontrolled set of examples (and the additional hope that counter examples will not turn up all too soon). However, one advantageous strategy to conquer the abundance of disciplines and subdisciplines is to assign them a group in which one aspect of systematicity is the same for all. If it is possible to show that the trait in question indeed promotes systematicity, then this result applies to all members of the group at the same time. I shall use this strategy whenever possible (see chapter 3).

Let me note at this point that all these complications in the argument for our thesis are a consequence of its primarily descriptive character (see section 2.1). If our thesis were primarily normative, our argumentative business would be much simpler, at least as far as the extension of the task is concerned (remember—if you can—the so-called good old days of philosophy of science when some normative deliberation plus some sketchy descriptive remarks about science on the level of high school physics were sufficient to establish or destroy some normative claim). We had to show that the higher degree of systematicity is a reasonable thing to posit, and we might make it additionally plausible by showing that it is indeed met in some real cases. But dealing with the entirety of science would be unnecessary.

However, I suggest that our descriptive analysis of science will have normative implications, even if "implication" should not be understood in the sense of logical or conceptual consequence in this context. If a descriptive analysis of science reveals something about the factual cognitive goals of science and some tendencies of how these goals are pursued, we may better understand why some candidate goals are more desirable than others. If, for instance, our analysis reveals in which sense and why the sciences have in fact a strong tendency in the direction of increased systematicity, we may conclude that the benefits of increased systematicity are desirable. Normative implications can therefore be suggested by a descriptive analysis, although, of course, they are not strictly implied. I shall come back to this topic in greater detail in section 5.3.

In concise terms, the sobering result of this section is this. In contrast to what I would like to do, namely, to present a fully systematic argument for my thesis, my argument will have to be very sketchy indeed. I cannot realize what is often an attractive goal in philosophy: that a book's content is completely consistent with its own way of proceeding. However, also in the sciences, one cannot always realize the desirable degree of systematicity.

The whole of science is nothing more than a refinement of everyday thinking.
—ALBERT EINSTEIN, 1936

3

The Systematicity of Science Unfolded

IN A SENSE, the subject of the present chapter is a refinement of Einstein's dictum as quoted above. At first, Einstein's dictum that the whole of science is nothing but a refinement of everyday thinking appears to be quite plausible (unless one believes in an epistemological break between everyday thinking and science). However, after a moment's reflection, one is left feeling bewildered about the precise meaning of "refinement" in this context. In what sense exactly is science "finer" or "more refined" in comparison to everyday thinking? As the literal, physical meanings of these terms have to be dismissed immediately, the intended sense must be metaphorical. A circumscription of "refined" by, say, "more sophisticated" is not really helpful, because it is then left open in what respect this higher degree of sophistication (whatever *that* is precisely) has to be sought. I propose that Einstein had a vision in mind that is similar to the one developed here. If one reads Einstein's "refinement" as "having a higher degree of systematicity," one ends up with roughly my thesis in a highly compressed form. That Einstein has indeed something like this in mind is revealed in another of his articles that was published in 1944. In this article, Einstein contrasts "the thinking in daily life" with "the more consciously and *systematically* constructed thinking of the sciences," and that is precisely the contrast I am using in this book as well.

As explained in the preceding chapter, I shall develop the thesis of the higher degree of systematicity of scientific knowledge in comparison to other kinds of knowledge by means of nine different dimensions. These dimensions are

- descriptions,
- explanations,

- predictions,
- the defense of knowledge claims,
- critical discourse,
- epistemic connectedness,
- an ideal of completeness,
- knowledge generation, and
- the representation of knowledge.

Before discussing these in detail one after the other, four remarks about the whole set of dimensions are necessary.

First, these nine aspects of science may look fairly unproblematic regarding their content and the contrast between each other, but they are not. For instance, in some areas of research like botany, the contrast between descriptions and explanations appears to be fairly sharp, and, correspondingly, the meaning of these concepts in these areas appears to be fairly clear. In others areas like history, however, the contrast is by far less clear, because particular descriptions (of processes) may constitute explanations as well. Whether this blurring of borders impairs our analysis will be discussed in the respective sections, with reference to the disciplines involved.

Second, the nine dimensions are not absolutely independent from one another. For instance, the ideal of completeness is strongly connected with the generation of new knowledge. In some areas, the ideal of completeness and the systematicity of knowledge representation both overlap with the systematicity of descriptions. In the course of this chapter, we will encounter more such connections. Therefore, the nine dimensions have some overlap and interrelationships and are not conceptually absolutely independent. This does not impair their use as analytical categories. All of the dimensions highlight a special aspect of systematicity, and as this is our main purpose, they are useful and appropriate.

Third, the category of predictions may seem problematic since they do not constitute a part of the program of all sciences, as the humanities demonstrate. The same is true for the historical natural sciences like cosmology, paleontology, paleoclimatology, or paleoceanography. However, this is not really a problem for the analysis presented here. It simply should be noted that the dimension of predictions only applies to a limited range of disciplines. Only for those disciplines is a difference in the degree of systematicity to other kinds of knowledge claimed.

Fourth, I have no systematic theoretical argument for choosing precisely these nine dimensions and whether this list is complete. Such a theoretical argument would probably consist of some principle that could be developed such that it yields just these nine dimensions. Lacking such a principle, my procedure to identify these dimensions is, broadly speaking, empirical. As a matter of historical fact, I started with three dimensions: the defense of knowledge claims, the structure of scientific knowledge, and an ideal of completeness. Later I realized that science was more systematic than other knowledge-seeking enterprises in more dimensions than just these three. Over the years, I added altogether six more, resulting in the nine dimensions mentioned above. This procedure may appear to be philosophically unsatisfactory, and perhaps it is. However, this is quite common in the natural sciences, for example. When studying new sorts of systems, there is no a priori way to find out what the relevant state variables are, or in another terminology, what the relevant dimensions are in order to describe and explain the systems in question. It is then a matter of trial and error, of "playing around" as scientists often put it, with different possibilities, and finally settling for some set of state variables without a systematic theoretical argument. It is the success of a set of state variables with respect to a comprehensive description and explanation of a system that finally counts. Lacking a deeper philosophical principle that could do the desired job, namely to single out the relevant dimensions in which science and other knowledge seeking enterprises differ, I do not see any alternative to this empirical procedure.

What I shall do in the following is this. In each of the nine dimensions, I will distinguish several means that are applied in different groups of sciences in order to increase their degree of systematicity. Where it is not entirely obvious, I will then shortly discuss the specific meaning of the pertinent notions of systematicity. Finally, I will show that in each of these cases, the claim of my general thesis is indeed fulfilled, i.e., that the degree of systematicity of scientific knowledge in the respective fields is higher than in the corresponding forms of nonscientific knowledge, especially of everyday knowledge. However, in the concrete execution of this sort of argument, I will often proceed a little more loosely than would be required by a very literal reading of the task, namely, comparing *corresponding* forms of knowledge. As I explained in the fourth item of section 2.1.2, that would mean to compare everyday knowledge about a particular subject matter with the corresponding knowledge about exactly the same subject matter. I will often use examples from science and then examples from everyday knowledge that do not precisely match in content but that are indeed less systematic. It will be obvious, however, that both examples will be representative for a whole domain and that in this way, I will have made plausible that the particular domain—scientific knowledge—is indeed more systematic than everyday knowledge.

3.1 DESCRIPTIONS

3.1.1 *Some Preliminaries*

Descriptions appear to be in some sense the basis of all fields of research. Whatever further aims of science there are—explanations, understanding, predictions, causal analysis, modeling, control, design, and so on—they have to refer not only to the phenomena themselves but also to descriptions of those phenomena that are the subject matter of the respective field. The necessity of descriptions originates from the need to communicate to other researchers and to preserve knowledge for future use. Of course, a researcher in a group may refer to a particular phenomenon present to all of his colleagues by pointing at it. This sort of reference to a phenomenon is possible without describing it in detail. If, however, the peers are physically absent, a description (or some other representation) of the phenomenon under investigation is required in order to refer to it.

Assuming that descriptions are the starting point for other activities of science is not entirely unproblematic. This assumption implies that there is a more or less strict division between descriptions and these other activities of science like explanations, understanding, and such, and that descriptions are suppositions for these activities. We shall later see that descriptions may indeed have some explanatory powers, thereby blurring an absolute distinction between these two scientific activities (section 3.2.7). However, what I will have to say about descriptions is not invalidated. They can still serve as our starting point.

Our main thesis contends that scientific descriptions of some set of phenomena are more systematic than everyday descriptions of the same phenomena. In this statement, the meaning of "systematic" must be a specification of the general meaning of the abstract term "systematic," possibly different from other specifications in other contexts like explanation, prediction, and so forth. According to section 2.2, it should also vary as we consider descriptions in different fields of research, and these variations should be interconnected by no more than family resemblances.

Before I begin to argue for this thesis, a few general remarks about descriptions are in order. This seems to be all the more important because in philosophy of science, descriptions are a much neglected topic. They seem to be taken for granted because they seem to be unproblematic and thus philosophically uninteresting (apart from the question of whether they are in principle theory-dependent or not). By contrast, much more philosophical ink has been spilled on, for instance, scientific explanation and prediction. These remarks will be useful when we discuss the differences between scientific and nonscientific descriptions.

First, *all* descriptions are *abstract* in the sense that they do not cover each and every aspect of the phenomenon described. For instance, the description of what happened to me yesterday when I went to the cinema (an everyday account) and a historian's description of events that finally led to the outbreak of WWI are both abstract. Each of them can be extended by innumerable details and in this way made more concrete. Therefore, there is no simple contrast between concrete and abstract descriptions, either in everyday use or in science. Rather, there is a difference in the degree of the abstractions. But not only may the degree of abstractions vary, but its focus also may vary. For example, chemists may include in the description of a gas its color and its smell, whereas physicists will omit these features. By contrast, physicists may include some information about the proportion of isotopes in the given gas sample, which feature may be entirely irrelevant for chemists. As we shall see below (third remark), a lowering of *or* an increase in abstraction may lead to an increase in systematicity, depending on the respective scientific field.

Second, it is often maintained that there is a (ontological) difference between reproducible events or processes as pertaining to the natural sciences and singular events or processes pertaining to historical disciplines (for simplicity, I shall speak here only of events, but I mean events *and* processes). A closer look reveals that the difference does not refer to the events themselves, i.e., the subject matter of the descriptions, but that only the events' modes of the descriptions differ. The reason is that all events are, by their very nature, never reproducible but always (historically) singular. The boiling of some water in a pot on my stove at some particular time is an event as historically singular as the drafting of the Declaration of Independence by Thomas Jefferson between June 11 and June 28, 1776. It is the typical denomination and the typical description of such events that falsely suggest that the events themselves are either reproducible or historically unique by their nature. A very detailed description of the boiling water, however, where and when exactly which steam bubbles were produced and how exactly they moved upward, changing their size and form and more, readily demonstrates that exactly the same process will probably never take place again. Add a definite description of that particular sample of water, i.e., a description that identifies it as a particular sample of water, the precise time, and the geographical coordinates of the event, and it becomes clear that this particular boiling of water perfectly constitutes a historical event, happening once and only once. By the same token, a more abstract description of the drafting of the Declaration of Independence, omitting proper names, yields an event that seems perfectly reproducible: "Someone writes something politically very important in a time span of less than three weeks." In this particular case, even the addition of an expression like "in the United States of America" would not demolish the seemingly inherent reproducibility of the event described.

The upshot is that the so-called reproducibility of some events, but not of others, is not at all an intrinsic property of these events. It rather constitutes an artifact (though sometimes an extremely useful one) created by a particular mode of description of these events. It holds in general that all events are essentially historical and therefore not reproducible in any strict sense.

What properties of a description make an event appear to be reproducible? Clearly, it is the abstraction of all potential descriptive elements that may lead to the identification of that event as a singular (historical) event. For instance, when physicists describe the state of some sample of gas under certain conditions, they will use only a few properties of the gas. Typically, its molecular weight, its amount, its temperature, its volume, and its pressure will be chosen (for gases in equilibrium). Even a quantitatively very precise specification of these parameters fits many different individual samples of the gas in many space-time locations. Under this description, the state of the gas thus appears to be reproducible.

Third, there is thus a contrast between descriptions intended for one single event only and those that are intended for a whole class of events. The first kind may be called "historical descriptions," the second one "generalized descriptions." As the first remark in this section implies, both kinds of descriptions are abstract in that they do not describe the entire concreteness of their subjects. However, the kind and degree of abstraction differs in these two cases. In the first case, the abstraction involved does not destroy the reference to a unique event or phenomenon. In the second case, the abstraction intentionally destroys all reference to a uniquely identifiable event or phenomenon, generalizing the description and making it applicable to a whole class of events or phenomena. In spite of them both being abstract, these two kinds of descriptions are associated with different concepts of systematicity. In other words, to say that a description has become more systematic means different things in the two cases. Roughly speaking, a historical description becomes more systematic by adding more details in an orderly way; the description then becomes more concrete. Similarly, a generalized description may become more systematic by an increase of its degree of comprehensiveness; the description then becomes more abstract. Therefore, an increase in or a decrease of abstraction may lead to an *increase* of systematicity of a description depending on the kind of description, which in turn depends on the pertinent field of research.

After these preliminaries, let us now discuss how the sciences (in the broad sense) contrive descriptions that are more systematic than descriptions belonging to other kinds of knowledge, in particular everyday knowledge. Several techniques exist to increase the systematicity of descriptions that are used across groups of disciplines. Typically, these groups are formed by neighboring disciplines, but sometimes even very distant disciplines belong to the same group. There is nothing, however, that is shared by all these techniques to increase the systematicity of descriptions: they are only regionally applicable and not universally. The techniques that I will discuss in turn are axiomatization (subsection 3.1.2); classification, taxonomy, and nomenclature (subsection 3.1.3); periodization (subsection 3.1.4); quantification (subsection 3.1.5); empirical generalizations (subsection 3.1.6); and historical descriptions (subsection 3.1.7). There are additional means to augment the systematicity of descriptions, namely various graphic forms of representation like tables, diagrams, and maps. I will postpone their discussion to section 3.9, where I will deal with the representation of knowledge in more general terms.

3.1.2 Axiomatization

Axiomatization of descriptions is most typically employed in the formal sciences of which mathematics and the formal parts of logic are paradigm cases. I take logic as an example. If a layperson is asked to describe logical inference, she will probably at best come up with examples like "All men are mortal, Socrates is a man, and therefore Socrates is mortal" and explanations like "yes, if you accept the premises of a valid logical inference, you will have to accept the conclusion, too." The descriptions of logical inference will be unsystematic; a few scattered examples, perhaps an example of an inference that is logically invalid, and some rather vague explanations. Compare this to the scientific treatment of logic. Already, Aristotle, the founder of logic, tried to compose a complete list of logically valid inference forms. The same holds for the modern treatment of logic. In one widely used approach, called proof theory, the goal is to write down a few axioms (or, to be more precise, axiom schemata) and a few rules on how to derive theorems from the axioms, such that a complete system of all logical truths results (which contains all the logically valid inferences). This way to describe logical truths is extremely systematic: it aims at a few logically independent principles that generate all logical truths. To those familiar with logic, a completely transparent body of knowledge emerges, unsurpassed by its degree of systematicity.

In the case of axiomatic descriptions, the abstract notion of "systematicity" is made more concrete in a very specific way. Instead of having some examples of the respective field of interest without apparent connection, and perhaps some unordered theoretical ideas about it, an axiomatic description of that field provides an extremely high degree of order and perspicuity. The axioms must be logically independent of one another, they must be simple, they must be as complete as possible regarding their logical consequences, and they must be consistent with each other. The posit of completeness of an axiomatic system regarding its logical consequences shall guarantee that all the theorems valid in the respective area can indeed be derived from the axioms. This posit already embodies another aspect of the systematicity of scientific knowledge, namely, its ideal of completeness, to which I shall return in section 3.5. Presenting a body of knowledge in an axiomatic manner is indeed extremely systematic, where "systematic" in this context just means "in the form of an axiomatic system." It is obvious that this is a powerful concretization of the element of order that is contained in the abstract notion of systematicity.

3.1.3 *Classification, Taxonomy, and Nomenclature*

Classification organizes an assortment of individual items. The order is established by arranging similar individual items into the same class and dissimilar ones into different classes. In addition to this establishment of order, for some research questions, the quantity of items to be considered is thereby much reduced because consideration of one representative item per class may suffice. The classification procedure can be iterated by putting similar classes into the same higher order class and dissimilar ones into different higher order classes. A hierarchy of classes, or *taxonomy*, results. Typically, a system of denominations for the classes at the different levels has to be devised that is called a *nomenclature*.

The principal idea of classification and taxonomy sounds simple, but in scientific practice, a number of severe difficulties may arise. To name but a few: Defining classes can be difficult as their boundaries may be fuzzy or controversial. Identifying the items' traits on which the classification shall be based can be controversial; after all any classificatory scheme depends on a choice of features deemed as relevant. Also, the principal nature of the classification can be controversial. Does the classification simply represent what can be found, or is the classification a human invention that imposes order on something that, by itself, does not exhibit that order? Fortunately, we do not have to deal with these difficult questions at this point since they do not directly influence our issue, i.e., the asserted increase of systematicity of descriptions by virtue of using classifications.

Items considered for classification or taxonomy in the sciences are extremely diverse. They comprise physical things like plants, animals, viruses, genes, elementary particles, chemical elements, chemical compounds, enzymes, and minerals, or physical conditions like diseases or nursing diagnoses, or abstract entities like mathematical objects, languages, literary genres, economical or political systems, or structures of societies. The Linnaean classificatory scheme for plants and animals and its nomenclature is an illustrative case in point. In its earliest full form for plants, published by Linnaeus in 1751, the core of his scheme is the division of the diversity of life forms into species at the lowest level and the collection of species into genera. This division is also the constitutive principle for the denomination of species. Linnaean naming for a species consists of an analogue of a family name and a first name, with the family name in the first place; it is a binominal nomenclature. The name of the genus corresponds to the family name. Any particular species within this genus is then identified by some specific trait of that species and designated by a name, an analogue of a first name. For example, *Bos taurus*, commonly known as the cow, is the species that belongs to genus *Bos*, or cattle, and is distinguished within this genus from, e.g., *Bos primigenius* (aurochs) or from *Bos grunniens* (domestic yak). Above the genus level, further taxonomic categories exist. Today, the number of levels in the taxonomic hierarchy is simply stunning. In one of its most elaborate forms, above the genus *Bos*, there are no less than twenty-six hierarchical levels, beginning with what is called a subfamily, the *Bovinae*, and ending at the top level with "cellular organisms."

In some cases, for scientific purposes, a large number of various items have to be denoted in a unique manner without being classified. For instance, the Geographic Names Information System (GNIS) contains information about more than two million physical and cultural geographic features in the United States and its territories (to be more precise, on March 10, 2010, it was 2,126,537 features). Similarly, planetary nomenclature is used to uniquely identify features on the surface of planets or satellites so that the feature can be easily located, described, and discussed.

Sophisticated forms of classification and an associated nomenclature are not restricted to the natural sciences, however. In the domain of the humanities, for instance, linguists classify human languages. In 2009, the authoritative system counted 6,909 living languages that are, at the top level, classified into 116 language families. They comprise between 1,510 (Niger-Congo) and 1 (e.g., Basque) family members. Below the family level, there are up to five additional levels. For instance, English belongs to the family of the Indo-European languages. The relevant level below are Germanic languages, and below that the languages of the West, and then the group of English languages of which English is one of two members.

It is obvious that an ordering of items in a multilevel hierarchy is an extremely systematic description of a given diversity. All the items subsumed have their place relative to the others, and the traits seen by the respective discipline as pertinent serve as the basis for the classification. It is no accident that the first book that Linnaeus published in 1735 on the subject of classification of animals, plants, and minerals bears the title *Systema naturae*, or *The System of Nature*. Everyday descriptions of some diversity are, in comparison, obviously much less systematic than the classificatory descriptions used in many branches of science and in the conscious systematic denotation procedures used where many individual items have to be uniquely identifiable.

3.1.4 Periodization

Periodization is the temporal counterpart to classification and sometimes even to taxonomy, involving several hierarchical levels of ordering. It is relevant both for sciences that deal with recurrent developmental processes like developmental psychology and for the historical sciences that typically conceptualize their subject matter as something unique. Contrary to a widespread stereotype, the historical sciences do not in their entirety belong to the humanities because there are also historical natural sciences. For instance, paleontology, the discipline that deals with the history of life on Earth, traces the history of a class of natural phenomena. Of course, political history or art history that are paradigm examples of historical disciplines do belong to the humanities.

Most, if not all, historical disciplines structure the historical development of their subject matter by introducing different "phases," or "periods," or "epochs." These phases are meant to subdivide the continuity of time into succeeding time intervals that comprise events and processes that are sufficiently similar, somehow related to one another, and sufficiently dissimilar to the respective preceding and later time intervals. A familiar example is the division of world history into antiquity, the middle ages, and modern times. This periodization, although not entirely uncontroversial, structures many history departments around the world. A more sophisticated example from political history is the division of ancient Egypt history into different dynasties and intermediate periods, totaling some thirty-three phases and three hierarchical levels. An extremely detailed example from the historical natural sciences is the periodization of Earth's history, the so-called geologic time scale. It features six hierarchical levels: supereons, eons, eras, periods, epochs, and ages. The Precambrian supereon, for instance, covers some four billion years, whereas the smallest units, the ages, are of the order of millions of years (small by geological measures but obviously very large by human measures). The order imposed upon the flow of history by periodization is analogous to the order imposed upon entities by classification and taxonomy (see subsection 3.1.3). In both cases, a variety of entities is structured by a partition or a hierarchy of partitions defining classes on every level.

Also, some of the generalizing empirical disciplines use periodizations. Disciplines that study regularities in the development of some individual entities typically also try to work out phase models. For instance, in developmental psychology, a variety of different life span theories have been proposed in order to schematically describe individual human development throughout life. These theories define certain stages of a human life that are assumed to be well defined and follow upon each other according to some pattern.

It is rare that periodizations are uncontroversial. For instance, even the well-entrenched periodization of world history into antiquity, the middle ages, and modern times has been called into question. Some historians think that instead of two major transitions, one should have only one, located roughly at the so-called saddle time around the turn to the nineteenth century. Controversies about periodization typically result from two problems. First, there are always continuities that bridge supposed discontinuities between any two successive phases, thus threatening their integrity. Second, it is often not clear on which specific traits a periodization should be based. Clearly, different periodizations may result from different traits.

These controversies, however, do not have to concern us here. My aim is to show that periodizations in the sciences add an element of systematicity to descriptions. And they do that indeed as they structure a flow of innumerable events by imposing an order upon them. In comparison to our everyday practice where we also use periodizations, for example in our life stories, the periodizations in the sciences are much more reflective. Typically, the basis of a proposed periodization is considered with some care because awareness of the at least potentially controversial character of any periodization exists. By contrast, periodizations in everyday life, like talk of different periods of one's life, are mostly done without much reflection. One may distinguish different periods of one's life by differences in residence, or partners, or jobs, or the like, and usually not much thought is spent on the appropriateness of these partitions. In comparison, scientific periodizations are more circumspect, more reflective, and often much more detailed, especially with respect to the number of hierarchical levels. This is the sense in which they are more systematic.

3.1.5 *Quantification*

It is obvious that for the last four centuries or so, there has been an increasing tendency toward quantification in many areas of scientific research. Quantification means the transition from the qualitative understanding and use of a concept to its quantitative understanding and use. For instance, the temperature of a bottle of white wine may be expressed qualitatively by stating "it is fairly warm," or quantitatively, by stating "it is 16°C." However, it should be noted that in both cases, a *quality*, or property, of the bottle and its content is expressed, once in qualitative and once in quantitative terms. It is a potentially misleading practice in the natural sciences to call quantitative *expressions* of qualities "quantities." This designation invites the misunderstanding that quantitative sciences deal with something called quantities; whereas qualitative disciplines deal with something completely different, namely, qualities. In this way, an ontological difference seems to be implied with respect to the subject matter of these groups of disciplines. Nevertheless, it is a fact that all disciplines deal with qualities of their objects (and their relations); they only use different modes of expression for their concepts. It is an interesting question under which conditions the transition from the qualitative use of a concept to its quantitative use is possible and makes sense to the respective discipline. It is by no means the case that every kind of quantification in some field is scientifically fruitful. However, I am dealing with a different question in this section. I am asking whether quantification, where appropriate and successful, leads, in a specific sense, to an increase of the systematicity of descriptions.

Quantification in science began in ancient times, especially in astronomy and harmonics. The most prominent ancient examples concern the quantitative description of the movement of the heavenly bodies. Physics was not a quantitative science in (Greek) antiquity. The quantitative treatment of local motion started in the fourteenth century, paving the way for the fuller mathematical treatment during the so-called Scientific Revolution. Chemistry had only begun to become a quantitative science in the late eighteenth century. Many more disciplines followed this path. A few decades ago, an additional boost toward quantification was triggered by the availability of computing power through electronic computers. In most areas, easy accessibility to computers did not foster quantitative descriptions directly. However, the impact of computers on this development was significant as large data sets could now be dealt with efficiently. In particular, if descriptions are quantitative, computers can be employed for various statistical forms of data analysis.

The tendency toward quantification in the sciences serves various purposes. At this point, however, the question is only whether quantitative descriptions are more systematic than nonquantitative descriptions where they are appropriate, and what the pertinent sense of systematicity is in this context. Let us first observe that quantitative descriptions are, where applicable, more precise than qualitative descriptions. For example, the quantitative description, "The temperature on this day at noon was 30.7°C," is more precise than "On this day at noon, it was really very hot." Due to their greater precision, quantitative descriptions also have many other epistemic advantages over qualitative ones regarding reproduction of data, intersubjective tests, and the like. Second, mostly as a consequence of their greater precision, quantitative descriptions allow for many more different and easily discernible descriptions than qualitative descriptions, typically not only in principle but also in practice. To stick with the example of temperature, using a household thermometer, one can easily distinguish and describe some five hundred different temperature states between −15°C and 35°C. Using our qualitative everyday language, we have perhaps two or three dozen descriptions at our disposal, ranging, for example, from "extremely freezing cold" to "extremely burning hot." Third, not only do we have many more expressions to describe temperature in quantitative terms, but also, as a consequence of the higher precision of these descriptions, they are by themselves uniquely ordered.

It is obvious that quantitative descriptions, where they are possible and appropriate, are more systematic than qualitative descriptions. If one considers the set of qualitative descriptions of outside temperature from, say, "extremely hot" to "extremely cold," and compares it with the corresponding quantitative description from −15°C and 35°C, it is evident that the latter set is more systematic. Not only are individual items better defined, but so is their ordering and their differences. They form much more of a system than the set of the somewhat vague qualitative descriptions. It is also clear that an increase in quantitative accuracy of determining temperature leads to a richer system of quantitative temperature values. In that sense, an increase of accuracy implies an increase of systematicity (although with respect to order and determinateness, the smaller system with fewer temperature values is as systematic as the bigger one).

Quantification leads to an increase of systematicity due to the system character of the set of all possible individual descriptions. In addition, quantitative descriptions, appropriately combined with a fitting classification, often admit empirical generalizations of a highly systematic character that are relevant in a large group of sciences to which we now turn.

3.1.6 Empirical Generalizations

In this subsection, I am referring to a group of sciences that can be called *generalizing empirical sciences*. Members of this group aim at, among other things, a particular mode of description of their subject matter. Whereas historical sciences aim at descriptions that make individual phenomena identifiable as such, generalizing empirical sciences do the opposite: they abstract away all features of phenomena that make them identifiable as individual events. The descriptions of these sciences are *generalized descriptions*, i.e., descriptions that reach beyond singular cases (compare section 3.1.1). These descriptions often express regularities or even laws, holding for whole classes of phenomena (or events, systems, processes, and the like). In the sciences themselves, such generalized descriptions are often called "phenomenological laws" or "empirical regularities." The terms "phenomenological" and "empirical" are here opposites of "theoretical": we deal with descriptions of things that are more or less directly observable and not the subject of theoretical speculation. "Laws" express a connection between events that holds (presumably) with necessity in contrast to mere "regularities" that may be accidental in character. I shall use the term "empirical generalizations" for this sort of descriptions. Of course, the generalizing empirical disciplines can and do deal with specific individuals, like a particular star or a particular sample of some protein or a particular middle-aged married engineer living in Pittsburgh. They typically treat these concrete objects as representatives of the whole class of these objects, however, and they aim at descriptions (and theories, explanations, and so on) of these objects that hold for the whole class. Typical generalizing empirical sciences in this sense include some parts of observational disciplines like astronomy, e.g., where generalized descriptions of stellar evolution are sought, or experimental disciplines like solid state physics or protein chemistry. Some of the social sciences also belong to this class, for instance the political sciences and, more precisely, the field of international relations, where the empirical regularity seems to hold that democracies are never at war against each other. I am not sure whether any of the humanities seek empirical regularities, let alone empirical laws.

Empirical generalizations describe regularities that hold for the objects in question. Often, these regularities are quantitative, and they may be deterministic or statistical. The regularities connect relevant observable features of the phenomena in question, often called state variables, by describing quantitative functional dependencies among them. A simple case in point is Boyle's gas law that, in its simplest form, connects the pressure p and the volume V of some sample of gas by the formula $p\,V = const.$, given that the temperature remains constant. Whether or not and under which conditions such regularities are called laws is irrelevant in our context, because we are only concerned with their role in increasing systematicity of scientific knowledge.

Three partially interconnected conditions must be fulfilled for such empirical generalizations to hold. First, to achieve the sort of order of the phenomena that is captured in empirical generalizations, an appropriate classification of the phenomena is presupposed. It is unlikely that classes of intrinsically heterogeneous phenomena will exhibit empirical regularities of some generality. Again, the identification of phenomena of the same kind, potentially exhibiting regular behavior, is a highly nontrivial task if the domain in question is not well known. For instance, in the chemistry before the early nineteenth century the nowadays well-established distinction between (physical) solutions and chemical compounds of two substances was not a part of the generally accepted body of scientific belief. As a consequence, there were no *general* restrictions on the proportions in which two substances could be combined (forming, in today's language, a solution *or* a compound). From today's vantage point, this is quite obvious as the class of compounds *and* solutions is too inhomogeneous to allow for general regularities. Only after the separation of solutions from chemical compounds, could an extremely powerful and consequential empirical generalization about the proportions of the latter be found: the law of constant proportions (that formed the basis for the determination of relative atomic weights and much more).

Second, for the phenomena in question, a set of appropriate state variables must be identified. This is not a trivial task at all in cases where the phenomena in question are hitherto poorly understood. For instance, in the 1730s, after more than one hundred years of research throughout Europe, a multitude of different and puzzling electrical effects were known. However, many of the effects were small, difficult to reproduce, and, above all, without apparent order. It proved impossible to devise empirical generalizations about electrical phenomena in which some regularity of their behavior was revealed. After extensive explorative experimentation involving many different materials and many different configurations of bodies, the French scientist Charles Dufay found the clue for the formulation of empirical generalizations. He abandoned the prevailing concept of one single electricity and replaced it by the idea of two different electricities that correspond to different materials. Armed with this concept, he was able to subsume hundreds of experiments under general empirical regularities. It is important to note that Dufay's research was not directed at the theory of electricity, i.e., at the hidden causes of electric effects or the nature of electricity. His goal was rather to establish order among the electrical effects by exhibiting their regularities. The clue to some order of the phenomena is therefore the identification of appropriate state variables. In Dufay's case, it was the splitting of one supposed state variable into two different ones.

Finally, in the dominant case of quantitative empirical generalizations, the relevant state variables must be in quantitative form.

These three conditions demonstrate that and how increased systematicity of descriptions is implied by the use of empirical generalizations. The necessary classification of phenomena into those that belong to the domain of an intended empirical generalization and those that do not contributes an element of order and hence of systematicity, as we saw in subsection 3.1.3. The increased level of generality of an empirical generalization, made possible by an appropriate choice of state variables, systematizes the description as opposed to separate descriptions of individual cases. The quantitative form of empirical generalizations further increases systematicity as successful quantification does in general (subsection 3.1.4). It is obvious that sciences that seek empirical generalizations, possibly of quantitative form, thereby increase the degree of systematicity in comparison to other kinds of knowledge. Whenever the empirical generalizations of these kinds of knowledge on some subject matter prove to be valid, they can be taken over by science. They may become quantified in science, adding to other empirical generalizations established by further research. It may be noted that the emergence of a keen interest in quantitative empirical generalizations is one of the characteristics of modern natural science that started developing in the seventeenth century.

3.1.7 Historical Descriptions

In all of the historical sciences, be it in the historical natural sciences or the humanities, descriptions of individual events and processes are predominant. These descriptions take on the form of narratives in which a particular sequence of events or processes is told. There are many interesting philosophical questions that can be asked about these narratives: for example, what constitutes their unity; whether they are in fact theory-free; whether they should contain theories from natural and/or social sciences; whether they are objective and if so, in which sense exactly; what is their exact difference to fictional stories; whether they are necessarily guided by interests. In our context, however, I am primarily interested in the relationship between these stories told by professional historians and the stories we tell in everyday life (stories about real events, not the other stories we may make up for various purposes!).

In principle, the structure of the two sorts of stories is the same. When I tell a friend the story about why I failed to turn up yesterday for our date at the expected time and place, I start at some earlier point in time, perhaps when I decided to leave home in order to get to the date. I will then continue to tell a sequence of expected and unexpected events that all contributed somehow to my failing to turn up, perhaps starting with an unexpected phone call that intervened, continuing with the malfunctioning of my car and the impossibility to get a taxi, my mistaken reading of the bus schedule, my turning up at the wrong street, and so on. When historians tell their stories—for example, about the decline of the Roman Empire, the emergence of modern science, the life of Napoleon Bonaparte, or the history of childhood and its relation to family life in a particular region during a particular period—the structure of their stories is essentially the same. They start at some point in time and continue to tell a sequence of events, situations, and processes that are relevant to the subject in question. The crucial difference between our everyday stories is that the historians' stories are in many respects stricter. For instance, in our everyday stories, things will be told that are, strictly speaking, irrelevant for the subject in question, things that may come to mind by association and are simply told. Historians are much more aware of the necessity to restrict themselves to events that are truly necessary for the story. They are much more consciously guided by what has often been called "the selectivity of historical judgment" or "the principles of historical relevance." Three such principles are discernible. Each of them selects material that should be included in the respective history, and often, but not always, the selections of these criteria overlap. The principle of *factual relevance* selects material that needs to be included in the story because of the story's subject matter. For instance, the history of some country must include information about its economy, about its political system, about its relations to its neighbors, and so on. Without such information, the story would be incomplete. The principle of *narrative relevance* selects material that must be included if the resulting text is to be a proper story. For instance, a story must provide some continuity in its course. Therefore, sometimes events must be told although they may not be particularly relevant for the story's subject matter, but they provide the necessary transitions to make the story continuous and thus intelligible. Finally, the principle of *pragmatic relevance* selects material that is necessary for the pragmatic goal of the story to be achieved. This includes material that makes the story intelligible for particular audiences, or references to other historical works that are defended or attacked, or especially careful steps of justification for particularly controversial contentions.

Of course, all of these elements are also present in our everyday stories even if in a much less stringent form. Our everyday stories tend to be much looser, and we may jump from one point to another; rather irrelevant details coming to mind are nevertheless told, previously forgotten pieces are injected at a later point, possible alternative interpretations are often not seriously considered, and so on and so on. So, the principal difference of these stories to the stories of professional historians is one of a degree of discipline, or methodicity, or, in a word, of—systematicity.

During past decades, another type of historical description has appeared on the scientific scene, mainly in the historical natural sciences. It is the computational reconstruction of certain complex historical processes or historical items to which we have only little access by means of direct historical sources. Such processes (or at least plausible scenarios of them) and historical items can sometimes be reconstructed by means of computer simulations or reconstructions. On the basis of the available data, factual assumptions about these processes or items and theories and models relevant for the possible course of events or the items in question, it may be possible to simulate these processes and their results or to reconstruct the historical items. This has been done for a variety of historical processes and items. Here are three examples from cosmology, Earth's history, and biological evolution.

The development of our universe during its early phase when the formation, evolution, and clustering of galaxies and quasars took place is a historical process of utmost importance for our understanding of the current structure of the universe. The theoretical basis for a recent simulation of this process was the current standard model of galaxy formation, the Cold Dark Matter model. According to this model, the initial sources of the hierarchical structure of our universe, which consists of a clustering of objects at different levels, are weak quantum fluctuations that arose shortly after the "big bang." Due to gravity, these fluctuations are amplified into the rich structure of our current universe. The recent computational simulation of this process starts at about ten million years after the big bang. As its initial condition, the simulation assumes a distribution of ten billion (!) mass points, each having a mass of one billion sun masses. Then, the temporal development of the mass distribution of the universe is calculated numerically. The results of the simulation are in remarkable agreement with current observations of the universe, and they describe, for instance, the early formation of quasars, the brightest objects of the universe. This is especially remarkable because it has been doubtful whether the Cold Dark Matter model could describe, for example, the formation of objects like an observationally confirmed quasar that emerged roughly 850 million years after the big bang and has a luminosity 10,000,000,000,000 (ten trillions) times that of the sun.

Here is another example. In its course over the last billion years, life on Earth underwent several serious disruptions, most prominently (for the public) the extinction of the dinosaurs. The largest extinction, however, took place some 251 million years ago, at the so-called Permian-Triassic boundary. Ninety percent to 95 percent of life in the oceans and some 70 percent of terrestrial life was extinguished. In recent decades, several hypotheses have been formulated to describe this event. One of the possibly significant factors was the presence of large magma streams, the so-called Siberian Traps, at about the same time the extinction took place. This volcanic activity would have released significant amounts of carbon dioxide, sulfur dioxide, and possibly methane. However, in order to describe the effects of this volcanic activity on life, one should know how it translated into climatic changes. A comprehensive climate model has recently been devised that couples land, ocean, and sea-ice using realistic paleogeographic and paleotopographic data. The simulation of climate development caused by the volcanic activity at the Permian-Triassic boundary describes inhospitable conditions for marine life due to disrupted ocean circulation and severe constraints on terrestrial life due to excessively high temperatures over land.

It should be noted, however, that such computer simulations do not intrinsically take into account that the resulting description is meant to be a historical one, i.e., that the described process is unique under that description (compare section 3.1.1, second and third remark). There is no fundamental difference between computer simulations of processes that can occur many times (under some description) and those that actually occurred only once, like the evolution of the universe (excluding at this point the idea of parallel universes) or some particular process in the Earth's history. However, this fact does not change the character of the resulting description as a historical one, which is due to the specificity of the factual information used in the simulation. For instance, the computer simulations of the historical processes mentioned above do not intrinsically represent unique events. It is rather the specificity of the initial and boundary conditions and of further factual assumptions entering the simulation that may have come true only once during history.

Unlike these two examples, the last example of a historical description does not concern historical processes but the reconstruction of extinct biological species. On the basis of the fossil record extending over more than three billion years, paleontologists try to reconstruct earlier life forms. In a new research field called computer assisted paleontology, researchers reconstruct fragmented and distorted fossil specimens in three-dimensional images so that their function, biomechanics, developmental changes, and evolutionary modifications can be determined. Roughly speaking, by means of computer tomography, the available fossils are analyzed in their three-dimensional structure and digitalized. This factual information is then combined with anatomical regularities in order to generate reconstructions of the species in question. Such reconstructions have proven to be relevant, for instance, for the controversial question whether *Homo neanderthalensis* and *Homo sapiens* represent morphologically discrete, separate species that belong to distinct evolutionary lineages.

It is obvious that such attempts at the reconstruction of certain historical processes or historical items by computer simulations are highly systematic. If, in our everyday practice of coming up with reconstructions of past processes, we employ something like models at all, these reconstructions are typically much sloppier, more superficial, or, in other words, much less systematic than the corresponding reconstructions in the sciences.

3.2 EXPLANATIONS

3.2.1 *Some Preliminaries*

Before discussing the main topic of this section—namely demonstrating that scientific explanations are more systematic than corresponding nonscientific explanations—some preliminary issues need to be addressed. They concern the difference between explanations and descriptions, the relation of explanation and understanding, my restriction to a discussion of explanations of phenomena and not of laws and the like, and the way I am going to proceed.

First, in many treatises, explanations are introduced as answers to a class of specific why-questions. These questions are identified as "explanation-seeking why-questions." For any given phenomenon x, prototypical explanation-seeking questions are "Why has x happened?" or "Why is x the case?" By contrast, prototypical description-seeking questions are "What has happened?" or "What is the case?" With this approach, the character of the difference between descriptions and explanations becomes transparent. It is a difference between two perspectives that can be taken regarding any given phenomenon: one may be interested in a description of the phenomenon or in its explanation. A perspective is a particular way of looking at a situation that singles out certain aspects. Therefore, any perspective is not simply given by the situation itself, but results from an active choice on the part of the observer. The choice of a perspective is well expressed by particular questions as they emphasize an activity on the part of the questioner. Hence, descriptions and explanations are different epistemic perspectives upon situations. The conceptualization of this difference as differing perspectives or questions will later prove relevant.

Second, I will employ the term "explanation" in a very wide sense. In particular, "explanation" is not to be taken as a contrasting term to "understanding," as has been a linguistic practice in many philosophical and humanist circles. I consider this an infelicitous practice, because innumerable misunderstandings have resulted from it. It originates from a particular philosophical discussion in a particular national context. This philosophical discussion took place in the second half of the nineteenth century in Germany in the neo-Kantian context. One of the main issues in this discussion was the supposed contrast between the natural sciences and the humanities (*Geisteswissenschaften*). Two fundamentally different procedures were identified that supposedly characterize these two groups of disciplines. To denominate these procedures, two *technical terms* were introduced: "explaining" (*erklären*) as the characteristic procedure of the natural sciences, and "understanding" (*verstehen*) as the characteristic procedure of the humanities. It is important to recognize that these terms were introduced as *technical terms*. This means that their sense *intentionally deviates from their everyday sense*. Explanations in this technical sense are what the generalizing natural sciences provide in their answers to questions of the form "why has x occurred?" whereby x is some natural phenomenon. In this technical sense, explanations make essential use of natural laws (so this fact was not at all an original discovery in the twentieth century by Popper, Hempel, Oppenheim, et al., as it is sometimes maintained). Understanding in the technical sense is provided by the social sciences and the humanities in response to someone's "not understanding y," where y is something of emphatically human origin, such as a concept, a literary or philosophical text, a piece of art, an individual or a collective action, an institution, a historical process involving human action, and the like. Understanding in its technical sense essentially refers to an "inner" dimension of certain phenomena that are unarguably part of the human sphere. Understanding attempts to grasp "meaning," a notion that is supposed to be entirely alien to the nonhuman natural world (and constitutes, if not explicated and exemplified, a rather ambiguous and unclear concept). In today's dominant worldview of the Western world, there is simply no such thing as an understanding *in the technical sense* of natural phenomena, because such phenomena do not embody a meaning in the same way as does a text, an action, a social institution, or a piece of art. Whether or not the dichotomy between the technical sense of "explanation" and the technical sense of "understanding" ultimately proves to be significant, it is at least a prima facie clear distinction. By contrast, the *common* words "explanation" and "understanding" (or the German originals of *erklären* and *verstehen*) do not really indicate this difference. In everyday language, it would be acceptable to say, "I do not understand the association of thunder and lightning; please explain it to me," whereas on the technical reading of these terms, such a request is totally incoherent. However, in the discussion about the relationship of the natural sciences on the one hand and the social sciences and the humanities on the other, the technical and the everyday senses of "explanation" and "understanding" have often been conflated, resulting in communication severely at cross-purposes.

I will therefore employ "explanation" largely in agreement with the everyday use of this word to denote all of those (adequate) answers given to someone asking not just for a description of some phenomenon, but for an additional something: *why* the phenomenon has occurred, *what* it means, *how* it could be understood, *why* it functions as it does, and so forth. In particular, I will not restrict the legitimate use of the word "explanation" to the natural sciences, nor will I restrict the legitimate use of "understanding" to the humanities. However, I take the philosophical motifs behind this infelicitous choice of technical terms very seriously, as shall be seen in section 3.2.7.

Third, one particular issue will not be treated in my discussion of scientific explanations. I shall deal only with explanations of phenomena (events, processes, states, and so on) and not with explanations of regularities, models, laws, or theories. Although the latter types of explanations are important in theoretically advanced branches of the natural and social sciences, there hardly exists any counterpart in our everyday practice of explanations. After all, I only have to provide an argument that the sciences are more systematic than *corresponding* everyday practices. Scientific explanations of regularities, laws, and theories can plausibly be viewed as a *systematic extension* of explanations of phenomena, and as such, they are in agreement with our main thesis. The reason is this: as we shall see in this section, many scientific explanations of phenomena draw on regularities, laws, and theories. Explaining these items therefore constitutes second-level explanations, i.e., explaining what is explanatory on the first level. By extending explanatory practice in this manner, a systematic extension of first-level explanations has been achieved.

Fourth, when subsequently discussing scientific explanations, I will not provide an exhaustive overview of all types of explanation, many of which have already been discussed in the literature. I shall choose a few prominent forms that demonstrate the plausibility of my main thesis. Unfortunately, there is no obvious or established order of the different types of explanation used in the various sciences. I therefore have no other choice than to discuss the different types of explanation in a somewhat arbitrary order, without stringent systematicity. I shall discuss explanations using empirical generalizations (subsection 3.2.2), explanations using theories (subsection 3.2.3), explanations of human actions (subsection 3.2.4), reductive explanations (subsection 3.2.5), historical explanations (subsection 3.2.6), explanation and understanding in the humanities in general (subsection 3.2.7), and explanations in the study of literature (subsection 3.2.8).

What may be perceived to be missing in this list are explanations using models and mechanisms, which are a type of explanation that is pervasive in many empirical disciplines. Especially mechanisms have been a recently much-discussed topic in the philosophy of science. No specific subsection has been dedicated to this type of explanation because the class of scientific models used for explanatory purposes is extremely heterogeneous. Without entering into a deeper discussion of models, I am assuming that explanations using models are of the same type as explanations using empirical generalizations, or theories, or reductive explanations. A model typically states something that is known not to be literally true, or even blatantly false, due to the idealizations and/or abstractions employed. However, a good model is used, in the appropriate context, as if it were a correct empirical generalization, or a well-confirmed theory, or the expression of real entities having certain properties and entertaining certain relations. Therefore, it functions in much the same way as explanations using exactly those resources. I shall deal more explicitly with models later when discussing different modes of predictions in section 3.3.

3.2.2 *Explanations Using Empirical Generalizations*

Let us first look at scientific fields in which the phenomena in question obey certain general regularities or phenomenological laws, summarily called empirical generalizations (compare section 3.1.6). If one is familiar with the pertinent empirical generalizations, explanations for changes of state may easily be provided. The explanation consists of a derivation of the value(s) of the changed variable from the empirical generalization together with some information about the concrete situation, demonstrating that the change is due to the regularity holding for the respective system. In its "deductive-nomological" form, the explanation is based on deterministic regularities, typically laws, and the derivation is a logical deduction from laws—hence its name (*nomos* is the Greek word for law). In its "inductive-statistical" form, the relevant regularities are of a statistical kind. Each of these types of explanation has been widely discussed in the literature. They have been called the "covering law model," or the "Hempel-Oppenheim" or "Popper-Hempel" schema of explanation. As I already pointed out in section 3.2.1, the latter labels are somewhat misleading because they imply that the twentieth-century authors lending their names had invented this scheme. In fact, already in the nineteenth century, it was well known that explanations belonging to the generalizing natural sciences are derivations from laws. At close examination, quite a number of problems can be identified with this type of explanation. However, as we are primarily interested in systematicity, we do not have to deal with these problems here.

As an illustration, let us consider how physicists, chemists, or engineers often explain the change of a system from one well-defined state to another due to a change of external conditions. The system is described by certain "state variables," i.e., some quantified properties of the system, for instance its mass, its temperature, or its volume. In its well-defined states, these state variables take on definite values, and the system's state can be characterized by the values of its state variables. For quite a few such systems, so-called state equations are known. A state equation expresses a regularity that the system in question obeys when changing from one well-defined state to another (usually induced by external influence). The state equation describes the functional dependence that the state variables of the system must fulfill in all of its well-defined states. For instance, fixed samples of gases in equilibrium can, for many applications, be sufficiently characterized by their amount n, their pressure p, their volume V, and their temperature T. The state equation of the gas expresses a functional dependence among these state variables. The simplest case of such a state equation is the so-called "ideal gas equation," which reads

$$pV = nRT$$

where R is some constant (the so-called universal gas constant). Under special circumstances, namely high temperature and low density of the gas, this equation holds for most gases.

When a system whose state equation is known is forced by an externally caused change of the value of some state variables to change to another state, the attainment of this new state can be explained (and predicted). By using the state equation and the known values of state variables, the new values of the state variables can be calculated. Hence, the explanation follows exactly the deductive-nomological pattern described above: the explanation consists of a derivation of the explanandum (the new state that has to be explained) from a law (the state equation) and some factual conditions (the externally changed state variables). For instance, let a sample of an ideal gas of amount n_0 be in a closed bottle. Its state can then be characterized by the values of the state variables, such as p_0, V_0, and T_0. These values have to satisfy the state equation of an ideal gas, i.e.,

$$p_0 V_0 = n_0 R T_0.$$

Now we heat the bottle such that it reaches a temperature T_1. As the gas is confined in the bottle, its amount and its volume remain the same. But due to the temperature increase, the pressure will go up as well. The new state will also satisfy the state equation, i.e.,

$$p_1 V_0 = n_0 R T_1.$$

From this equation, we can calculate the pressure of the gas in the new state as

$$p_1 = n_0 R T_1 / V_0.$$

We can thus explain why the pressure of the gas has taken on the new value p_1, and the state equation has provided the necessary resources for this calculation.

The distinction between a description and an explanation may appear artificial in the case above. The reason is that the *explanation* for a change of state is deduced logically from the generalized *description* of the system, the state equation, as well as the values of some state variables. Nevertheless, I would maintain that the distinction between descriptions and explanations remains because, as explained previously, they answer different sorts of questions. The *description* of the new state includes stating the new pressure value p_1 that may be determined empirically. In this case, p_1 must be accepted as given. The derivation of p_1 from the state equation, however, *explains* why the pressure has taken on this value and not another one.

This type of explanation whose most important explanatory source is some regularity, is, of course, well known from our everyday explanatory practices. For instance, we explain somebody's being late by maintaining that the person is "always" late, or the failure of our car's engine to start in the morning by pointing out particular weather conditions in which the engine often refuses to start, or the apparently strange behavior of someone by his belonging to some other culture, or the perhaps surprising weather by being almost typical for this particular month, or someone's falling ill by her preference for light clothes also under unfavorable temperature conditions, and so on. In all of these cases, we refer to empirical regularities that are at the heart of the explanation. Of course, these regularities are not quantitative; very often, they are not even mentioned explicitly, and their epistemic status is often questionable. Wherever such explanations have scientific counterparts, it is obvious that they are less systematic then the corresponding scientific ones. The latter are typically quantitative, the regularities are made explicit, and their epistemic status must be sufficiently robust in order to make the proposed explanation acceptable. Otherwise, in science, the proposed explanation is refused and competing explanations are sought.

3.2.3 Explanations Using Theories

Science is very closely associated with theories. What is understood by "theory," however, varies widely across the whole range of the sciences (remember, in the broad sense!), and it is also not very clear. A mathematical theory like number theory, a physical theory like the general theory of relativity, the biological evolution theory, a theory of mind in cognitive ethology, a philosophical theory of truth, feminist theory, or literary theory represent very different types of theories that have little in common. Unfortunately, the Standard English philosophical reference works are not very helpful with respect to "theory" because usually they do not feature that term as an entry of its own. In this subsection, I shall restrict myself to explanations provided by theories as they are typical for the natural and social sciences. Another and somewhat looser use of "theory" can be found in literary theory with which I shall deal in subsection 3.2.8.

Typical examples of theories in the natural and social sciences are classical electromagnetic theory in physics, evolutionary theory in biology, valence bond theory in chemistry, the theory of plate tectonics in geology, the theory of transformational generative grammar in linguistics, structural-functionalist theory in sociology, Gestalt theory in psychology, or neoclassical theory in economics. What do these and other theories in these and similar empirical disciplines share?

The property common to all these theories has become a part of the everyday meaning of the term "theory." Theories are hypothetical, i.e., they are not supposed to be the last word on the matter, or, in other words, the existence of theoretical rivals to any given theory is always at least conceivable. In many cases, of course, theoretical rivals to any given theory actually exist, especially in the social sciences. Note, however, that the hypothetical character of theories is a part of the meaning of "theory," but only if applied to theories of the empirical sciences. What is called a theory in mathematics, such as "number theory," is not a priori assumed to be hypothetical. Note also that the degree to which theories are hypothetical may vary greatly. A theory may be so hypothetical that it can even count as pure speculation because there is little (or even no) supporting evidence for it. String theory in high-energy physics can serve as an example. By contrast, a theory may be so well established that it borders on being perceived, or is actually considered, to be a fact. An example of the latter case is the theory of common descent, i.e., of the evolution of all life forms out of a single or a few primordial forms (this theory only concerns the *factual interrelationship* of all life forms, not the *mechanisms* that produce this connection).

There are two main sources for the hypothetical character of theories. The first is the generality of theories, i.e., the scope of their possible applications. Of course, theories differ tremendously in their degree of generality. They range from mini-theories that are tailored to a special class of cases to theories of breathtaking generality like evolutionary theory applying to all life forms, or quantum theory, in principle applying to just everything material. But whatever their range, the generality of theories implies that the set of their intended applications is open and does not only comprise a number of known cases. In other words, theories are designed to be applicable to yet unknown cases that may turn out to be significant counterexamples.

A second source for the hypothetical character of theories is their essential reference to so-called theoretical entities. Theoretical entities are not directly observable and are therefore posits. They are either too small to be directly observable, like electrons in physics or chemistry, or unobservable by their nature, like fields in physics, or fitness in evolutionary theory, or social status in social science, or by their nature as idealizations, like the *homo economicus* in economics, or by their nature as counterfactual posits, like the assumption of the atomic nucleus to be a liquid drop with surface tension. These entities are called "theoretical" entities for three reasons. First, insofar as "theory" is opposed to "direct observation," they are attributed to the theory side. Second, these entities always figure in the context of some theory. Third, insofar as the term "theoretical" carries with it the association of "hypothetical," these entities are indeed hypothetical. They are hypothetical either in the sense that they are posits believed to be real but not proven to be real, or they are hypothetical in the sense of a consciously counterfactual assumption. At any rate, the hypothetical character of these entities immediately adds to the hypothetical character of the theories in which they figure. Either it is uncertain whether the postulated entities exist or whether the consciously counterfactual assumptions are fruitful.

Explanations based on theories do not differ in principle from explanations based on empirical generalizations that I discussed in the last subsection. The main difference in the latter case is, of course, the recourse to unobservable entities. This fact, together with the usually wide scope of the theory's intended application, conveys the impression that theoretical explanations are somehow deeper or more fundamental than explanations by empirical generalizations. Let me illustrate by means of an example. The free fall of a body near the surface of the Earth can be quantitatively explained by Galileo's law of free fall. This law is an empirical generalization connecting s, the distance fallen, and the elapsed time t. It involves a constant g with the complicated name "acceleration due to gravity on Earth." This constant can be measured and is a characteristic parameter for free fall near the surface of the Earth. The formula for free fall is $s = 1/2\ g\ t^2$. But the free fall of a body may also be quantitatively explained by Newton's gravitational theory. This theory involves an unobservable entity, a postulated gravitational force, and the law of gravitational attraction that is claimed to be universally valid. The universality of the law means that the law holds for each and every case of gravitational attraction, anywhere at any time. This law states the strength and direction of the gravitational force F, depending on the masses m_1 and m_2 involved and their locations ($F = Gm_1m_2/r^2$). This law also involves a constant G, the universal gravitational constant that is also measurable (although this is more complicated than measuring the acceleration due to gravity on Earth). In order to quantitatively explain the fall of a particular body, the numerical values of the mass of the Earth and its radius are also necessary, but then the explanation proceeds along the same lines as the explanation by means of the law of free fall. The fundamental difference between these two kinds of explanations is the intended range of application of the two laws involved. The particular formulation of the law of free fall is only locally applicable, near the surface of the Earth, and only to free fall; whereas the law of gravitation is supposed to be universally applicable to all situations in which gravitation plays a role. Theoretical explanations involving such laws are therefore much more comprehensive; they bring order to a wide range of phenomena that may appear to be totally unrelated. The universal law of gravitation provided, e.g., explanations for such diverse phenomena as free fall, the generation of the tides, and the movement of celestial bodies. It has unified these phenomena as cases of one and the same fundamental physical interaction. In that sense, explanations based on theories are more systematic than corresponding explanations based on empirical generalizations.

Also, in everyday life, we use assumptions we call theories. As their scientific counterparts, these theories are hypothetical and often involve unobservable entities in order to explain the course of events. In our social interactions, we often ascribe to people certain character traits or interests that are obviously not directly observable. Nevertheless, these features function in many explanations of their behavior. This holds both for the private and for the more public sphere, for instance in the explanation of actions of politicians. In addition, people believe in all sorts of supposedly explanatory theories about natural, social, and even supernatural events, all involving nonobservable entities. These theories include religious and pseudo-religious convictions (whatever their precise difference) as well as superstition of all sorts and conspiracy theories. Whatever the cognitive or existential merits or other of these theories are, they certainly differ from the sort of theories discussed above by their looseness. Their explanations are often vague, sketchy, certainly not quantitative, and stand more often than not on dubious epistemic grounds. In a word, they are much less systematic than the theoretical explanations supplied by various sciences. In spite of these properties, we should not simply deride them because some of them seem to be necessary for us to carry on with our ordinary life.

3.2.4 Explanations of Human Actions

There is a type of explanation that exclusively occurs in the social sciences and the humanities. It differs crucially from the two kinds of explanation discussed before. The explanatory potential of the former types of explanations is rooted in empirical generalizations and theories as they apply to concrete situations. By contrast, a typical explanation of individual human actions does not refer to any general statements, i.e., to empirical generalizations or theories. Rather, what is referred to is an actor's intentions and beliefs about the current situation. Schematically, this form of explanation answers the question, "why did someone do p," by stating that this person had a certain intention and that she believed that in the given situation, she would realize her intention by doing p. To take an everyday example, John's taking the cup and lifting it to his mouth may be explained by John's intention to quench his thirst and his belief that drinking the content of the cup will be an appropriate action to achieve this goal. No general statement like a theory or an empirical generalization is needed to offer the explanation; an appropriate intention (the element of will) and certain beliefs about the current situation (a cognitive element) suffice. Upon closer inspection, however, many interesting problems connected with this form of explanation present themselves, but in view of our goal, they do not have to concern us.

Here is a somewhat more elaborate example from the historical sciences displaying the same explanatory pattern. Why did President Truman decide in 1945 to drop atomic bombs on Japan? This complex question has been the subject of numerous studies. Many details of the answers to this question are controversial. The standard story is that Truman's intention was to end the war with Japan at the earliest possible opportunity. Given the military situation, atomic bombs, dropped without prior warning, were thought to be the most efficient means to force Japan into the capitulation, minimizing American and Japanese casualties. An alternative interpretation is that the bombs were used for purely political reasons, such as intimidating the Soviet Union, anticipating the postwar situation. Or were there motives of revenge for Pearl Harbor or even anti-Japanese racial biases involved in the decision? Whatever is best supported by the evidence, any answer to the question, "Why did President Truman decide to drop atomic bombs on Japan?," will involve Truman's intentions and his reasoning why dropping the bombs would be the best way to realize his intentions. The historians' main task will therefore amount to reconstructing President Truman's intentions and his judgment of the situation. They will have to use all available sources pertinent for this question, including not only official documents but also President's private letters and diary entries and reports by others. Despite all background information, the actor's intentions or beliefs about the situation may remain questionable or so strange that they are hardly believable.

Then they are in danger of losing their explanatory power. Especially in this case, the reconstruction of the development of the actor's strange intentions and beliefs is also part of the historian's task in order to contribute to the intelligibility of the action. President Truman's decision to use the bomb against Hiroshima may remain perplexing on consultation of his diary entry of July 25, 1945:

> This weapon is to be used against Japan between now and August 10th. I have told the Sec. of War, Mr. Stimson to use it so that military objectives and soldiers and sailors are the target and not women and children....
>
> He and I are in accord. The target will be a purely military one and we will issue a warning statement asking the Japs to surrender and save lives.

Obviously, Truman had changed his opinion sometime during the following days because no warning statement was issued before August 6, 1945, when the bomb was dropped, and Hiroshima was not a purely military target. Why did he finally decide to forgo issuing a warning statement and to attack a city, i.e., why did his intentions expressed above change? These questions need to be answered for Truman's decision to become intelligible.

It is obvious that this form of explanation of action is used both in everyday life and in disciplines belonging to the social sciences or the humanities. The crucial difference between an explanation's everyday use and its scientific use is that in everyday life, the ascription of intentions and beliefs about the situation to the actor is often done without much argument. By contrast, in a scientific context, the ascription must be carefully argued in most cases, and a variety of ways exist to do that. Documents of all sorts and circumstantial evidence often make the existence of particular intentions and beliefs about a situation of an actor plausible, and also empirical generalizations and theories may play a role. In other words, the sources of the potential explanation are much more systematically identified, and their explanatory power for the pertinent action is much more systematically scrutinized although the general pattern of the explanation is the same. In the case of explanations for actions, our main thesis is exemplified again: namely, the scientific mode of this kind of explanation is more systematic than its everyday mode.

3.2.5 Reductive Explanations

In a wide range of research fields, one particular pattern of explanation is extremely prominent. In this particular pattern, the explanation of the features or of the dynamics of some system does not rely on elements that are part of the description of the system itself, as, for instance, in explanations involving empirical generalizations (see subsection 3.2.2), but on other elements. This explanatory pattern is called "reductive" or "reductionist," derived from the original meaning of the Latin *reducere*, which is "to trace back." The explanation's elements are located on another level than the system itself. The most popular case of such reductive explanations draws on the *components of the system* as explanatory resources and the way they are composed and they interact. This form of explanation has also been called "microexplanation," referring to the system as located at a "macro" level and its components at a lower "micro" level. However, reductive explanations may also draw on levels "above" the system in question, for instance, when individual actions are explained by social factors alone that are, somehow, located at a societal level above the individual. Correspondingly, a research strategy aiming exclusively for such explanations because they are judged to be solely legitimate is called "sociological reductionism."

Let us first take a brief look at some examples in order to see how widespread reductive explanations are in the sciences. Large areas of physics are reductive in character: the properties and dynamics of solids, liquids, and gases, of atoms, nuclei, and composite particles are explained by recourse to their component parts. The same holds for quantum chemistry with respect to molecules. Classical biology explains the functioning of organisms drawing on their organs or other parts, the functioning of organs by recourse to their constitutive parts, and the functioning of these parts by recourse to their constitutive cells. Molecular biology and biochemistry try to elucidate all sorts of processes in living beings referring to their molecular constituents. Engineering sciences explain the functioning and malfunctioning of artifacts referring to their parts and their configuration (this, of course, is also true for many technological explanations not belonging to science proper). Macroeconomic and other social processes are often explained as aggregative effects of individual actions. Large-scale natural or social historical processes are explained by interplay of various constituent factors. The meaning of a linguistic expression is explained by the meanings of its parts and the way they are syntactically combined, and so on.

Reductive explanations have been widely and controversially discussed, both in various scientific fields and in the philosophy of science. Without going into too much detail, I should at least mention some of these disputes. First, in various fields of the *natural* sciences, controversies exist with respect to the scope of reductive explanations (or even reductive research strategies in general). For example, in biology, is the only type of legitimate explanation reductive, for instance, in terms of the molecular composition of the respective biological phenomenon in question? Can, in principle, all explanations in chemistry be reduced to physical explanations? Are the really fundamental explanations in physics all reductive, e.g., in terms of elementary particles and their interaction? Those who do not answer these questions in the affirmative typically claim that there are types of explanations in these fields that do not refer to some lower level. These types of explanations are claimed to be fully legitimate and even scientifically necessary because they cannot be reduced to some lower level. Thus, the claim is that some levels possess what can be called explanatory autonomy despite the fact that entities of that level are indeed composed of parts from a lower level. Typically, in the natural sciences, it is beyond dispute that reductive explanations are often useful and legitimate. The controversial question is whether reductive explanations exhaust the set of all legitimate fundamental explanations.

Second, in the *social* sciences, the analogous controversies are sometimes more pronounced. Some defenders of more holistic approaches deny the legitimacy of *any* reductive approach, considering them totally misguided; whereas others affirm that *only* microexplanations are legitimate (of course, others defend less extreme approaches). These topics have also been discussed in philosophy of science.

Third, concrete details of microexplanations have been a matter of extended and controversial philosophical discussion. For instance, it is clear that the vocabulary of the reducing level, the microlevel, must somehow be connected to the vocabulary of the level to be reduced, the macrolevel, as far as it is specific to that level. However, the nature of this connection is not clear. Are comprehensive general definitions of macrolevel terms needed? Or are partial reconstructions of these macroterms sufficient, applying to a limited range of particular phenomena? What is the epistemic status of such sentences that connect the different vocabularies? Are they empirical or analytical? Do they have the character of natural laws? Are they in need of some sort of explanation, or do they express brute facts and therefore simply have to be accepted as given? Another issue for philosophical discussion concerns the question, what are the justificatory resources for antireductionist positions?, i.e., what are the arguments that microreductive explanations are impossible for some class of phenomena? This is an interesting question because arguments for impossibility claims are notoriously tricky.

In recent decades, the terms "reductionism," "reductive," "reductionist," and similar ones have often been used with a pejorative connotation. This may be due to epistemic claims often connected with these terms that were perceived in some quarters to be exaggerated, misleading, unproductive, or narrow-minded, at the expense of a more holistic approach. Subsequently, I intend to abstract completely from these value connotations. I am not aiming at ultimately assessing the scope and potential of reductive explanations, but presenting this explanatory pattern and its use in the various sciences as well as comparing it with similar patterns of everyday life. Of course, I hope to show that the scientific form of this explanatory pattern is more systematic than its everyday form.

It should be noted at this point that there is a substantial overlap of reductive explanations with other types of explanations—first and foremost with explanations using theories or models. The reason for the overlap is obvious. In the social and natural sciences, theories are entities that are developed mainly for explanatory purposes. For this goal, so-called theoretical entities, i.e., entities that are not directly, or even not at all, observable are introduced. Very often these entities are located at a level below the systems to be explained, for example, as their component parts. For instance, in atomic and molecular physics, all quantum mechanical explanations of atoms' and molecules' properties refer to the component parts of these particles and their interaction, thus being reductive. Similarly, models used for explanation are typically models about the system's composition, together with assumptions or laws of the interaction of these components. Examples abound, from solid state physics, biochemistry, climatology, or economics.

Obviously, reductive explanation presupposes a difference of certain "levels," and it must be admitted immediately that this notion is somewhat problematic. How is this notion to be understood? There are at least three areas of problems. First, is a difference in levels something that is rooted in nature itself, independently of us, i.e., is this an ontological difference? Or is a difference in levels a difference of descriptions such that we, and not nature, are the primary creators of this level structure? Or is this, perhaps, the wrong way to approach the level problem because levels contain both originally subject-sided (the subject being the epistemic subject) and originally object-sided elements? Second, from the extended discussion about reductive explanation and emergence in the last decades, it became clear that the postulate of the existence of an unambiguous and universal level structure of the world—whatever its origin may be—is highly contentious. Rather, the different levels used in the sciences seem to be context dependent. In biological research, for instance, there are many different possibilities of distinguishing levels below the level of the organism. Biologists deal with various levels depending on the respective research questions. Finally, in some concrete situation, it may even be unclear to which level a given entity belongs. Fortunately, these complicated and important questions are of no concern here because levels are fairly well defined in many situations in the sciences. Hardly any problems regarding the identification of the levels or their nature present themselves in the preceding list of examples.

Microexplanations often coexist with other forms of explanations of the same phenomenon. For instance, microexplanations often supplement explanations based on empirical generalizations because the latter leave completely unexplained how the phenomenon under investigation is produced. By contrast, microexplanations aim at elucidating the "mechanism" of its production, i.e., the way the phenomenon is produced by its component parts, their mutual interaction, and their interaction with the environment. To illustrate, let us take up the main example from subsection 3.2.2. Explanations of a state change in a given system with recourse to the system's state equation may be supplemented by explanations that refer to constituent parts of the system, their composition, and dynamics. In the case of the ideal gas, an explanation for state changes refers to the constituent molecules of the gas and their dynamics. State change is then explained by reference to changes in the molecules' motion due to the external variation of state variables, which in turn translate into changes of other state variables.

Similarly, the existence of an explanation for an action as described in subsection 3.2.4 does not exclude the possibility of a reductive explanation. Let us assume for the sake of argument that a neurophysiologic explanation could also be given. Again, it seems sensible to say that this explanation uses resources situated below the level of actions. Actions are then seen as highly complex facts involving as constituents, among other sorts of facts, especially neurophysiologic facts. These facts may indeed be explained by recourse to other neurophysiologic facts. On the other hand, explanations of actions may also involve levels above the level of individual action. Imagine that in the example mentioned above, where John is taking the cup and drinks, that the cup is in fact a chalice, and drinking from it is part of a religious ceremony. John's action may then be explained by reference to his intentions and his belief about the situation he is in, i.e., by his will to perform the ritual and by his belief that performing the ritual in the given situation is the appropriate thing to do. However, this explanation is only illuminating for someone who is familiar with the ritual and the circumstances under which it is appropriate. For someone lacking this information, the reference to John's intentions and beliefs is of no value because she will still not understand the meaning of the action. In order to obtain that meaning, a description of the ritual and the context in which it is embedded must be provided. In other words, the full explanation for the action must refer to facts involving societal (or social group) values, rules, roles, and practices. Again, in a fairly clear sense, such explanations involve a level above the level of individual actions.

Reductive explanations or, more generally, reductive research programs, are very systematic in character. Once the relevant levels have been determined, it is in principle clear what route reductive explanations and research programs have to take. First, one must find out the properties of the entities on the component level and construct or find out the laws/models/theories these entities obey. Second, one must determine the configuration of the components that make up the system of interest. Finally, one must derive the features of the system from the application of the laws/models/theories to the configuration at hand. The larger the scope of the laws/models/theories in question is, the stronger the unifying achievement of the reductive explanation or the reductive research program is, at least potentially. Especially physical theories like electrodynamics or quantum mechanics or the biological theory of evolution display an extremely strong systematizing power due to their unification of an enormous range of heterogeneous applications.

Also, in our everyday life, we use reductive explanations. Why does the car not start? Because there is no gas in the tank, or the carburetor is clogged. Why was the football team not successful today? Because one member of the team who has a strategically important position was unusually weak. Why was the party such a success? Because so many different people cooperated fruitfully. Why did the cake not rise? Because you forgot the baking powder. Why did she behave as she did? Because of her family background. And so on. In all of these cases, we refer to the components of the respective object, or to a level above it, blaming or lauding them or it for success or failure of some property or behavior of the object. The configuration of these components can often be inferred from the context, as well as the regularities that are invoked (even if only implicitly), or an explanatory higher level. By now, we are encountering a familiar pattern. Everyday explanations of the reductive variety are much more fuzzy, much less articulate and confirmed, much more prone to prejudice, and much less explicit than their scientific counterparts where they exist. Both kinds of explanation have basically the same structures, but the scientific ones are more systematic.

3.2.6 Historical Explanations

In the historical sciences, several types of explanations are used; some are specific for the historical sciences; others are not. Let me first briefly discuss two nonspecific types before proceeding to two specifically historical types of explanations.

1. Historical sciences dealing with natural history also use the types of explanations that are based on empirical generalizations or theories that I discussed in sections 3.2.2 and 3.2.3. This is the case when a known theory, law, or empirical regularity covers the historical development of some entity or system, and no intervening, contingent, and—from the viewpoint of the developing system—therefore unforeseeable factors play a role. For instance, the historical development of the universe is reconstructed entirely on the basis of lawlike developmental patterns. Obviously, for the development of the universe as a whole, no external contingent factors necessitate an explanatory story in order to understand its course. Similarly, long stretches of the geological history of the Earth or of the Earth's climate can be reconstructed on the basis of laws and models—as long as the cosmic environment behaved regularly and there were no contingent intervening events. By the same token, some stretches of the history of biological evolution can be accounted for in a purely lawlike manner—that is, when the environment was stable and no additional contingent factors played a significant role.

2. Historical sciences dealing with the human realm also use the pattern of explanation of human actions that I discussed in section 3.2.4. In fact, the main example I gave there was a historical example, the question about why President Truman dropped atomic bombs on Hiroshima and Nagasaki.

3. The next type of explanations belongs exclusively to the historical sciences. Its explanandum, i.e., the event, process, or the state to be explained, is a singular thing, and the explanation has essentially a narrative form. Such a historical explanation consists of a story about a chain of happenings that lead to the explanandum. The story makes the explanandum intelligible by reporting why it has occurred, i.e., in which way it was produced. Various elements may enter these explanatory stories, including laws, regularities, and even theories, but the main explanatory burden lies in the narration of a sequence of earlier events or of converging parallel sequences of earlier events. The principal structure of a historical explanation is as follows. We start with a state A that developed into state B. But then C happened, which was unforeseeable in state B and must therefore be told. B and C then led to the development of D. Then an unforeseeable E happens or becomes relevant, leading to F, and so forth. There is no overall regularity connecting A and F because there are intervening events that—from the expected course of things—are external and contingent. Such intervening events are essential ingredients to the explanatory story that need to be reported.

Take as an example the outbreak of World War I. In order to explain this event, the situation of various European countries before 1914 must be presented first. Then, the assassination of Franz Ferdinand, Archduke of Austria, on June 28, 1914, by Serbian nationalists is to be told. This event—as the culmination of a series of diverse parallel threads involving several European countries—triggered the outbreak of war on August 1, 1914.

Such a historical explanation can obviously not be brought into the form of an explanation whose main ingredient is a theory, an empirical law, or some regularity. By far too many independent events happening at different times are involved that identify the given configuration of events and factors as a historically unique combination. There simply is no general law known to us connecting the various events and factors such that an explanation of the outbreak of World War I could be given on the basis of that law. Rather, in cases like the outbreak of World War I, the narrative form of the explanation is essential and unavoidable. In a narrative explanation, all of those more or less independent events are told which, appropriately connected, bring about and thus explain the event in question.

Although sometimes perhaps less evident, many historical explanations in the natural sciences have the same narrative structure. For example, when in paleontology the extinction of some species or the emergence of new species is explained, first a description of some earlier state of the evolution of life has to be given. Typically, no natural law exists that leads from this initial state to the state that is in need of explanation, i.e., the extinction or emergence of a species. This holds at least if the initial state is temporally not too close to the explanandum event. The reason is that in most explanations of the extinction and emergence of species, factors whose existence and dynamics are not covered by evolutionary theory itself play a causally important role. For instance, all of the environmental factors and their consequences that play a role in the evolutionary story at different points in time have to be narrated.

4. There is another type of explanations belonging exclusively to the historical sciences, although it seems that, as far as the everyday business of historians is concerned, it is not as common as the previous type. In this case, the explanandum is not a singular event, process, or state but rather a variety of events, processes, or states that are puzzling and cry out for a unifying historical explanation; the whole variety of events, processes, or states is seen as an effect of one singular major event, process, or state in the past. However, what exactly the nature of this historical event, process, or state in the past is may be a matter of considerable dispute; even the unity of the events to be explained may be controversial.

Here are two examples. Alfred Wegener's continental drift hypothesis, advanced in 1912, was designed to explain a variety of puzzling features on the surface of the Earth regarding the Atlantic coasts of Africa and South America. These coasts exhibit complementary shapes, similar geological formations, and fossil records as if no gap extending several thousand miles was between them. As is well known, Wegener's hypothesis was not accepted in his own time because physically, continental drift appeared to be impossible; the continents were supposed to move only vertically. However, the continental drift hypothesis exemplifies, especially in its later accepted form of plate tectonic theory, how a variety of events, processes, or states can be seen and unified as traces of an earlier event, process, or state, and how they are thereby explained.

A second example concerns human history. In Western culture, it is a deeply entrenched persuasion that at least in European history, the time period that is called the Middle Ages is very different from what is called modern times. If one believes this, then all of the historical items of modern times that are characteristically different from historical items belonging to the Middle Ages must somehow be explained in their modern character by tracing them to the events, processes, and states of the transitory period. For instance, historians of science stressing the difference of modern science from medieval traditions usually assume the existence of the "Scientific Revolution" that is basically responsible and therefore explanatory with respect to the characteristics of modern science. Again, we see the same explanatory pattern: a somehow compact event or process in the past explains and unifies a variety of later events, processes, or states by identifying them as traces of the earlier events or processes.

With respect to explanations, the most common situation in the historical sciences is thus the following (type No. 3). We are interested in an explanation of a historical event. Its existence can be understood by connecting it to earlier events. No law is known to us, however, that would allow for a direct connection of the historical event with those earlier events. At various stages in the process, we are thus forced to refer to contingent events and factors. The resulting explanatory representation of the process is therefore a story. Again, the structure of such narrative explanations is identical to the structure of most of our everyday explanatory stories, for instance when we explain a complicated case of having been late. We may relate the chain of all of the intervening events that we could not foresee and that disturbed the planned order of events resulting in our being late. Even for historical explanations that have the unifying character (type No. 4), there are analogues in layperson's explanations. For instance, people may describe their life as having two very distinct periods, divided by a decisive event that changed their life, for instance a religious conversion. The character of a diversity of events in the second period is traced to the events during the cut and thereby explained. Although the structure of these layperson's explanations is the same as their scientific counterparts, in various respects, the former are less systematic than the latter. For instance, historians are typically much more careful to exclude possible alternative explanations than we are in our everyday explanatory narrative practice. For every factual contention, they aim to present sources or other argumentative backing. They are careful to avoid narrative gaps in order to have as continuous a flow of the story as possible, and so on. In comparison, our explanatory everyday stories are much more unordered, less disciplined, less circumspect, or, in a word, less systematic.

3.2.7 Explanation and Understanding in the Humanities in General

The objects of study in the humanities are cultural products such as all sorts of texts, images, traffic signs, paintings, sculptures, installations, theater plays, movies, operas, songs, symphonies, dances, performances, gestures, rituals, juridical laws, archaeological remains, buildings, etc. Cultural products are distinctly different from purely natural objects like some anonymous stone on the bottom of a lake. They are specifically related to human life by being outgrowths of it. Cultural products do not have to be artifacts, however. A sacred mountain is as much a cultural product as is a religious interpretation of solar eclipses. This special relationship to human life is the basis for the characterization of cultural products as having *meaning* (or as being bearers of meaning). I am aware of the fact that the concept of meaning just employed is neither very clear nor unambiguous. Indeed, as there are many ways how things can be "outgrowths of human life"—the open list of cultural products above demonstrates that—, any term covering this immense variety cannot be very specific. But when confronted with the choice whether any of the items on that list rather resembles a text or some anonymous stone on the bottom of a lake, the answer seems to be clear. It is their meaning—whatever that is!—that makes them analogues of texts rather than of stones. Very abstractly and equally roughly, to state that something has a meaning (in the intended sense) is to say that this thing has a special relationship to human life, that it plays a role for humans. Although physical appearance and its use are key to the meaning of any given cultural product, its meaning is by no means identical with any of its purely physical properties or physical aspects of its use.

One of the principal activities of the humanities is the investigation of cultural products with respect to their meaning(s)—these disciplines want to explain, or understand, what these products mean. This is the basis for the popular description of the humanities as being "united by a commitment to studying aspects of the human condition" because the human condition, as experienced by humans, is encoded in cultural products. As meanings are not physical, they are not readily available to inspection (in the literal sense)—in fact, meanings are completely hidden from any sense experience. But if meanings are in no way observable, which human faculty allows exploration of meanings and enables humans to grasp them? This is of course a highly complex question which cannot be answered here in any detail. But basically, in one way or another, meanings are the subjects of thought, or, more to the point, of *reflection*. Roughly, reflection about observable cultural products and their use (although their being *cultural* products may not be available to inspection) enables us to form hypotheses about their meaning. It is exactly here where the *technical* use of the word "*Verstehen*," or of "understanding," has its place (I have briefly discussed the technical use of "understanding" in section 3.2.1). Understanding, in the technical sense, refers to the grasping of meaning, whereas "reflection" is the way to achieve understanding. Again, as in the case of "meaning" and other key terms in this section, it is not a trivial task to circumscribe what the precise meaning of the concept of reflection in the given context is (it is much easier to explain it in optics!). I shall try to elucidate it by illustrating how it works in practice.

Let us begin with the arguably most common, or at least the most traditional, object of study in the humanities: the study of written documents. Such records are studied with very different aims in different branches of the humanities. The same document, especially when it belongs to the so-called "classics," may be studied even in one and the same discipline with very different objectives in mind. (In fact, the status of "classic" is derived from a document's property acquired over time from being considered a worthwhile object for study from different angles.) Frequently, the aim of the study of some text is to understand it, to grasp its meaning, if put very schematically. In the case of a historical document—for example, an old deed—this means something that is different from the case of a philosophical treatise, and again something different from the case of a fictional text, such as a novel. Different contexts will have to be considered in order to enhance our understanding of written documents from different literary genres. Once we concentrate on the document itself, we aim at understanding "what the document itself says." Some terms may be unknown or unfamiliar; there may be obscure transitions between different parts of the document; or the overall thrust of the document may be unclear or even entirely unintelligible. When such difficulties present themselves in the humanities, the methodological situation is often described as the "hermeneutic circle." *Hermeneutics* is both the art of interpretation and the theoretical discipline that studies the process of interpretation. One is often confronted with a particular *circle* when trying to make sense of a given document. Roughly speaking, the hermeneutic circle consists in this. In order to understand the whole document, one must understand its parts, but in order to understand the parts, one must understand the whole. Put in this unhelpful form, the hermeneutic circle seems to suggest the paradox that improving one's understanding of yet poorly understood documents is impossible. This, however, is not and should not be the message of the hermeneutic circle. Rather, the metaphor should suggest that improving one's understanding of the document implies a movement between its parts and its presumed overall message. By moving back and forth between parts and the presumed whole, one should be able to increase the understanding of both in a stepwise process. This process comes to an end once one has reached a reflective equilibrium, i.e., a balance and mutual support of the supposed meanings of the parts and of the whole. Some writers have therefore rightly suggested that the metaphor of the hermeneutic circle should be replaced by the metaphor of the "hermeneutic spiral," indicating that an upward move in understanding involves a repeated feedback loop bringing the parts and the whole into ever greater harmony and mutual support.

This process is mainly kept in motion by reflection. Based on a first reading, or on knowledge of its context, or on secondary literature, one may form initial ideas about some of the document's parts and about its overall thrust. A lack of understanding of parts or the identification of implausible tensions between some of them will lead to their being read more carefully, to questioning the meaning of key terms, to comparisons with other parts of the document, and so on. Drawing on the tentative meaning of the parts, a hypothesis is formed about the overall meaning, which is then followed by attempts at understanding the roles of the parts in relation to the whole text. Implausible tensions demand going back to some of the parts, devising and testing other potential interpretations that can then be evaluated with respect to the attempted increased coherence of the overall meaning. Or, one may identify tensions between the parts as essential elements of the overall meaning of the document.

Throughout this activity, all relevant data are at hand in the written document. Any progress in understanding is based on a fixed data set, and it is the result of reflection, i.e., considering and reconsidering, these data. Of course, for various reasons, one may wish to compare the given document with other documents, or to adduce other sources of information. For instance, one may want to identify congruities, or incongruities, or developments in comparison with other texts. Or, gaps in understanding may be bridged by putting the document into a larger historical, social, economical, political, or literary context. In this case, new data do play a role for productive work in the humanities. However, "new data" in this context are usually unlike "new data" in the natural or social sciences. In the latter disciplines, "new data" usually means data that are absolutely new because they were, for instance, produced by newly designed instruments. Of course, an analogous process can also happen in the humanities when, for example, a historical document that is yet unknown to the relevant scientific community is detected in some library. Whatever constitutes progress in the humanities, it is certainly not driven primarily by discoveries of new data in this quasi-absolute sense. It is rather reflection that promotes the humanities by producing new insights or, at least, new possibilities of interpretation. Indeed, reflecting on known data may trigger new questions that, in turn, may call for a reevaluation of the importance of those data and the inclusion of data as yet not considered. This may then lead to new interpretations of the issue at interest, or even generate new issues.

The hermeneutical approach that is so prominent in the text-based humanities is by no means restricted to an understanding of written documents. Especially during the last decades, the character of cultural products as being text-analogs has been stressed in many disciplines of the humanities. Such cultural products can be treated very much like written documents. Correspondingly, the hermeneutic approach to texts has been transferred in one way or another to other subjects that are texts only in the figurative sense. One example must stand for many. In pictorial semiotics, i.e., the study of pictures from the point of view that they are signs, it has been stressed that an understanding of a picture often also involves a hermeneutical circle. In fact, even a hierarchy of nested hermeneutical circles may be involved. Some dots and strokes on a painting may be entirely meaningless when looked at in isolation, separated from their context. They may only be perceived as standing for different facial traits, for instance, when they are integrated into a local whole, say, the sign of a person. On the other hand, that sign of a person may consist of nothing but those parts that signify different pieces of that person. One level above, the same structure may be repeated. What that sign of a person means in more detail may depend on a higher order local whole, e.g., the group of people with which that person is associated in the picture. Again, the sign for this group of people may be constituted by the signs for the different individuals, and we encounter the same circular structure.

This reflective practice that is characteristic for the humanities in their attempt to understand cultural products is not alien to our everyday practices. In fact, we are constantly interpreting our physical and human environment with respect to their meanings. In our everyday life, we see practically every object as serving possible human aims, and we constantly try to read other people's behavior from their intentions, i.e., we try to understand what they are doing and what they are up to. Even more obvious, we try to understand what other people say. On many or even most occasions, these processes run so fast and without a conscious effort that we are hardly aware of them. From time to time, however, we encounter things the purpose of which is unknown to us; we see behavior of people that does not seem to make sense; we hear sentences whose meaning eludes us. In cases like these, we immediately start reflecting in order to grasp the meaning we are lacking. We look or listen again in order to be sure to get the data right; we think of possible interpretations and consider alternatives; we try to assimilate the unintelligible phenomenon to familiar phenomena; and so on. In the humanities, we find in principle the same activities but typically with a higher degree of systematicity, which is often reached by a stricter commitment to reflection: by a more complete exploration of possibilities; by consistently exploring manifest and hidden presuppositions; by trying to spell out more consequences than those that come immediately to mind; by confronting one's own way of thinking about a certain subject with alternative attempts; by submitting our hypotheses to the professional discourse; and so on. Of course, all of these activities can go terribly astray—by exaggeration, biases of all sorts, ignorance about immediately pertinent or relevantly contrasting subject matters, and so on. Certainly, they do go astray at times, as there are developments in all fields of research that cannot be lauded unconditionally. However, it is not my goal in this subsection to evaluate these and related matters. Rather, my goal has been to illuminate that also the humanities are more systematic in their explanatory endeavors than the corresponding everyday thinking and that the appropriate concept of systematicity especially includes the aspect of a higher intensity and methodicity of reflection.

3.2.8 *Explanations in the Study of Literature*

Finally, there are types of explanation that are only applicable to the study of literature. The main characteristic of this subject matter is that the texts under study are fictional in character. Of course, there are many questions that can be asked about any text, irrespective of its content and its character as fictional or factual: When was the text composed? Who was its author? What are its linguistic peculiarities? Is it representative of a certain genre? And so on. But fictional texts admit of a broad range of additional questions, including explanation-seeking questions, deriving from their fictionality. Again, this is an extremely broad and heterogeneous area, and different schools profoundly disagree about what the important questions about fictional texts are.

As I will be unable to cover this field in any degree of completeness, I shall first pick out, almost entirely arbitrarily, one particular explanatory device that is specific for some literary theorists' work. The specificity of this device for literary studies derives from the fact that the author of a fictional text has, in comparison to an author of a historical text, additional freedom in his choice of elements of the story, and it can therefore be asked why the author chose a particular given element. At this point, the concept of the "poetological difference" comes into play. This concept is central to a thorough understanding of literary texts (if this can ever be achieved). The poetological difference becomes relevant when one asks questions about inner-fictional facts, which are part of the story told; they would appear in a description of this story. For instance, it is an inner-fictional fact of Shakespeare's play *Hamlet* that in the third act, Hamlet accidentally kills Polonius. With respect to any given inner-fictional fact, the question may be asked why this fact is part of the story. There are two fundamentally different viewpoints from which this explanation-seeking question can be answered, and their difference is the poetological difference. The first viewpoint is inner-fictional. It explains the inner-fictional fact by reference to other inner-fictional facts that happened before. In our example, the inner-fictional explanation for Hamlet's killing of Polonius is that Hamlet wanted to kill Claudius but mistook Polonius for him. The extra-fictional viewpoint explains what role the given inner-fictional fact plays in the intended overall message of the text in question. Of course, questions about the overall message of some literary text are always bound to be extremely controversial, but that is not relevant here. In our particular case, why did Shakespeare arrange the train of events in his play such that it came to the accidental killing of Polonius by Hamlet? Broadly, the answer may be that by killing Polonius, Hamlet becomes similar to Claudius in that he killed a father whose son will avenge the killing. And it was Shakespeare's objective to equate Hamlet to Claudius, because the main message of the play was the approximation of the avenger to the one on whom he takes revenge—this is at least one possible explanation.

Also, a layperson occasionally reading a novel or watching a play may dimly be aware of what has been called the poetological difference above. When asking specific questions about the composition of a literary text, she may even use it more or less explicitly. Children in their early teens may be aware of the difference between the inner logic of a fictional story and the author's intentions concerning a certain bit of the story with respect to future developments in the story or even the purpose of the story as a whole. But it is only in literary theory where this difference is clearly articulated and developed and systematically applied to the analysis of literary texts. Again, we see what should by now be a familiar picture. The explanatory pattern used in literary theory that makes use of the poetological difference is basically the same as the one used by an educated reader, but it has a higher degree of systematicity in its articulation and is applied in a more systematic fashion.

Finally, explanations involving so-called literary theory need now be addressed. It should be noted from the beginning that what is commonly called "theory" in this context is of a very different type and plays a very different role from what I discussed earlier under the rubric of "explanations using theories" (subsection 3.2.3), where theories from the natural and the social sciences were at issue. In spite of some vagueness that a closer look reveals both in the natural and the social sciences, theories in these areas are fairly well-defined entities, at least when contrasted with "theory" in literary studies. Here, "theory" denotes an extremely heterogeneous and large variety of works that typically do *not* directly deal with the subject matter of literary studies, namely, pieces of literature. Rather, it includes "works of anthropology, art history, film studies, gender studies, linguistics, philosophy, political theory, psychoanalysis, science studies, social and intellectual history, and sociology." It is characteristic of these works that despite their apparent different subject matter, they can be put to use in the discussion of literature. In this context, their primary role is to undermine the common-sense notions of meaning, writing, literature, and experience that influence us when dealing with literature. More to the point, theory reveals that these common-sense notions are not simply given and unquestionable but that they are the products of typically long historical processes involving various social factors. Which factors are singled out to be of particular importance depends, of course, on the subject and on the theory adduced. It is plain that in contrast to the use of theories in the natural and social sciences, the primary purpose of theory in literary studies is not explanation in the standard sense, which is: here is a phenomenon, given and indubitable, and an explanation for its existence and its characteristics is sought. Rather, what is presented as literature and in literature should be the subject of sustained reflection in the light of writings from various disciplines. The purpose of this exercise is a critical examination and possible dissolution of common sense notions that take for granted what should not be taken for granted because of an underlying process of historical genesis of what appears to be given and immutable. However, explanations do come into play, but only at a later stage of the game and not so clearly separated from other elements of this "discourse." Whatever the detailed changes are that the reflection and dissolution of common-sense notions amount to, after this process, many things appear in a new light, which, in turn, leads to new ways of explanation and understanding of the phenomena in question, i.e., of literature.

Because of the extreme heterogeneity of the subjects and the theoretical elements involved, there seems to be no schematic way to describe this process in general terms. Furthermore, literary studies are permanently permeated by controversies of all sorts that make an unbiased account of this discourse even more difficult. But whatever the details, this process does not look completely unfamiliar when compared with processes when lay people think and talk about literature. An occasional thought of potentially subversive character concerning the fundamental ingredients of literature may also cross their minds. Literary theory develops such moves by more sustained reflection, by a wider range of other subjects involved, by a confrontation with alternative views, and so on. In that sense, the activities in literary studies involving literary theory are just more systematic than the comparable activities in everyday life.

3.3 PREDICTIONS

3.3.1 *Some Preliminaries*

Before discussing predictions in science, I should make a terminological remark. Literally, predictions concern future events (or states or processes and the like). In the sciences, however, a looser usage of "prediction" also exists. In that looser sense, a prediction of a theory or model is simply a consequence of that theory or model about some presumed fact that needs not be in the future. Thus, to say in this sense that a certain description of an event is a prediction of some theory means that this description can be derived from that theory, irrespective of whether and when the event has occurred or will occur. In the following, however, I shall use "prediction" always in the literal sense, meaning a reference to future, not yet observed events.

Many natural scientists believe that the ultimate test for a field to be truly scientific is its ability to successfully produce predictions. From this conviction they infer the intellectually lower status of disciplines that are unable to predict. However, this view is badly biased.

First, the formal sciences like mathematics are completely excluded from this perspective.

Second, there are successful natural sciences whose main epistemic goals do not include producing predictions. I am here mainly referring to the historical natural sciences like paleontology, cosmology, paleoclimatology, and the like. Their task is the reconstruction of some particular historical natural process, and their ways to realize this task bears strong similarity to the practice of historical disciplines dealing with human affairs, like political history, or history of mentalities, or art history, or history of science. It is not to be denied that the results of the historical natural sciences may be very relevant for disciplines that do indeed predict. For instance, the results of paleoclimatology provide an extremely important testing ground for climate models that are designed to predict climate behavior. In the domain of the historical human sciences, however, this spin-off for predictions in the literal sense is rare.

Third, disciplines not aiming at predictions are, by this very fact, in no sense intellectually inferior to (experimental) disciplines directed at them. Though it is true that predictions have an intrinsic value and that they provide particularly severe testing opportunities for the disciplines involved, nonpredictive disciplines have developed their own tools in order to launch effective self-criticism. Each discipline must develop tools for critical evaluation of its knowledge claims, appropriate to its given subject matter, its specific perspective, and its procedures, and these tools obviously differ significantly across the range of all of the sciences. This subject, however, will not be treated here in any detail but in section 3.4.

Fourth, there are several arguments that make it plausible that reliable longer-term predictions concerning human affairs are impossible in principle. One well-known argument stresses the possibilities of "self-destroying prophecies." Because human beings can take note of predictions concerning them, they will be able to behave in such a way that the prediction will not come true. Another argument states that longer-term developments of human affairs will depend on future knowledge, i.e., knowledge that we do not possess today. As a consequence, it is in principle impossible to forecast such developments as we cannot predict the content of future knowledge. The final argument to be mentioned stresses that cultural change can be so profound that we may presently lack the concepts that might become necessary to describe our future. To believe that the prediction of future cultural states is possible is equivalent to the belief that we have a theory that exhaustively describes all possibilities of the human condition. This, however, is definitively not the case. Therefore, the humanities are bound to be sciences that in the long run at best understand ex post, but certainly do not predict.

We should therefore be aware that what will be dealt with in this section does not apply to all of the sciences in the wide sense but only to a subset of the natural and social sciences.

Scientific predictions can be sorted into several classes, which I will discuss in turn. However, I will not be able to claim completeness for this set of classes, nor is the assignment of a particular mode of prediction to one of these classes always unequivocal. In addition, some modes of prediction combine elements from different classes. Given the objective in this section, namely the demonstration that scientific predictions are more systematic than predictions based on other kinds of knowledge, this somewhat sketchy overview will suffice. I shall discuss predictions that are directly based on empirical regularities of the previous data of the phenomenon to be predicted (subsection 3.3.2), predictions that are based on correlations with other data sets (subsection 3.3.3), predictions that follow (more or less) directly from (fundamental) theories or laws (subsection 3.3.4), predictions that are based on models (subsection 3.3.5), and predictions using so-called Delphi methods (subsection 3.3.6).

3.3.2 Predictions Based on Empirical Regularities of the Data in Question

This subsection deals with the simplest and historically oldest case of predictions where, on the basis of the historical data for the phenomenon in question, a prediction of its future occurrence(s) can be made. This opportunity arises if the historical data exhibit temporal regularities. The most obvious and stringent regularities, accessible and of interest to all cultures of all ages, have been and still are astronomical regularities concerning the (apparent) motion of celestial bodies, most notably the sun and the moon. Although knowledge about the regular change of day and night, for instance, can certainly not count as scientific knowledge, knowledge about the more sophisticated regularities of phenomena like solar or lunar eclipses is at least a candidate for scientific knowledge. This is due to the fact that these events are too rare for their regularity to be obvious to common sense. In many cultures, records of the observations of eclipses were kept, and attempts at their predictions were made. It would, however, be premature to classify all systematic records and even predictions of eclipses in some culture as scientific, as in some cases, the means for predictions are of a particular kind, and the predictions themselves are fairly unreliable. For instance, in Assyrian and Babylonian letters and reports, some eclipses were foretold by liver- and oil-divination, by halos, by the untimely appearance of the new moon, or by the occurrence of fog.

Other prediction methods, however, resemble much more what we understand as science, and some belong to the Western heritage such that even a historical continuity with today's science exists. The earliest of such methods, used at least by 600 BC, was the so-called Saros period. Already in the middle of the eighth century BC, Babylonian astronomers seemed to have a more or less complete record of observed eclipses. Several regularities were discovered by analysis in these observations, among which a period of 223 (lunar) months (approximately eighteen years) was most important, the "Saros period." In addition, within any Saros period, slightly more complicated regularities for the possibility of eclipses were discerned. Simple extrapolation of these regularities into the future resulted in predictions of the potential occurrence of eclipses.

This exemplifies one of the simplest prediction methods used in science. Its general outline is: collect past data of the phenomenon to be predicted, analyze them with respect to temporal regularities, and employ these regularities for predictive purposes. Such regularities exist in different forms, for instance simple constancy of the data, or permanent increase or decrease (qualitatively), or a quantitatively specifiable pattern of increase or decrease, or simple periodicity, or a superposition of different periodicities, and so forth. This prediction method has various applications in many fields. It goes without saying that it is not always successful. For instance, the above-mentioned example of the prediction of possible eclipses on the basis of the Saros cycle was fairly successful for a couple hundred years but began to fail later. By the end of the fourth century BC, small eclipses that were not foretold were beginning to be visible, thus forcing revisions in the predictive scheme.

Another form of application of this prediction method occurs in economics. One temporal regularity used there is known as the "pig-cycle," because it was first investigated in this particular segment of the meat market. It can also be observed in other markets such as the labor market. It is an empirical fact that the supply for pig meat undergoes periodic fluctuations with roughly a four-year periodicity. The economic explanation for this periodicity appears simple at first glance, but the intricacies need not concern us here. Once this simple regularity of the supply is known, it is possible to predict the development of supplies in the future—so long as the regularity really holds.

Of course, in everyday life, we also make innumerable predictions of various events by extrapolating regularities found in previous data. We expect people to behave in a certain way because in the past they have behaved in that certain way; we expect, in those regions of the Earth where it applies, the same change of seasons every year; we expect, in some areas, quicker weather changes in April than in other months because that has happened in past Aprils; we are afraid of delays in air traffic because we have experienced them often before. Obviously, this sort of forecasting differs substantially from the corresponding scientific one. In the everyday cases, we never have carefully written records of data nor statistically analyzed data to find out whether the regularity is really supported, nor have we attempted to quantify the regularity, nor critically tried to understand the supposed regularity in order to determine its validity. Furthermore, we are using heuristics that are useful but sometimes lead to severe and systematic error. Therefore, it is evident that the everyday mode of prediction on the basis of regularities of past data is much less systematic than its scientific counterpart.

3.3.3 Predictions Based on Correlations with Other Data Sets

An important technique to derive predictions for some variables exploits their correlation with other variables. There are two main cases of how correlations of one set of variables with another one can lead to predictions. Let us call P the set of variables to be predicted and C the set of variables that are correlated with it. For the sake of simplicity, let us assume that P and C consist each of only one variable. For the first case, let us also assume that the correlation of P and C consists in a known functional dependence of P on C, i.e., $P = f(C)$, and that the temporal development of C obeys some known law $C = C(t)$. Under these circumstances, the temporal development of P can be predicted as $P = f(C(t))$.

Here is a concrete example that was not, however, enthusiastically received by the relevant community. The economist William Stanley Jevons (1835–1882) developed the sunspot theory of the business cycle. He was persuaded that the business cycles and the sunspot cycles were of the same length, approximately eleven years. Moreover, he believed that this coincidence was not accidental but rather causal, namely, that the sunspots causally influenced the economy. This influence was not immediate but mediated by effects of the sun's activity on the weather and the weather's effects on harvests. The lawful temporal pattern in the sunspot activity allowed the prediction of sunspot cycle peaks that, in turn, licensed the prediction of commercial cycle turning points. It also allowed the retrodiction of a peak in sunspots from an observed peak in corn prices. However we assess the validity of this approach, it perfectly exemplifies the forecasting technique based on dynamic laws for correlated variables.

In the second case, there are also correlated sets of variables P and C, but contrary to the first case, no laws for the time evolution of P are known. However, a specific regular correlation between $C(t)$ and $P(t)$ is known. In the simplest case, C anticipates the changes of P. Or, in other words, P covaries with C with a certain constant time lag. With appropriate units and with Δ denoting the time lag, this can be expressed as $C(t) = P(t + \Delta)$. Graphically speaking, the function $C(t)$ has the same shape as $P(t)$, but $P(t)$ is shifted by Δ toward the future. In other words, if one can measure the behavior of C at time t, one can predict the behavior of P at time $t + \Delta$. The behavior of C is thus an indicator for the future behavior of P. Here is a qualitative everyday example: the forecast of weather changes based on barometric pressure changes. Falling barometer readings indicate weather changing for the worse, rising barometer readings portend a weather change to the good, because the respective barometric pressure changes precede the respective weather changes.

I have only sketched the very simplest case of this forecasting technique, but more sophisticated scientific cases are in essence the same. This technique has been and is still widely used in many kinds of economic forecasting, and it is called "forecasting with leading indicators." The earliest attempt was the so-called Harvard A-B-C barometers introduced in the early 1920s. The idea was to first collect time series of many economic variables. These variables were then put into three classes. Most of them were indicators of the current situation. They were attributed to class B and called *current indicators*. Some variables showed the same movement as the current indicators, but a bit earlier. They were attributed to class A and called *leading indicators*. Finally, some variables showed the same movement as the current indicators, but somewhat later. They were attributed to class C and called *lagging indicators*. For obvious reasons, the mean of the variables of class A was called the "A barometer" because it could be used—analogously to a barometer reading—to forecast economic activity. For several years, the A barometer was quite successful in predicting the movements in the stock market. However, as it failed to predict the 1929 crash, it lost much of its reputation (in parallel to many people losing a lot of money).

Obviously, the identification of leading indicators presupposes a serious amount of data collection and analysis. In everyday life, we do not get involved in these sorts of activities. If we use any type of leading indicators at all, they are certainly not based on the sort of systematic sampling, recording, and evaluation of data. Again, the scientific practice is much more systematic than its everyday analogue (if that exists at all).

3.3.4 Predictions Based on (Fundamental) Theories or Laws

The class of predictions we discuss in this subsection cannot sharply be distinguished from the predictions treated in the following subsection, i.e., predictions that are based on models. The reason is that even predictions most directly derived from theories or laws involve additional elements. Typically, they comprise information about the system under survey and other "technical" elements, such as simplifications, abstractions, approximation procedures, limiting processes, and the like. The greater the number of these additional assumptions and the stronger their role is, the more one is inclined to speak of a model instead of a straightforward application of a theory. Thus, there is a continuous transition from a prediction "purely" based on theory to a prediction based on a model (I will deal with predictions based on models in the following subsection 3.3.5). Furthermore, predictions based on theory do not stand in opposition to those based on models with respect to their principal use of theory. The latter type of predictions also makes use of laws or theories, although typically in simplified form, but the predictions themselves are usually strongly dependent on the assumptions characteristic of the model. At any rate, scientific predictions exist that are clearly based on theory only.

A classic case is the discovery of the planet Neptune in 1846. Earlier, discrepancies between calculations and data for the orbit of the planet Uranus had been detected. These calculations were based on Newton's theory of gravitation. A possible explanation for these discrepancies was the existence of a yet undiscovered planet beyond Uranus that gravitationally influenced its orbit. The British mathematician John Couch Adams and the French mathematician and astronomer Urbain Le Verrier independently worked out this hypothesis. On the basis of the observed discrepancies and Newton's gravitational theory, they were able to predict the position of this yet unknown planet. On September 23, 1846, the new planet was indeed detected by the German astronomer Galle at the Berlin Observatory, almost exactly at the position predicted by the calculations.

Another famous case of a scientific prediction that was not mediated by a model but was more or less directly derived from a theory is the bending of light by gravitation. On the basis of his general theory of relativity, Albert Einstein predicted in 1916 that a ray of light passing a massive object should be bended by some particular small amount. The British astrophysicist Arthur Stanley Eddington then set out on an expedition to observe this effect during a total eclipse of the sun in 1919. Light from stars in the immediate vicinity of the sun should be bent, and during a total eclipse of the sun, these stars would become visible. It was then possible to compare the position of these stars when photographed during an eclipse of the sun with photographs of these stars in absence of the sun, i.e., in the night sky some six months earlier or later. If light bending takes place, there should be a shift in the apparent positions of the stars. On May 29, 1919, an ideally appropriate eclipse occurred with a very long duration and in the immediate vicinity of several fairly bright stars. Photographs showed indeed the predicted slight bending of the star light, not only qualitatively but also quantitatively as Einstein had predicted. It was a great triumph for the general relativity theory and its creator.

This scientific prediction technique has no real analogue in nonscientific thinking because there are usually no analogs to scientific (fundamental) theories or laws. At best, in our normal thinking, we predict events by using regularities that we believe are present in previous data of the respective phenomenon. We have discussed this case in subsection 3.2.2. By the invention of theories, the sciences systematize large portions of experiences that often appear to be entirely unrelated. By using these theories for predictions, science displays a higher degree of systematicity in this area, too.

3.3.5 Predictions Based on Models

Models (in the sense relevant here) are used when systems are too complex to be treated by theories or general laws alone. For instance, the global climate system involves a large number of variables interacting in various and complex ways such that it is impossible to establish a set of equations from fundamental physics describing the system, let alone to solve them. Similarly, an economic system should be described in terms of measurable variables like prices, costs, incomes, savings, employment, and so forth. The relationships among these are complex and derive from the interaction of millions of households, millions of firms, and thousands of governmental units, producing and exchanging millions of products. A complete representation of these relationships would involve trillions of equations, each of them as complex as human behavior. Again, this is intractable.

For any theoretical treatment of such hyper-complex systems, simplifications in the representation of their real structure are necessary. In our context, a set of such simplifications that somehow captures at least aspects of the behavior of the system is called a model of the system. Stripped down to its bare bones, a model is often a set of interlinked (or "coupled") equations. In the typical cases in the natural sciences, it is a set of time-dependent, coupled partial differential equations whose solution, given some appropriate initial and boundary conditions, should describe the behavior of the system. There are many conceptual, empirical, and technical problems that obstruct setting up such a system of equations, of getting appropriate initial and boundary conditions, and—last but not least—of solving the equations, even if only approximately, but this need not concern us here.

The simplifications constitutive of model building are introduced at various locations. Different variables are lumped together resulting in aggregated variables; known but mathematically difficult relations are simplified (e.g., by linearization); submodels are developed for particular processes; unknown relations are treated by reasonable assumptions or even neglected; ad hoc assumptions are introduced in order to correct systematic prediction errors; and so on. It should be noted that more often than not, these additional simplifications and assumptions are literally false. In quite a few cases, their falsehood may not even be euphemistically covered by ornamental supplements like "but they are approximately true." For instance, there are meteorological models in which even the law of energy conservation is violated—and this is one of the most fundamental laws of physics altogether. Of course, the scientists making such contra-factual assumptions are fully aware of them, and they do not even have a bad conscience about it. The reason is entirely pragmatic. If some assumption makes possible or improves the predictive power of a model, it is considered legitimate. The main goal of these models is not to get the fundamental working of the system in question right, but to make predictions. Of course, getting aspects of the fundamental working of the system right may be an additional asset for the predictive purpose, but even that is not necessarily so. In order to achieve good predictions, getting one part of the system right may require getting another part right as well, but the latter may turn out to be impossible. A literally false assumption about the net effect of the two parts together may be the best solution available to reach the set goal, namely, to get good predictions about the system as a whole.

Let us now look at the case of meteorology for a concrete example of a predictive model. A global meteorological model consists of a grid, dissolving the whole of Earth's atmosphere into a discrete set of points, a set of variables, and a set of dynamic equations involving these variables. At each point in the grid at some time t, each variable has a definite value, describing the actual weather state at this point at that time. The dynamic equations simulate the time evolution of the weather system; they are the core of the model. To get an idea of the size of such models, take the global model of the German Weather Service known as GME as a representative example. It consists of sixty vertical layers, each layer containing 1,474,550 grid points; its grid length, i.e. the horizontal distance between two neighboring grid points, is 20 kilometers. The weather state at each grid point is characterized by variables describing pressure, temperature, wind components, water vapor, cloud water, cloud ice, and content of rain and snow. The global weather state at one particular time is thus described by the numerical values of the variables at a total of 88 million grid points. Regional meteorological models like the COSMO-EU model for Europe are embedded in such global models; its grid length is 7 kilometers on 40 layers, totaling 17.5 million grid points. A high-resolution model like the COSMO-DE for Germany with a grid length of only 2.8 kilometers and fifty layers uses a total of 9.7 million grid points; it is also embedded in the larger models.

On the basis of the set of dynamic equations, the initial weather state at each grid point at some time t_0 and some further initial conditions concerning the Earth's surface, it is in principle possible to predict the future weather state of the system. This process is called numerical weather forecasting. In order to determine the initial weather state, a worldwide observation system has been installed, consisting of ground stations, ships, drifting buoys, land and ship radio probes (fixed on balloons to make measurements in the vertical dimension), aircraft reports, and satellites. In spite of these efforts, the determination of the weather system's initial state at time t_0 by measurements is always incomplete, because at many grid points, especially in the oceans, there are no measuring devices. For these grid points, the values of the variables at time t_0 have to be inferred (a process that is called data assimilation) and then fed into the model. The forecast range for the routine forecasts is up to seven days, up to three days, or eighteen hours for the global, the regional, or the high-resolution model, respectively.

Of course, there are many more areas where models are used for predictive (and other) purposes, and they may differ considerably in structure from the case discussed above. For instance, ever since the pioneering and controversial work of Jay Forrester and Dennis Meadows in the 1970s, so-called world models have received much public attention. Their models mapped important interrelationships between world population, industrial production, pollution, resources, and food. The models forecasted a collapse of the world's socioeconomic system sometime during the twenty-first century given the continuation of certain trends. Today, for similar reasons, global climate predictions on the basis of sophisticated models encounter high interest by the public. Again, these models' forecasts often describe devastating future scenarios. For instance, a model calculation of the consequences of long-term fossil fuel consumption was performed on a supercomputer that was the world's fastest machine in November 2005. The result was summarized as follows: "If humans continue to use fossil fuels in a business-as-usual manner for the next few centuries, the polar ice caps will be depleted, ocean sea levels will rise by seven meters and median air temperatures will soar to 14.5 degrees [Fahrenheit, or 8 degrees Celsius] warmer than current day." The truly dramatic predicted increase of sea level is due to the temperature increase in the polar regions of more than 20 degrees Celsius. Nowadays, many institutions around the world deal with global predictions that are all based on models.

In everyday life, we do have analogs of these scientific models, namely, sets of simplifying assumptions that permit predictions. Especially in the social world, we operate on such assumptions. Predictions that Ann will be the most suitable partner for Peter, that the current president will be reelected, that the unrest in country X will not cease soon, and so on rest on sets of simplifying and, to say the least, possibly literally false assumptions. For better or worse, we have to constantly make decisions in daily life that seriously affect our future—we are dependent on such predictions. Very rarely are we aware of the whole set of assumptions guiding these predictions; very rarely do we critically investigate these assumptions; very rarely can we really distinguish them from mere prejudices. In other words, despite their structural similarity to scientific models, these everyday models lack the sophistication in many dimensions that the scientific models own.

3.3.6 Predictions Based on Delphi Methods

An entirely different type of scientific predictions is based on so-called Delphi methods. As the name indicates, there is a certain similarity of these methods with the ancient oracle located in the Greek city of Delphi. The method, which is also categorized as "judgmental prediction," was developed in the 1950s in the Rand Corporation, surprisingly enough by three philosophers. Basically, the procedure is as follows. Locate a number of experts on the phenomenon to be predicted. Ask these experts individually about their prediction regarding the phenomenon. If all their answers are roughly the same, then this is the final prediction. If their answers diverge significantly, feed these divergent answers back to the experts in order to give them a chance to change their individual predictions in light of their colleagues' predictions. If the updated predictions then converge, this is the final prediction. If they do not converge, try repeating the process, or give up: the attempted prediction failed.

Delphi methods are predominantly used in areas where all of the other prediction techniques fail. This holds especially for areas where qualitative changes that do not admit of any quantification predominate. Forecasting long-term developments, i.e., concerning the next twenty-five years, for example, in technology, science, society, and warfare provide cases in point. In retrospect, it is obvious that such predictions partly fail very badly, but they may also catch some trends quite accurately.

Of course, what has been cultivated here to be a respectable means of prediction in science is again well known from our everyday practice, at least in principle. In our daily life, we are used to asking people about their guesses about the future, and we may confront the guess by one person with the guess by someone else. Also, in many rather important areas of life, several experts try to build up a consensus in an informal way. Take the prediction of the future course of a disease that afflicts a particular patient. In such a situation, health professionals from different medical disciplines may try to find a consensus about the most likely course of the disease and about the best possible therapy. The discussion is usually informal, i.e., there are no explicit rules guiding the discussion and the formation of a consensus. A similar case is the prediction of criminal recidivism in court. Often, various experts are heard and interrogated on the basis of which the judge forms his or her opinion about the case at hand. Important court decisions may depend on this judgment. Clearly, these procedures of prediction on the basis of expert opinions are typically fairly cursory and in many respects much less systematic than a scientific Delphi study.

3.4 THE DEFENSE OF KNOWLEDGE CLAIMS

3.4.1 Some Preliminaries

The high esteem that science enjoys almost everywhere derives from its reputation to produce a superior form of knowledge. Scientific knowledge is supposed to be less prone to error than other forms of knowledge; it is supposed to have a higher quality in being more reliable. The central insight that science takes more seriously than most other comparable human activities is that human knowledge is constantly threatened by error. Error may arise as the result of individual or collective mistakes, by false assumptions, entrenched traditions, dogmatic indurations, belief in authorities, superstition, wishful thinking, prejudice, bias, reliance on error-prone heuristics, and even fraud. Of course, in principle, we all know of these possibilities also in everyday thinking, but science is typically more careful and also more successful in detecting and eliminating these sources of error. To be sure, it is by no means invariably successful, but science appears to be the human enterprise that is *most systematic* in its attempt to eliminate error in the search for knowledge. It is thus evident that the defense of knowledge claims is an absolutely indispensable dimension in science's systematicity. This is, of course, no surprise, because ever since the reflection on science has begun, both in the sciences themselves and in philosophy, this aspect of science has always been in focus. Systematicity theory continues this tradition.

Before discussing the defense of knowledge claims in the sciences in detail, some preliminary remarks are in order. They concern the terminology used in this section, the differences among the sciences in their ways to defend knowledge claims, the relationship of the questions treated in this section with the so-called context of justification, and a distinction between empirical sciences in a narrow and in a wide sense.

First, a few words on my terminology in this section. In the philosophy of science, there are a host of expressions designating the ways by which the sciences attempt to secure high quality of their knowledge claims. These expressions have been taken over from ordinary or scientific language and have been endowed with a more precise meaning. Typical expressions of this kind include the following: proof, verification, empirical or inductive support, justification, certification, confirmation, corroboration, validation, critical test, disconfirmation, falsification, refutation, organized skepticism, and the like. These expressions belong to two groups, depending on two conceptions about how the sciences operate regarding the enhancement of their knowledge claims. On the first conception, science is seen to improve its knowledge claims primarily by positive measures regarding hypotheses. The details of these measures vary widely, but their common claim is that they positively support a hypothesis. On the other conception, the higher quality of the sciences' knowledge claims is reached by negative measures. The underlying idea is that a positive support of empirical knowledge claims is in principle not possible. The only possibility left is a sustained attempt at the diagnosis of error and its subsequent elimination.

However, in this book I do not want to take sides in this controversy because it is not relevant for my concerns. Whether the pertinent activities in the science are more positive regarding their direction, or negative, or an inextricable mixture of both such that their opposition only raises pseudo-problems is of no concern here. I am interested in their degree of systematicity when compared to similar procedures that are used in other forms of knowledge. Therefore, I am trying to use a vocabulary that is as neutral as possible with regard to these different directions. Already in the title of this section, I am speaking of "the defense of knowledge claims," leaving it completely open whether this defense moves in the more positive direction of confirmation or in the more negative direction of criticism.

Second, with respect to the defense of knowledge claims, there are vast differences among the sciences. On the one hand, there are very different things that must be defended against error: there are data, singular hypotheses, empirical generalizations, theorems, models, theories, explanations, interpretations, classifications, and the like. On the other hand, there are very different procedures by which the putative correctness of such things may be defended: by proof, observations, experiments, statistical analyses, comparison with sources, and the like. It is obvious that procedures that are successful in one field may be inapplicable in other fields, without the implication that the latter fields are inferior. However, it is an empirical fact that the higher the reputation of one's own field is regarding error elimination and the demonstrated stability of knowledge claims, the stronger the tendency to look down on other fields. The stereotype of the "hard" versus the "soft" sciences is very common, sometimes even suggesting that the "soft" disciplines are not really intellectually respectable enterprises at all. Similarly, even within the natural sciences, there is a tendency for those working in the experimental sciences to depreciate the historical natural sciences because the latter are notoriously unable to subject all their claims to controlled laboratory experiments. This attitude is, however, definitely unjust. Different fields admit of different procedures, of different degrees of rigor, and different success in error elimination, and very often the very best scientists in a given field represent the respective *realistic* gold standard of critical scrutiny. This standard reflects what can realistically be achieved in that field in the given historical situation. Scientists from higher reputation fields invading lower reputation fields typically do not successfully import higher standards for the defense of knowledge claims into their new field. Rather, they will have to learn that the standards that are valid in a particular field are typically not so easily improved as they may imagine. Therefore, we will have to strongly differentiate between different fields.

Third, it should be noted that the questions discussed in this section seem to belong to what in analytical philosophy of science has traditionally been called "the context of justification." However, some caution is recommended regarding this concept. The expression "context of justification" has been used in philosophy of science in a rather specific sense. In this (dominant) sense, it was not philosophically neutral by just denoting the evaluation of knowledge claims, i.e., the question of whether some knowledge claim is justified or not. Rather, in its standard usage, the concept of the "context of justification" was heavily loaded with key assumptions of the logical empiricist and the critical rationalist tradition. More specifically, the context of justification was believed to be sharply distinguished from the "context of discovery," i.e., from the processes and procedures that led to the initial discovery of some hypothesis or theory. Furthermore, the consideration of questions belonging to the context of discovery was banned from the domain of the philosophy of science. Finally, it was believed that the questions belonging to the context of justification, i.e., questions of epistemic justification or test, had to be exclusively treated by logical means. All of these assumptions turned out to be highly controversial. I will not follow up this controversy in any detail here, but only note that in some of the examples that follow, the application of the traditional context distinction is impossible. These are scientific episodes in which there is just one process with regard to which it is impossible to distinguish discovery aspects from justificatory aspects. In the relevant respect, these processes are just like finding out the product of two numbers, e.g., 349 times 981: the process to "discover" the result is exactly identical with the process to "justify" that result, namely, by carrying out the necessary steps of the multiplication. We will see that some controlled experiments, to be discussed below, are exactly of this type. Thus, one should not approach this section with a sharp (traditional) distinction between the context of discovery and the context of justification in mind. What I shall not doubt, however, is that questions regarding the defense of knowledge claims make sense, or, in other words, that questions in a normative mode regarding the legitimacy of knowledge claims can and must be posed.

Fourth, an ambiguity in the use of the expression "empirical sciences" should be noted if "science" is here taken in the wide sense, including all research disciplines. Typically, "empirical science" means natural sciences like astronomy, physics, and crystallography and social sciences like psychology, sociology, and economics. These sciences are called empirical because they produce and use empirical data when attempting to justify knowledge claims (in fact, empirical data seem to be the final arbiter regarding justificatory questions in these fields). By contrast, it sounds a little odd to call humanities like art history, Slavic literature, or musicology "empirical" disciplines. Upon closer inspection, however, they are also empirical in the sense that their ultimate basis for all justificatory claims is data that are empirical. Of course, most of the relevant data are of human origin, predominantly texts, pictures, and the like. Undoubtedly, the only access to those data is through our outer senses, and they can therefore be called empirical. In another respect, however, they differ from typical empirical data. Because of their human origin, i.e., by being texts or text-analogs, they embody meaning that is not accessible to our outer senses (compare our discussion of the concept of meaning as it is pertinent for the humanities in subsection 3.2.7). Without the necessary reading (in the literal sense) of these data or their interpretation, they are tacit regarding all justificatory purposes in the humanities. Thus, their empirical quality as affecting the senses is only necessary but by no means sufficient for their quality as providing relevant data for justificatory purposes. By contrast, the same sort of interpretation of data is totally absent in most of the natural empirical sciences (except, perhaps, some areas of ethology) and also often absent in the social sciences.

The upshot is that one should distinguish a narrower and a wider sense of "empirical science." In the *wider* sense, sciences are empirical if their justificatory procedures depend on any sort of data that are empirical in character or, in other words, on empirical evidence. I take "empirical evidence" to include textual sources and text-analogs of all sorts when used for justificatory or critical purposes regarding knowledge claims. Therefore, the text-based disciplines are also empirical in this wide sense. The contrasting class to the empirical sciences in this wide sense is the class of the formal sciences (whose justificatory procedures do not depend on empirical evidence). In the *narrow* sense, disciplines are called empirical if their justificatory procedures depend on empirical data that are not themselves texts or text-analogs.

In the following, I shall first discuss procedures to defend knowledge claims that do not depend on empirical evidence. This mainly concerns the formal sciences but also some considerations found in the theoretical parts of empirical sciences (subsection 3.4.2). In the following subsections, considerations of empirical evidence will always be involved. I will start discussing procedures most often found in the empirical sciences in the narrow sense: the defense of empirical generalizations, models, and theories (subsection 3.4.3). In subsection 3.4.4, I will continue with a discussion of procedures that allow the identification of factors that are causally relevant for some phenomenon. Subsection 3.4.5 will investigate the so-called *verum factum* principle that is important in some natural and engineering sciences. The role of mathematics with regard to the defense of knowledge claims in various sciences is considered in subsection 3.4.6. Finally, in subsection 3.4.7, I shall discuss some peculiarities of the historical sciences.

3.4.2 Nonevidential Considerations

Nonevidential considerations are considerations that are not based on some sort of evidence, "evidence" meaning any kind of empirical data. Clearly, nonevidential considerations concerning the defense of knowledge claims play a preeminent role in the nonempirical, that is the formal, sciences like mathematics and logics. In these disciplines, the most rigorous form of a defense of a knowledge claim has been practiced since antiquity, namely to provide a proof for any statement that is not an axiom, a definition, or a convention. Basically, a proof consists of a derivation of the statement in question from axioms and definitions. However, this is a fairly flexible formulation so long as it is not precisely specified what is and what is not allowed to be part of the "derivation." A proof is believed to establish cognitive certainty for the statement in question, at least relative to the axioms. Although this idea seems not to have changed in the course of history, the means that were deemed adequate to constitute proofs have changed, and they have also varied across mathematical disciplines. In fact, both components of proofs have undergone historical change: the axioms and the admissible rules of derivation. To cut a long and complicated story short: in the course of the nineteenth century, the status of axioms that from antiquity onward were believed to be necessarily true was downgraded to the status of mere assumptions whose truth could not be a matter of dispute. The rules admitted for derivation in a proof was more and more restricted, culminating in the most rigorous demand for rules that are mechanically executable. Very recently, a new controversy about admissible proofs has emerged due to the invention of mathematical procedures that claim to be proofs but involve the indispensable use of computers. It should also be noted that for the longest time of its history, genuine scientific knowledge in general was posited to be absolutely certain. As we have seen in section 1.1, up to the middle of the nineteenth century, this was the epistemic ideal not only for the formal sciences. It was believed that this ideal could be realized by the application of proofs (up to the seventeenth century) or of the scientific method. However, at least from the end of the nineteenth century on, it was realized that proofs were only available in the formal sciences, and even there, they could not establish absolute truth of statements but only truth relative to some set of axioms.

Of course, this way of error elimination displays an unsurpassed degree of systematicity. Hence, regarding the dimension of the defense of knowledge claims, for the formal sciences, the thesis of the higher systematicity of scientific knowledge in comparison to other kinds of knowledge is trivially fulfilled.

Nonevidential considerations also play a role in the empirical sciences, and they have been extensively discussed in recent decades in the philosophy of science, usually in the context of the theory choice situation. When scientists have to choose between competing theories, it is not only the available empirical evidence that is relevant for their choices. Other properties of the competing theories also will be considered, for instance, their internal consistency, their unifying power, their relationship with other theories, their simplicity and elegance, their scope, and their perceived potential to guide future research in a fruitful way. Thus, in the history of science, scientists have indeed defended their particular choice of a theory with recourse to these so-called epistemic values. However, in the context of this section, the question suggests itself whether such a defense can really count as the defense of a knowledge claim. Typically, it cannot. The reason is the nonevidential epistemic values mentioned usually play a role, for some scientists even a decisive role, in theory choice situations in which not enough empirical evidence is available on which to base the choice. In spite of this shortage, scientists have to make up their mind about which theory to work with. Nonevidential values can then play an important *heuristic* role. We shall therefore not deal with these heuristic nonevidential considerations in this section because we are here concerned with defenses of knowledge claims.

3.4.3 *Empirical Generalizations, Models, and Theories*

Empirical generalizations, models, and theories differ from one another in important respects. With respect to procedures of error elimination, however, they can often be treated similarly. The basic idea here is quite simple: to confront these theoretical constructs with empirical data. If the data fit the theoretical construct, no error has been detected. If the data considerably disagree with the theoretical construct, something is wrong. However, in most practical cases, this confrontation turns out to be quite complicated. There are three different steps involved in this confrontation, and their order is not necessarily the same as they are presented here. First, one has to produce relevant empirical data. Second, one has to manipulate the theoretical constructs such that a confrontation with the data is possible. Third, one has to interpret the result of the confrontation.

In the most simple cases, all steps are straightforward. Today, one can find these simple cases only in undergraduate physics or chemistry classes. In much of seventeenth-century science, however, they were cases of real productive research. Let us take Boyle's law, discovered in 1662, as an example to illustrate the three steps. Boyle's law reads

$$pV = const.,$$

where p is the pressure and V the volume of a gas sample contained in a vessel. Given constant temperature and a constant amount of gas, the law states that the product of pressure and volume of the gas is a constant, no matter how you vary the volume of the gas sample. Of course, there is no way to *directly* confront this equation with any empirical data. However, in a trivial step, one can derive a set of equations for different values of p and V from Boyle's law, namely,

$$p_1V_1 = p_2V_2 = p_3V_3 = \ldots$$

This is the second step from above. It is fairly simple to set up an experiment in which the volume of a gas sample can be manipulated and the respective values of p and V can be measured. This is the first step from above. The confrontation of the law with the data consists in the comparison of the products of the different measured values for p and V. If these products are constant, the law has been defended, if they are not, the law has not been defended. This is the third step from above.

Although this really is the general pattern of the confrontation of empirical generalizations, models, and theories with empirical data, in most cases, a number of substantial deviations from this simple example can be found. There are at least four ways in which things can be and typically are more complicated.

1. In Boyle's case, it is unequivocal which empirical data are relevant to defend or criticize the law: it the pressure and the volume of gas samples of constant mass and temperature. This can be directly derived from the law in question. In many other cases, however, the theoretical construct's hints about which empirical data exactly are relevant for its defense or criticism are much weaker. For instance, in Schrödinger's equation for quantum mechanics or in Einstein's field equations for gravitation, most of the variables contained in these equations cannot be directly measured because they denote so-called theoretical entities, i.e., entities that are theoretical posits that cannot be directly observed or measured. Thus, it is not at all clear in which way exactly the contact with empirical data can be made. As the gap between theoretical entities and observables may be fairly wide, for most models and theories, much creative work is needed in order to derive consequences from them that are at least in principle empirically testable. Furthermore, the values of the variables to be measured must not only be in principle empirically measurable, but they must be in the technical reach of real measuring instruments.

There is a further difficulty for the comparison of theoretical constructs with empirical data that is located at the theoretical side. The derivations from a theory or a model that shall connect it with the data typically involve other theoretical elements. This includes assumptions of various sorts like (mathematical) approximations, simplifications, or idealizations, or other theories, laws, models, or fragments of them, and so on. The involvement of these additional elements in the confrontation of the theoretical constructs with empirical data has the unpleasant consequence that the result of the confrontation becomes equivocal, at least in principle, and very often also in practice. Neither a fit nor a disagreement between theoretically derived and measured values can be unequivocally traced to the law, model, or theory. A fit may occur because of a cancellation of errors from different sources, and a disagreement may be due to errors in the additional assumptions while the original theoretical construct may be correct. This is the essence of the so-called Duhem-Quine thesis. A deviation of empirical values from values derived from a theory can usually not immediately be directed against the theory in question but only against the whole of theory, auxiliary theories and assumptions, and anything else that is involved. Of course, this does not exclude further procedures of error localization. It just rules out simple-minded ideas about theory confirmation and falsification.

2. Also on the data side itself, things are typically not as simple as in Boyle's case. Usually, data that are measured are "raw data," indicating that some sort of processing is necessary in order to obtain data in the form in which they are needed.

First, very often raw data have to be corrected. For instance, in the case of temperature measurements with thermometers that come in physical contact with the entity to be measured, the temperature value read off the thermometer is not truly the temperature of the measured entity. Rather, it is the temperature of the system that consists of the original entity now combined with the thermometer, once that system has reached thermal equilibrium. Thus, the temperature shown by the thermometer must be corrected; it is only a raw datum. The correction involves the temperature of the thermometer before the measurement, its heat capacity, and the pertinent part of thermodynamics. Its result is the temperature of the entity before the measurement was done.

Second, very often the value of the variable of interest cannot be directly measured. Rather, some other quantity is measured from which the value of the quantity of interest can be derived. For instance, in most cases, the velocity with which astronomical objects move cannot be directly measured. The radial component of the velocity (the velocity toward us or away from us) is often measured by means of the Doppler effect, i.e., by measuring the shift in the electromagnetic spectrum of the source. This involves knowledge of relevant spectral lines and how a shift in their frequency translates into velocities. The transverse component of the velocity (perpendicular to the observer–object–axis) is often measured by its angular velocity, which involves measurements of angles and time differences. If the distance of the object is known, the transverse velocity component can be calculated from the angular velocity.

Third, data may have to be interpreted. This is, of course, not the same meaning of "interpretation" that we have used when discussing the specifics of the humanities, for instance in section 3.2.7. Interpretation of data here means, in the context of the natural sciences, the establishment of a definite relation to one or several hypotheses, especially of consistency or inconsistency. By that sort of interpretation, the data become "meaningful" in that their relevance for the hypotheses becomes explicit, whereas before the interpretation they were mute with respect to them. Such interpretations may be far from unique and consequently controversial. For example, the 1976 Viking Lander missions to Mars carried out several independent robotic experiments on its surface designed to determine whether there was extant and/or extinct life on Mars. Of course, these experiments were meant to produce an unequivocal positive or negative result regarding the (former) existence of life on Mars. Yet the experimental results were far from unequivocal. On the contrary, the combination of the results of all experiments was completely unanticipated and led to a variety of alternative theoretical explanations. The main problem was and is that there is no possibility of performing further experiments on Mars at this point in order to assess the various assumptions on which these theoretical explanations were based. In spite of that, even thirty years after the Viking mission, the discussion about the interpretation of its results has not ended.

3. In the examples discussed so far, the access to the relevant data was in the hands of the scientists because they were experimentally produced. Of course, this does not imply unrestricted and unlimited access to such data. Many contingent factors like available finances, available technology, ethical restrictions (in the case of the biomedical and environmental sciences), and other limitations of all sorts restrict the possibilities of the production of experimental data. However, in many natural sciences, the access to data is largely, or even exclusively, observational and thus, at least to some degree, dependent on lucky circumstances. This does not only concern the historical natural sciences, like paleontology or cosmology. It is obvious that the unearthing of a specimen of a yet unknown fossil species cannot be experimentally forced, just as the explosion of a particular kind of supernova cannot. However, there are many other sciences in which the relevant data can only be observed and not experimentally produced. For instance, astronomy, cultural anthropology, and ethology (the study of animal behavior) strongly depend on data that are exclusively gained in the field. Sciences that are concerned with natural disasters are dependent on the occurrence of such events. For instance, since the 1990s, theoretical research on tsunamis has developed computer models for tsunami propagation through the open ocean. However, investigators had few observations to compare against their models. The effects of the tsunami-generating earthquake of December 26, 2004, were recorded by three earth-monitoring satellites that happened to orbit the relevant region between two and nine hours after the earthquakes. This coincidence allowed making the first radar measurements of a tsunami propagating across the open ocean and thus checking theoretical models. In fact, these models were pretty much validated.

4. At least in the case of theories, especially foundational theories, it may be the case that the typical situation of a confrontation with empirical data concerns more than one theory. More to the point, it is a competition among theories that is the typical context of the confrontation of theories with data. Theory comparison has been a much discussed subject in the philosophy of science because some theorists have claimed that particular difficulties are raised if the theories in question are "incommensurable." This is not the place to follow up this discussion at any length. The central element is that with respect to incommensurable theories, relevant data may not be neutral in the sense that exactly the same data play exactly the same role for the theories in question. Instead, not exactly the same data are seen as relevant for an assessment of one or the other theory. To be sure, according to most defenders of incommensurability this does not make an achievement-based comparison of incommensurable theories impossible; it only makes it more complicated by involving some weighing and judging. But we can leave this point at that.

The result of this subsection is this. Checking scientific empirical generalizations, models, and theories for their empirical correctness is a fairly complicated affair, although the basic pattern is simple. In fact, the basic pattern is the same as in daily life when we check, if we do, generalizations for their correctness. Clearly, we check whether the generalization applies to particular instances that it should cover. Fundamentally, we do the same in the sciences, although in a much more sophisticated way. One way of putting this, appropriate in our context, is to say that science is much more systematic in defending the knowledge claims associated with empirical generalizations, models, and theories.

3.4.4 Causal Influence

A very specific kind of scientific task concerns the identification of causally relevant factors for a given phenomenon. It has been commonplace since the time of David Hume (1711–1776) that it is impossible to identify causal factors by pure observation. Even if an event A is always followed by an event B, this fact alone does not establish a causal connection between the two. For instance, it could be the case that another event C, taking place before A, invariably causes first A and later B. For instance, an illness C could first manifest itself by symptom A, and only later by its characteristic feature B. The symptom A would certainly not count as the cause of the characteristic feature B of the illness C. How then can genuine causal connections be identified and tested?

In the experimental sciences, the answer to this problem is a specific experimental arrangement, commonly called a "controlled experiment." Already in the nineteenth century, John Stuart Mill (1806–1873) had clearly described this arrangement in his

System of Logic under the label "method of difference." Here is the simplest case. The question is whether or not some factor A is causally necessary for the occurrence of a feature B in some well-defined situation S. For example, take the question whether the presence of a particular substance (A) is a necessary condition for the occurrence of a specific chemical reaction (B) in a particular situation S. Two experiments are needed in order to decide this question experimentally. In the first experiment, the situation S that includes the putative causal factor A is generated. If, in this situation, B is regularly produced, then the question can be asked whether A is a part of the causal mechanism of S that brings about B. A second experiment, the so-called control experiment, is then set up to answer this question. In that experiment, the situation S is generated again, but without the occurrence of A. If A is not causally necessary for B, then B will nevertheless occur in the control experiment; if A is causally necessary for B, B will not occur.

This sort of experimental setup is of utmost importance in many branches of chemistry, pharmacology, medicine, biology, education, criminology, and other areas. For instance, the question of whether a particular gene G is causally relevant for some disease D can be investigated in this way, using modern techniques of molecular biology. In order to prove the causal role of the gene G for the genesis of the disease D, genetically modified organisms are produced that lack the gene G but are otherwise completely intact (so-called "knock-out organisms," because the gene G has been "knocked out"). If these organisms still develop the disease D, the gene G is not causally relevant for D; if they do not, G is causally relevant for D.

In the preceding case, the question is a qualitative one: is something a causally necessary factor (in a given situation) or not? However, there is a quantitative, i.e., statistical, variant of the controlled experiment that is extremely important in medical research. Here, the question is not simply whether or not A causes B, but whether the occurrence of A causally increases the probability of the occurrence of B. This is a relevant question in situations in which many variables play a role that cannot be completely controlled. This is exactly the situation in those areas of medical research where the efficacy or the side effects of some treatment are to be assessed. For the recovery from some illness, or the prevention of its recurrence, or the palliation of its symptoms, or the question whether some treatment has side effects, a host of factors play a role, many of which are unknown or uncontrollable. In situations like this, the best one can do is to assess whether in a statistical average some treatment has better results than no treatment or a different treatment. Stripped to its absolute essentials, the procedure of these "treatment-control studies" (or "randomized trials") is this. First, two statistically approximately identical groups of patients are formed by randomly distributing the patients into two groups. One group is given the treatment to be assessed. The other group, the so-called control group, is either not treated at all or is given a treatment with which the new treatment is to be compared. Differences in the development of the disease or the occurrence of putative side effects between the two groups can then be ascribed to the treatment as its causal effect. For instance, in advanced cancer treatment, the question often arises whether a combination of two pharmaceuticals is more effective than treatment with one of them alone. In order to study this question, two groups of patients are formed. One group is treated with only one pharmaceutical; the other group gets the combination treatment. The progress of the two groups is monitored, and their difference, if existent, is assessed.

This procedure of identifying causal effects by controlled experiments is also used in the social sciences. For instance, it is an important question what the different effects of different legal sanctions are with regard to the rehabilitation of convicts. Because the goal of legal sanctions is, roughly speaking, not revenge but rehabilitation, we would like to choose those sanctions that have an optimal rehabilitating effect. But which legal sanctions have what rehabilitating effect? For instance, does community service rehabilitate better than short-term imprisonment? Such a question can be reliably answered only by means of controlled experiments, and the procedure is principally the same as in the medical case discussed above. Again, one has to form two statistically roughly identical groups by randomly distributing convicts among the two groups. Members of group 1 receive sanction 1, and members of group 2 receive sanction 2. Roughly speaking, statistically significant differences in the short-term and long-term behavior of the members of the two groups are then interpreted as causal effects of the difference in the sanctions. It is very interesting to see that such controlled experiments may have very counterintuitive results: interventions that were initially seen as probably having positive effects may turn out to be harmful in the long run. It is important to note that probably only sophisticated controlled experiments could discover these harmful effects that would otherwise have gone unnoticed. An interesting example of this kind is the famous Cambridge-Somerville study, a pioneering longitudinal study of delinquency prevention that was initiated in the 1930s. More than six hundred boys at high risk for later delinquency were randomly assigned to two groups. With respect to a number of relevant variables like physical health, mental health, social status of parents, delinquency prediction scores, and so on, these two groups were equal. Members of the first group were intensively supported by social workers with whom they could build up a personal relationship while members of the second group were left untreated. Roughly thirty-five years later, more than five hundred participants of the program could be traced and investigated. The surprising result was that the two groups did not significantly differ with respect to their criminal records but that members of the group that had received treatment had a significantly lower mental and physical health status and, correspondingly, a significantly lower life expectancy. These findings suggested that contrary to all expectations, the treatment intended as delinquency prevention was eventually harmful. This fact could not have been detected without the controlled experiment because, in retrospect, both the treated subjects and the social workers involved had very positive recollections about the program.

There are many variants of this principal scheme of treatment-control studies, especially nonexperimental (or "quasi-experimental") ones, which have their own problems. There are many details that I have not touched at all, like further methodological and statistical issues or the ethical problems involved when controlled experiments are carried out with human subjects. I shall not delve into these ramifications, but rather will follow our main objective, namely, to compare these procedures with similar ones from ordinary practice. In fact, sometimes in everyday life, people try to assess causal factors in the same way as in the simplest case of the above-mentioned "method of difference." Here is an anecdote. The philosopher J. T., who had a brilliant career both in the United States and Europe, was known to join in and contribute to discussions on almost any subject. Some of his colleagues at Harvard grew suspicious that he might do that regardless of his actual knowledge state about the subject in question. In order to investigate the case, they started, while he was present, a discussion about the doctrine of soul in the work of Bertrand of Hildesheim, who was supposedly a middle-age scholastic intellectually located in between Thomism and Scotism. After a while, J. T. joined the conversation, apparently impressing everybody with his deep knowledge of Bertrand's work. Unfortunately, Bertrand of Hildesheim never existed so there was nothing to know about him—his existence was made up by J. T.'s colleagues. This was a decisive empirical test that J. T.'s joining a conversation was not causally dependent on J. T.'s knowing anything about the subject of the conversation, as his colleagues had suspected. The upshot of this anecdote is that the basic line of reasoning concerning the experimental identification of causally relevant factors of some phenomenon is a part of common sense. In the sciences, however, this basic line of reasoning is developed into highly sophisticated, or systematic, methods in order to generate and defend knowledge about causal connections.

The drive to increase systematicity in the sciences does not stop at single randomized trials. The two main reasons for this are that one cannot tell whether the results of an individual study are sufficiently robust against the effects of chance (i.e., whether they are statistically reliable) nor whether there is so-called external validity (i.e., whether their results can be transferred to contexts different from the original study). In order to investigate these problems, different randomized trials conducted under different circumstances should be combined in order to yield results with higher statistical robustness and a higher degree of external validity, or an identification of yet unnoticed relevant causal factors. This is the task of a huge collaboration in the medical sciences called the Cochrane Collaboration, named after the British epidemiologist Archie Cochrane and established in 1993. In 1972, Cochrane had published an article in which he drew attention to the collective ignorance about the effects of health care, which turned out to be the initial spark of the collaboration. Today, this collaboration is supported by hundreds of organizations around the world, and it is engaged in the production and maintenance of what is called "systematic reviews." A systematic review tries to combine a number of randomized trials in as reliable a way as possible; hence "systematic." It tries to avoid preconceived opinions and outright prejudices by using a predefined, explicit methodology. In this way, bias in all parts of the process should be minimized. The relevant studies that are identified and selected for inclusion are, of course, sought regardless of their results. The methods to be followed in the identification and selection of studies and the collection and combination of their data are set forth in detail in the *Cochrane Handbook for Systematic Reviews of Interventions*. The main product of the Cochrane Collaboration is the *Cochrane Database of Systematic Reviews* that contained a total of more than 4,600 records as of March 2007; the *Cochrane Central Register of Controlled Trials*, i.e., the subject matter of the systematic reviews, contained almost half a million records!

In the area of the social, behavioral, and educational sciences, there is a sister organization called the Campbell Collaboration that cooperates closely with the Cochrane Collaboration. Founded in 1999, its aim is, quite similarly to the Cochrane Collaboration, to help "people make well-informed decisions by preparing, maintaining and disseminating systematic reviews in education, crime and justice, and social welfare." Evidently, these two organizations produce knowledge that is even more systematic than the already highly systematic empirical studies that they combine into systematic reviews. Of course, these reviews are much more systematic than anything we encounter in everyday knowledge when we want to know whether some intervention will lead to the desired effect. What is particularly nice in our context is the fact that the description of the reviews as "systematic" is not my invention but the characterization of these reviews by the creators of the collaborations themselves.

3.4.5 The Verum Factum Principle

The aspect of the defense of knowledge claims I am discussing in this subsection bears a venerable name, the *verum factum* principle, also called Vico's principle. This is because the principle was introduced and discussed by Giambattista Vico (1668–1744) in the context of his philosophy of science. It roughly states that true knowledge of a thing presupposes that we know the origins of the thing as a result of human actions. Acceptance of this principle implies that true knowledge is only possible in the realm of human products, but not regarding nature. Today's main application of the principle, however, is not in the humanities, despite its Latin denomination, but in the natural and engineering sciences—even if not always under its original name. The idea is that a test for the presumed insight into the functioning of some system is the ability to re-create it, either as a physical model or at least as a computer model of it. Of course, strictly speaking, the test can only be negative—if your re-creation doesn't work, you have certainly missed something of the original. It cannot be definitively positive because the re-created function could be realized in the original by other means than in the model. The model would then only represent a functional equivalent of the original.

Here is an interesting example from biology. Desert ants of the species *Cataglyphis* follow a circuitous path of several hundred meters when foraging. However, once they find food, they run back to their starting point, the nest, that is not visible for them, in a straight line. How do they determine the direction in which to run? The brain that performs the trick weighs only one-tenth of a milligram and has a couple hundred thousand nerve cells. Observations and sophisticated experiments have revealed that the main ingredients of the ant's navigation system consist in its ability to perceive the polarization of sunlight and in a specific processing of this information by comparatively few neurons. These elements also explain why the ant performs peculiar bodily movements when leaving the nest: they serve to calibrate its perceptual system. Even the ant's specific neural realization of its orientation capability by particular nerve cells could be discovered. However, the ultimate test for the correctness of all these hypotheses was a physical model of the ant, a ten-kilo robot in which an electronic equivalent of the ant's neural network was implemented, together with appropriate electromechanical devices. Indeed, this robot exhibited the same sort of behavior as the ant, thus showing that the neural elements identified in the ant were capable of delivering the sort of behavior that was observed.

In much more embryonic form, we use the same sort of reasoning in our normal life. For instance, imagine someone shows you a simple magic trick. You may not discover the secret yourself, but after having been told how it works, the first step usually is to try out the trick oneself. Having successfully rehearsed the trick, one is certain to have understood how it works. Clearly, as I mentioned often, the everyday practice is much less systematic than its scientific counterparts.

3.4.6 *The Role of Mathematics in the Sciences*

So far, we have seen in this book at various places that mathematics plays a major role in the sciences, for instance, when I discussed the quantification of descriptions in subsection 3.1.5. In this subsection, I shall discuss the specific roles that mathematics can play when it comes to the defense of knowledge claims. From the start, one should be aware that the expression, "the use of mathematics," should not be equated with "the use of quantitative procedures." The latter is an important part of the former but by no means exhausts it. Let us nevertheless begin with quantification. But I must add a word of caution. In quite a few of the social sciences—particularly in psychology, social anthropology, and sociology—there is a debate raging between people preferring quantitative research designs and people defending qualitative research designs. By discussing quantitative approaches and their relationship to systematicity, I stress that I am not taking sides in this debate. This is not because I am a coward but because it is irrelevant for my present purposes. Nor do I discuss the conditions under which a quantitative approach is more desirable than a qualitative one or vice versa. I assume in this subsection, without specifying the preconditions, that in a particular situation, a quantitative approach is possible *and* appropriate in order to bring to light the consequences of this approach for the systematicity of science.

Let us start with an example of a hypothesis that sounds quite plausible in qualitative terms but may get in trouble once it has been quantitatively formulated. In the mid-nineteenth century—well before the acceptance of continental drift—various hypotheses were formed in order to explain the existence of mountains. A family of explanations tried to relate the formation of mountains and other geological features to contraction of the earth due to cooling. The cooling of the earth appeared to be a well-established fact, and the suggestion that cooling may lead to a shrinking Earth with a surface wrinkling like a dry apple sounded very plausible. However, in the 1870s and early 1880s, this hypothesis was put in quantitative terms. Various possible physical mechanisms were discussed that lead to contraction. However, "whether solid, partially liquid, previously liquid, or partially gaseous, the earth simply could not contract sufficiently to do the work required of it. At best, contraction would produce elevation differentials of eight to nine hundred feet," clearly in gross contradiction to the facts observed. Thus, while the hypothesis was plausible when articulated qualitatively because it resulted in an effect of the right quality, it fared much worse in quantitative terms because it missed the empirical data by a maximum factor of 30. This can be expressed in terms introduced by Karl Popper into the philosophy of science, which are now also quite common among scientists. A quantitative hypothesis has a higher degree of falsifiability, or is easier empirically testable, than a qualitative one, and its falsifiability is higher the more precisely it is formulated. In other words, the more precisely a false hypothesis is formulated, the easier it can be eliminated.

Also in situations of theory comparison, quantitative data often play a decisive role. Two hypotheses or theories may both be capable of deriving and thereby possibly explaining a given phenomenon in qualitative terms. Assuming that only this phenomenon is at stake, a decision between the two theories according to their empirical merits is then not possible on the basis of qualitative considerations alone. However, when aspects of the phenomenon can be quantified and measured, the situation may change drastically because one theory may get the numbers right and the other not. The following is a well-known example.

The phenomenon in question concerns an anomaly of the orbit of the planet Mercury. The anomaly consists in the fact that Mercury does not orbit the sun in a stationary ellipse, as expected from Kepler's first law. Kepler's laws were published in 1609 and became empirically well established in the course of the seventeenth century. By the end of the century, they were also theoretically well understood by their derivation from Newton's dynamics together with his law of gravitation. In the middle of the nineteenth century, however, French mathematician and astronomer Urbain le Verrier very accurately calculated the orbit of planet Mercury on the basis of the best observational data and concluded that there was a problem with Mercury's elliptical orbit. This orbit rotates very slowly. The technical expression for this peculiar movement is "advance of Mercury's perihelion." This expression means that the ellipse's closest point to the sun, the so-called perihelion, also moves around the sun, it advances—and that is an indicator of the ellipse's rotation. The effect is rather small, seen from the Earth only 5599.7 arcseconds per century (today's value), meaning that it takes Mercury's perihelion roughly 23,100 years to move full circle. Verrier was able to account for the perihelion's advance on the basis of Newtonian physics in a remarkable way. However, he didn't get the numbers quite right. He was able to calculate more than 99 percent of the effect, but a tiny amount was missing: 38 arcseconds per century (today's value is 42.7 arcseconds per century). Verrier considered this to be a "serious difficulty," and he discussed several hypotheses for its cause. For instance, it could be a yet undiscovered additional planet between Mercury and the sun. However, Verrier was skeptical of this possibility because given that no trace of this planet had ever been seen, it was unlikely that it existed. He thought it more likely that analogously to the asteroid belt between Mars and Jupiter, there might be a number of objects circling between Mercury and the sun.

I am skipping the details of the unsuccessful search for an explanation of the unaccounted part in Mercury's perihelion advance between 1859 and 1916. In 1916, the story took an unanticipated turn. Albert Einstein applied his recently finished general relativity theory, a novel theory of gravitation, to Mercury's orbit and could derive an additional value of 43 arcseconds per century for Mercury's perihelion advance. It was extremely important in this case that the general relativity theory got the value *just right* within the margins of error and not somewhere in the vicinity of the right value. In the latter case, relativity theory would not have come out so superior because to get *some* advance in the *vicinity* of the right value was also an achievement of Newton's theory. So this is a clear case where the quantitative value of a derivation from a theory played a decisive role that could not have been achieved by any merely qualitative consideration.

The power of quantification also becomes evident by the possibility of using statistics. Data that can be subjected to statistics do not necessarily result from quantitative measurements of certain variables; they may also result from simple counting. One of the most important applications of statistics is the investigation of correlations. Correlations describe the dependencies between variables, or the covariation between them. It is well known that positive correlations should not be mistaken for causal dependencies (because the dependence may be generated otherwise), but in the analysis of causal relations, they play an important role. For example, I can refer here to section 3.4.4, where I sketched treatment-control studies in which the statistical element is absolutely essential. This is, of course, only one particular experimental setup in which statistics plays a role; there are countless others, also pertaining to purely observational data. However, I don't need to go into any further details here because my argumentative task is only to show that science is, also in this respect, more systematic than our everyday life. In order to realize this, just compare scientific practice that makes use of statistics (here not further elaborated) with something similar from ordinary life. Suppose someone states that the food and service in a certain restaurant you know well and value highly has always been bad whenever she visited it, thus implying that it is a bad place. You may be puzzled and ask how often the person frequented the restaurant. If the answer is "twice," you may be satisfied, because it could have been just two bad days of the restaurant. If the answer were "certainly a dozen times," you would start wondering what the case is, considering possibilities like whether the person has a taste very different from yours, or that you are not talking about the same restaurant, or that she is pulling your leg or lying, and so on. It goes without saying that such procedures lag far behind regarding their systematicity in comparison to science.

Finally, mathematics plays an important role in the generation and control of deductions. There are many situations in which scientists are interested in the (logical) consequences of certain statements. I only mention what I discussed above in section 3.4.c, the test of empirical generalizations, models, and theories. In these (and many other cases), logical consequences must be generated from certain statements. Logical consequences can be generated in a highly controlled way if the pertinent statements have a mathematical form: mathematical manipulation fundamentally consists in the generation of logical consequences. Thus, although drawing logical consequences is of course also possible from nonmathematical statements, it is much facilitated in the case of mathematized statements. Again, the advantage regarding systematicity in the comparison with routine drawing of logical consequences is obvious.

3.4.7 Historical Sciences

What is the historian's basic task? It is "to choose *reliable* sources, to read them *reliably*, and to put them together in ways that provide *reliable* narratives about the past." Although this quote refers to the historical *cultural* sciences like political history, art history, or archeology because it speaks of sources that can be read, in principle, it also applies to the historical *natural* sciences like cosmology, Earth history, or paleontology. However, in these disciplines, the relevant data are not sources in the same sense as above because they are not artifacts but remains not produced by human beings. At any rate, structurally, all historical sciences have the same task. They start from some data (called sources for the historical cultural sciences) that are somehow accessible in the presence. These data must in themselves be reliable. In order to make these data relevant for the desired knowledge of the past, they must be interpreted; that is, some information about the past must be extracted from them. It is as if the data allow one to jump from their physical presence into the past, like a signpost that stands somewhere but points to somewhere else. Finally, the information obtained about the past must be composed into a story about the past that tells us how something developed or came about.

Let us begin with the data. The historical cultural sciences have developed a broad spectrum of so-called historical auxiliary sciences, devoted exclusively to the securing of the quality of historical sources. To name just a few: sigillography (the study of seals), papyrology (the study of writing on papyrus), heraldry (the study of coats of arms), or numismatics (the study of coins). In addition, written sources must be critically evaluated both with respect to being genuine (are they really what they claim to be, e.g., regarding the author or the issuing institution?) and with respect to the information they contain, which may be inaccurate or even wrong, both unintentionally and intentionally. Countless questions can and must be asked and answered in order to evaluate the sources, to compare, interpret, and weigh them and finally to weave them into a story for which they are the basis. A visible sign for these activities are the countless footnotes, each often containing several or even many references, which usually accompany historical texts. Each footnote bears testimony to the attempts to have a responsible narrative as the end product.

Work in the historical natural sciences does not differ in principle from work in the historical cultural sciences. Again, a narrative should be constructed on the basis of available data. The main difference is that the relevant data are not artifacts but natural remains of all sorts. Also, these data have to be critically analyzed, interpreted according to what they tell us about the past, and finally woven into the story to be told. Later, in section 3.8, I will discuss in more detail the example of ice cores, some of them more than 3,000 meters deep, that have been taken from continental glaciers by hollow drills. A particular layer of ice may be assigned to a particular year, so we have physical remains of that year. However, to use the available information from the core, that data must be interpreted in order to be informative about the past, for instance about the climatic conditions or about a volcano eruption. The same term "interpretation" is used both here and in the cultural historical sciences because the function of the "interpretation" is the same: to extract information about the past from what is currently present in front of our eyes. However, regarding the content of the interpretations, there is typically a significant difference: in the cultural historical sciences, the human origin of the remains plays a significant role in that they are bearers of the specific meaning that is completely absent in objects of nonhuman origin. I have discussed this specificity earlier in section 3.2.7 and shall not go further into it here. Rather, what is important at this point is the comparison of the activities of the professional historians to procure the reliability of their stories with our corresponding activities when telling stories. Clearly, we also rely here on certain data that we interpret and then process into a story. But it is equally clear that immeasurably less care is usually exerted when these stories are composed. Just imagine someone telling that story about why, against all his intentions, he was late again, or someone telling the story of his life to a potential new lover. As in professional historiography, these stories are in danger of being somehow bent toward the goal that they are supposed to reach. However, with respect to the sources, their interpretation, and their composition into a story, they are far less systematic than the professional historian's efforts.

3.5 CRITICAL DISCOURSE

3.5.1 Some Preliminaries

The dimension of systematicity that is the subject of the present section is somewhat different from the other ones. It does not concern scientific knowledge itself but rather the peculiar social organization of science that bears on the specificity of its product. In the previous section, we saw various ways and techniques how science defends its knowledge claims in a much more systematic way than we do in our life.

However, these ways and techniques can only function properly and effectively if they are realized by people who are socialized in a certain way and further embedded in fitting social organizations. Roughly speaking, the social norms and the social institutions that constitute the *social* organization of scientific communities must be conducive to the exertion of the *cognitive* norms that must be operative for the enterprise to reach its institutional goal. In the present perspective, this goal comprises the maintenance and even increase of the systematicity of scientific knowledge in various dimensions. In particular, the dimension that I discussed in the previous section, the defense of knowledge claims, is of utmost importance here. Science must, in order to maintain a high quality of its knowledge, be constantly attentive to avoid errors that may have various origins. With respect to the social structure of science, the scientific community must be organized in such a way that all knowledge claims are scrutinized by its members from as many possible different points of view. We are thus looking for the social reflection of something epistemological: the highly systematic defense of knowledge claims.

Thus, in the present section, I am only interested in those particular aspects of the social organization of science that are related to the systematicity of scientific knowledge, in particular the defense of knowledge claims. The social organization of science has other functions as well, of course—for instance the procurement of junior scientific staff or of financial resources, but they are of no concern here. The aspect of the social organization of science I will focus on here is institutionalized critical discourse. The term "critical" signifies the goal of probing claims, the term "discourse" signifies the involvement of various members of the community, and the term "institutionalized" signifies some sort of social organization and order, hence some sort of systematicity. This is the reason this social aspect of science can figure here as one of the relevant dimensions of systematicity that characterize science, according to the main thesis of this book.

When developing the systematicity of critical discourse in science, one must be careful in two respects. First, one should not uncritically assume that the sciences behave constantly in a self-critical fashion. How far the self-criticism of the sciences goes or should go is, or at least was, a controversial issue in the philosophy of science. Whereas Popper declared that good science is constantly in a self-critical mood, Kuhn described so-called normal science as quasi-dogmatic, meaning that it does not usually, let alone constantly, question its foundations, and this for good reasons; Lakatos similarly followed suit. According to Kuhn, the specific efficacy of normal science is a (desirable) result of this attitude. Be that as it may: because our focus is a comparison of science with others forms of knowledge, we do not have to fathom how far science's self-criticism ultimately goes; we have only to argue that science is, also in this respect, more systematic than the enterprises it is compared with. The second thing we have to be careful about in this section is with which institutions science should be compared regarding critical discourse. For it is clear that also many nonscientific institutions have established social structures whose purpose it is to secure the reliability of the knowledge (and the information) that is vital for the institutions' functioning. Think of banks, insurances, the legal system, the military, or the administration. I am not claiming that the sciences' social organization is more systematically structured with respect to securing knowledge quality than these societal fields. Neither do I know this, nor is this really relevant here. I am only emphasizing that also with respect to its social organization, science displays a high degree of systematicity, certainly higher than anything comparable in our day-to-day lives.

I shall discuss the social realization of scientific critical discourse on a more abstract and a more concrete level. On the more abstract level, certain norms can be identified that are relevant in scientific communities (subsection 3.5.2). On the more concrete level, we can identify certain standard practices or institutions that foster critical discourse (subsection 3.5.3).

3.5.2 Norms and Institutions

The norms relevant for the specific functioning of scientific communities have been discussed in the sociology of science starting with Robert Merton's classic paper "Science and the Social Order," first published in 1938, and especially in Merton's "The Normative Structure of Science," first published in 1942. Merton identified four norms, also called "institutional imperatives," that are binding for scientists because they implement the institutional goal of science. The institutional goal of science is, in Merton's words, "the extension of certified knowledge." The four norms are universalism, "communism," disinterestedness, and organized skepticism. The first three are less interesting in our context so I will not discuss them further. The fourth factor, however, is relevant to our discussion. Merton's "organized skepticism" is usually seen as the imperative to emphasize "primarily an institutionally enjoined critical attitude toward the work of fellow scientists." Of course, *organized* skepticism already entails a lot of systematicity, in preestablished harmony with my thesis. However, what does "organized skepticism" mean in more concrete terms?

Although not with reference to Merton's norm, philosopher Helen Longino has fleshed out what "critical discursive interactions" are. They "are social processes of knowledge production. They determine what gets to remain in the public pool of information that counts as knowledge." In particular, Longino identifies four features that are "necessary to assure the effectiveness of discursive interactions," i.e., critical discourse:

- Venues: "publicly recognized forums for the criticism of evidence, of methods, and of assumptions and reasoning";
- Uptake: "community members pay attention to and participate in the critical discussion taking place";
- Public standards: "There must be publicly recognized standards by reference to which theories, hypotheses, and observational practices are evaluated";
- Tempered equality: "equality" here means that criticism of all members of the community is admitted, and "tempered" means that there is some sort of weighing according to intellectual authority. What that precisely means is difficult but without further relevance for us.

This description is, I think, correct, but it is still somewhat abstract. What we have to discuss now is the scientific practices that embody these (idealized) features of scientific communities. We will then see how science achieves a high degree of systematicity in the implementation of these features.

3.5.3 *Practices in Science Fostering Critical Discourse*

Let us start with publications. Publications are an obvious precondition for an open scientific discourse in which every member of the community can participate. Thus, there is a sort of imperative for scientists to publish their results. Typically, the journals and presses that publish academic work have a reviewing system in which the quality of the submitted work is assessed—the so-called peer review system. This system has two functions. First, it acts as a critical filter by eliminating work that is assessed as unsuitable by the reviewers and the editors because of either quality issues or an unfitting subject matter. In many cases, authors receive a review report in which the refusal is argued. In the most prestigious journals, the rejection rate is rather high, typically more than 90 percent. Second, the peer review system helps to increase the quality of scientific work in cases of "conditional acceptance" or "suggested resubmission." In these cases, changes in the original manuscript are suggested and argued in order to improve it. The reviewing process is mostly "blind" in the sense that the author does not know who the reviewers are. Often, it is even "double-blind" in the sense that also the reviewers do not know who the author is. Clearly, the latter practice is designed to avoid positive or negative bias due to known authorship. Academic presses use similar procedures as the journals for the publication of books.

For discussion of the work of others, there are many established channels. Conference talks or invited talks are invariably followed by a discussion period, often introduced by a commentator who studied the paper beforehand. For special topics,

topics, panel discussions are organized at conferences. Journals often have a particular section of shorter discussion papers that take up issues typically published earlier in the same journal. Book reviews that not only inform potential readers about the work but also critically assess it are a regular part of many journals; there are even journals that exclusively publish book reviews. In addition, the current state of play in some fields is critically summarized and assessed in review articles that are published in special sections of regular scientific journals or in specific review journals.

This sketchy overview makes clear that there are many platforms established in science on which critical discourse may and indeed does take place. One additional feature is remarkable. As already emphasized by Robert Merton, there are no taboos regarding subjects to be discussed in science. Of course, there may be quite touchy subjects, for whatever reason, be it political, ideological, scientific, or otherwise, and people without a permanent position at a university are well advised to exert great caution when approaching such subjects. However, in science, one will hardly find the rejection of some contribution for the officially declared reason that such a subject must not be discussed. Officially, there are no such taboos.

In the natural sciences, the institutions of critical discourse just discussed have been in place, at least in rudimentary form, since the seventeenth century. From the nineteenth century on, also the humanities and the then emerging social sciences have developed similar mechanisms. From the mid-twentieth century on, novel institutions have been developed for a novel brand of research, so-called big science. "Big Science is characterized by large-scale instruments and facilities, supported by funding from government or international agencies, in which research is conducted by teams or groups of scientists and technicians. Some of the best-known Big Science projects include the high-energy physics facility CERN, the Hubble Space Telescope, and the Apollo program." Big science started in the 1940s with the United States' Manhattan program, the development of atomic bombs. Big science today is really big. This can be seen, for example, in a 2011 publication by the so-called ATLAS Collaboration. The paper reports research results in high-energy physics done at the Large Hadron Collider (LHC) at CERN near Geneva (Switzerland); one of its detectors is called ATLAS. This paper lists no less than 3,172 authors affiliated with some two hundred institutions. Clearly, collaborations of this size require innovations regarding their social organization in comparison to "little science." Sociologist of science Karin Knorr-Cetina has investigated the specifics of different scientific communities, including how they organize critical discourse, especially in high-energy physics and in molecular biology; she calls the specific organizational form the "epistemic culture" of that community. In the largest experiments in high-energy physics, the intellectual input of thousands of people must be critically coordinated. Physicists were quite creative regarding the invention of social means designed to master this task. Not only are there many informal discourse occasions (as in other scientific institutions), but also a host of more formally arranged "meetings" and workshops of all sorts that constitute the main channels of information and critical discourse. There are research and development group meetings, working group meetings, detector meetings (divided according to subdetectors), panel meetings, institute meetings, steering group meetings, collaboration meetings, technical board meetings, editorial board meetings, referee meetings, accelerator meetings, fixed committee meetings, special workshops, and so on and so forth. The sequential order of these meetings is important: it "suggests a passing of knowledge and technical decisions from the expert group where the responsibility lies to wider and wider circles that take note of these details and play them back—through discussions, questions, and comments." Clearly, this kind of dense communication is absolutely essential in order to establish and maintain a collaboration that results in joint publications of thousands of authors.

It is obvious that we have here highly systematized forms of information flow and critical discourse. As I said at the beginning of this subsection, such a highly systematic organizational structure regarding information transfer and critical discourse is not a unique feature of science; it also exists in other institutions of society. However, it is a feature of the social organization of science that is continuous with its cognitive organization, and for this reason it should be seen as an additional aspect of science's systematicity.

3.6. EPISTEMIC CONNECTEDNESS

3.6.1 Preliminaries: The Problem

So far, scientific knowledge has been characterized in five dimensions by a higher degree of systematicity in comparison to everyday knowledge: its descriptions, its explanations, its predictions, its defense of knowledge claims, and its critical discourse are just more systematic than comparable aspects of everyday knowledge. It may appear that these features are already sufficient in order to delineate the realm of scientific knowledge and to demarcate it from other kinds of knowledge. However, this is not the case. There are areas in which knowledge is produced professionally in ways that are at least analogous to scientific knowledge production, i.e., with the same sorts and degrees of systematicity as discussed above, and still this kind of knowledge does not count as scientific knowledge. I shall first introduce these areas of knowledge production using five examples, and then show why the most obvious ways to demarcate this kind of knowledge from scientific knowledge do not work. In subsection 3.6.2,

I shall introduce the concept of epistemic connectedness in order to deal with the problem. In subsection 3.6.3, I shall come back to the examples given and show how their nonscientific character can be understood by means of the concept of epistemic connectedness.

Example 1. Two firms, A and B, both produce the same product and compete in the same market. The management of firm A tries to find out its potential customers' preferences and wishes in a very informal way by asking some people of the sales department, by talking to friends and acquaintances, and by pure imagination. The management of firm B tries to find out its potential customers' preferences and wishes with the help of a questionnaire, designed by a market research institute, presented to a representative sample of their potential customers, and finally evaluated statistically. Clearly, the knowledge of firm B about the market is much more systematic than the knowledge of firm A, which was gained by common procedures. Therefore, according to our main thesis, the knowledge of firm B should qualify as science. Nevertheless, firm B's knowledge usually does not qualify as such, although it may be much more reliable than firm A's knowledge and it was gained by what may be called scientific procedures.

Example 2. In automobile development, nowadays one of the most important goals is the decrease of fuel consumption. Engineers working for automobile manufacturers who are involved in the development of a new car model may get the task to improve the fuel efficiency of the new model in comparison with the previous model. They may pursue this task by trying to modify the previous model's engine in appropriate ways, among other measures. They will perhaps slightly change the design, they may use other materials at some places, they may try to improve tuning, and so forth in order to optimize the engine's efficiency. After some experimentation, some calculations, and some modeling, they may know how to increase fuel efficiency by several modifications of the old engine. This is a typical case of product development, and it may be carried out in a highly systematic fashion. All of the conceptual tools engineers may use for the task may be borrowed from science. Thus, the knowledge gained in the process presumably shares all the aspects of systematicity so far discussed. Yet the knowledge gained in the process will typically not be a part of engineering *science*. It will be preserved in internal documents of the company but will usually not be published as it would if it were a part of science.

Example 3. Let us consider chess theory. Chess theory usually comes in three branches, corresponding to the three phases of the game: opening theory, middlegame theory, and endgame theory. Opening theory systematically deals with chess openings, middlegame theory refers to principles and rules concerning the middle phase of the game, and endgame theory mainly concerns specific types of positions arising in the last phase of the game. There is an enormous literature on chess theory, estimated at several ten thousand volumes, beginning in the fifteenth century. For example, the *Encyclopedia of Chess Openings* contains examples from more than 150,000 games and analyzes them in over three thousand pages; the *Encyclopedia of Chess Endings* features five volumes. Clearly, chess theory is very systematic in several of the dimensions so far discussed—for instance, descriptions: opening theory features a sophisticated classification system of openings. Regarding predictions or explanations, endgame theory can predict and explain certain outcomes. Regarding defense of knowledge claims and critical discourse, openings may be "refuted," and there are lively critical discussions going on in the chess community. Certainly, chess theory is much more systematic than anything the average chess player knows or is using. Shouldn't chess theory qualify as science according to systematicity theory? Clearly, it should not, and the question is "why?"

Example 4. Consider now high-quality political journalism and ask how it is different from certain parts of contemporary history or political science. What I have in mind is a situation in which the journalist deals with the same subject matter as the contemporary historian or the political scientist, for instance, with the political development in one of the Earth's crisis regions within the last twelve months. Now compare the two typical kinds of publications resulting from these professionals' work. Political journalists typically publish their articles in daily newspapers (or weekly or monthly journals), whereas scientists publish in scientific journals (unless they act as journalists by publishing at more popular locations). Regarding their information gathering about the situation in the country in question, there may not be much of a difference between the journalist and the scientist. Both will consult all sorts of official and unofficial sources, and both may conduct field interviews in the region or talk to other informants. Certainly, the scientist is more obliged to also consider the existing scientific literature on the subject, but a journalist going for an in-depth analysis may also do so. The degree of systematicity regarding descriptions, explanations, possible predictions, the defense of knowledge claims, and critical discourse may be roughly the same. The basic messages conveyed by the journalist in the newspaper and by the scientist in the scientific journal about the events and developments in question may be nearly identical. However, the newspaper article will not count as science, whereas the article in the learned journal will. Why?

Example 5. Look now at a hobby genealogist, Ms. Miller. She may try to reconstruct her family tree out of a number of remaining documents, even for several generations. We may grant her that this is done with the diligence of a professional historian, i.e., with all the systematicity in all the pertinent dimensions that is required of professional historiography. Still, the result will usually not qualify as being part of historical science. Somehow, the Miller family tree does not seem worthy of being a scientific subject; it is not "relevant enough." Why not?

3.6.2 Failing Answers

Here is the first attempt at answering the question about why the obviously nonscientific knowledge domains discussed in the previous subsection are indeed nonscientific. Scientific knowledge may be distinguished from nonscientific knowledge on sociological grounds, namely, by relating knowledge to the site of its production. Scientific knowledge is produced at scientific institutions like research universities or government-funded research laboratories. Nonscientific knowledge is produced at nonscientific institutions like development labs of industrial companies or editorial offices of newspapers. However, this move is unsatisfactory for two main reasons. One reason is that even if this characterization-by-localization worked, we would not understand what the intrinsic differences of the two sorts of knowledge are. Without that understanding, the difference of knowledge production sites could be a purely contingent fact saying nothing about different qualities of the knowledge produced. In more technical terms, even if the characterization-by-localization worked extensionally, i.e., if it correctly distinguished two kinds of knowledge, it would not work intensionally, i.e., with respect to the different intrinsic features of the two kinds of knowledge that would be left in the dark. The other reason the sociological characterization-by-localization approach is unsatisfactory is that although it may work for some cases, it certainly does not work for all cases. Sometimes, in industrial laboratories, work is done that unequivocally qualifies as foundational research that could just as well be pursued at research laboratories. Correspondingly, researchers may easily move back and forth between jobs at such industrial labs and labs located at research universities. The sociological characterization-by-localization approach, however, would categorize knowledge produced at one location as different from knowledge produced at the other location, although there is, in some cases, no intrinsic difference between the kinds of knowledge produced.

The second attempt at answering the question of why the obviously nonscientific knowledge domains discussed in the previous subsection are indeed nonscientific refers to the aims of science. Knowledge gained for and tuned to immediate concrete application usually does not belong to the body of scientific knowledge because it serves nonscientific goals, similar to the chess theory or the Millers' family tree. However, what precisely are the goals of science? Answering this question turns out to be extremely difficult, if not impossible. Certainly, the answer must be historically variable and is certainly also discipline dependent. It is worth looking back a little in order to appreciate the size of the problem.

Some 150 years or so ago, it may not have been too difficult to state the specific goals of scientific knowledge production, as distinct from the goals of knowledge production in other domains. Very roughly, scientific knowledge was mostly sought for its own sake. The natural sciences had very little practical applications; on a larger scale, practical application of the natural sciences only emerged in the last quarter of the nineteenth century with the upcoming electrical industry and the chemical industry producing organic dyes. Thus, the main goal of the natural sciences was cultural: to inform us about the wonders of the natural world. Engineering sciences and social sciences did not really exist 150 years ago. Predominantly, the humanities also served cultural functions. Perhaps only historiography was a partial exception, because historiography could be used and was used in politics when it came to the legitimacy of certain claims about property and borders, and to more firmly establish the identity of some historical entity.

Today, however, the situation is much more complicated with respect to specifying the goals of science (in the all-encompassing sense). In many countries, one of the main arguments in the political debate about the financing even of the fundamental natural sciences is future economic profit. Thus, at least the institutional goal of scientific knowledge production is widely seen as the future application of that knowledge for practical purposes. In the public arena, even the humanities sometimes try to demonstrate their usefulness in economic terms in order to secure their financing. Thus, within the last 150 years, the sciences have been more and more strongly integrated into the economic domain, which makes the old determination of the goals of science as knowledge production for its own sake obsolete. Not only has the contrast between "curiosity-driven research" and "product development" become very fluid by intermediate research areas like "application-oriented research," but also has "curiosity-driven research" been functionalized for long-term goals of the economic sector. A well-known example from the past is the putatively pure research in number theory, especially the theory of prime numbers, which was supposed to be absolutely remote from any potential application. This mathematical discipline is now a cornerstone of cryptography, the theory and practice of encoding messages such that they become unintelligible for those who do not know how to decode them. Cryptography on the basis of prime numbers is practically relevant in a variety of areas, from banking to military applications. The sobering result is that scientific knowledge cannot be distinguished from nonscientific knowledge by a particular relation to application: scientific knowledge may also be tuned for application as in the engineering sciences; it may be extremely remote from any conceivable application, like cosmology; and it may change its status in this respect, like number theory. Furthermore, even if we found a way to characterize certain areas of science by means of a particular relation to practical application, this strategy would certainly not work for other examples in our list above. Chess theory (example 5) will probably have a similar relationship to practical applications as some scientific theories. Or, to turn to example 4, a well-researched newspaper article on some political development will probably have the same range of possible applications, for example, in politics, as a scientific article with roughly the same content.

The result is that I shall not pursue the project of distinguishing nonscientific areas as exemplified above from genuine scientific knowledge by reference to any aims of science because I believe that it does not work.

The third attempt at answering the question of why the obviously nonscientific knowledge domains discussed in the previous subsection are indeed nonscientific refers to the degree of generality that some body of knowledge exhibits. The basic idea is that scientific knowledge must have some degree of generality. By contrast, the knowledge gained, for instance, in a process of product development is usually so specifically tuned to that product that this knowledge lacks the required degree of generality to count as scientific. This is certainly a promising idea for *some* areas, but its potential application is certainly severely limited. The criterion of higher generality may be suitable to distinguish knowledge gained in the area of product development from genuinely scientific knowledge belonging to engineering science, but it does not apply to analogous situations in, say, historiography. Compare here, for example, example 5, above. The difference between the family tree of the Millers and that of, for example, the Habsburgs that a professional historian may investigate, is not a difference in generality. Similarly in example 4, the difference of an article belonging to political journalism and an article about the same subject matter belonging to contemporary history is also not a difference in generality. So at least for these cases, we need a different criterion.

The result so far is this: in order to distinguish the sort of nonscientific knowledge exemplified in examples 1 to 5 from genuinely scientific knowledge, we need a criterion that somehow encompasses "generality" for the appropriate cases but also applies to the other examples, especially the historical ones. For that purpose, I want to suggest the concept of "epistemic connectedness" as a criterion.

3.6.3 *The Concept of Epistemic Connectedness*

So far, the concept of epistemic connectedness is not a standard notion. Therefore, I will have to introduce it carefully. I shall do so in three steps; in two additional steps, I shall make comments on the concept.

First, in an abstract characterization, epistemic connectedness means the existence of manifest connections of knowledge to other pieces of knowledge; the nature of those connections, however, is left unspecified. They comprise all sorts of purely logical relations like logical equivalence, implication, dependence, or consistency; or more epistemic relations like confirmation, disconfirmation, verification, falsification, generalization, extension, modification, amendment, extrapolation, interpolation, specialization, reduction, criticism, reflection, interpretation, citation, and so on. This variety of possibilities may convey the impression that on an abstract level, the concept of epistemic connectedness covers almost anything and is therefore almost empty—and indeed, this is the case.

Second, the poorness of the abstract characterization of epistemic connectedness is a consequence of the fact that this concept has similar features as the concepts of systematicity or refinement, which I discussed earlier (see section 2.2). These features were:

- On an abstract level, i.e., with no particular area of application in mind, the attempt to clarify these concepts does not lead very far. The reason is that there simply isn't very much that can be clarified; in the abstract, these concepts have very little content.
- In certain contexts, i.e., in certain areas of application of the concept, such concepts do gain more content and a much clearer contour. Thus, further clarification has to presuppose a context.
- A comparison of the use of such concepts in different contexts reveals that only family resemblances exist among them. This implies that no universal criterion for their application exists, which agrees with the first feature given: on an abstract level, these concepts have little content. It further implies that two arbitrary members of this set of concepts that are united by family resemblance only may display very little similarity indeed.

"Epistemic connectedness" shares these features. On an abstract level, epistemic connectedness only states that there are manifest connections to other pieces of scientific knowledge, but the kind of connections is unspecified. In order to concretize the concept, one must give contexts of application. I shall do so in the next subsection.

Third, there is a fairly straightforward conceptual relation between epistemic connectedness and systematicity. This is due to the fact that epistemic connectedness is related to the older notion of a "system of knowledge." A system of knowledge in the strict sense is an axiomatic system with logically independent axioms as its base, specified rules of deduction, and theorems that can be proven on the basis of the axioms and rules of deduction. Clearly, this idea of a system of knowledge is still alive and kicking in large areas of mathematics (and in some pockets of empirical science). However, as an executable idea of how the whole universe of scientific knowledge could be ordered and represented, it was abandoned a long time ago. Nevertheless, there is a successor idea that generalizes (or weakens) the older idea of a system of knowledge in the following way: science as whole must not consist in isolated pieces of knowledge. There should be something like the unity of science, though certainly not in the form of a rigid axiomatic system, but generated by epistemic connections existing between different research fields, thereby providing some coherence between them. Empirically, as far as I know, there is no research field that is completely independent from all the rest of science; there are always some epistemic connections to other research fields. Thus, the whole of all the sciences, being a whole due to these epistemic connections, has the property of epistemic connectedness. "Epistemic connection" is thus the successor relation to "deductive connection" in the old idea of a system of knowledge, and "epistemic connectedness" is the successor property to the "system character" (in the strict sense of axiomatic system) of the old idea of a system of knowledge. Thus, the predecessor concepts are special cases of the successor concepts; the predecessor concepts are weakened and generalized. This reflects the general tendency of systematicity theory: it weakens and thereby generalizes aspects of older conceptions of science, thus keeping the older conceptions as special cases.

Given that all pieces of scientific knowledge have some epistemic connections to other pieces, it may be said that the whole of science forms a system, though of course *not* in the old sense of an axiomatic system of knowledge. It is a system in the sense of a rather loose assembly. Thus, a "higher degree of epistemic connectedness," meaning the existence of more or stronger relations to other pieces of scientific knowledge, can be said to be a "higher degree of systematicity" in the sense of being integrated more strongly into the (loose) system of scientific knowledge.

Fourth, epistemic connectedness overlaps considerably with other dimensions of this chapter. For instance, any theory that produces more and better predictions or explanations than others has also a higher degree of systematicity in these dimensions. At the same time, this implies a higher degree of systematicity regarding epistemic connections because every prediction or explanation consists in an epistemic connection of the theory to certain phenomena. Another example is given by any theory that is in a relevant sense more general than others. This theory has more epistemic connections than the others because of more applications; it is therefore more systematic in this sense. At the same time, it is more systematic in the sense of the defense of knowledge claims because due to its greater number of applications, it can be subjected to more tests.

Fifth, epistemic connectedness leaves room for a continuous transition area between scientific and nonscientific areas, for instance, between scientific and more applied work like product development. In section 1.2, I have already discussed the existence of such a transition area and its consequence with respect to this project. However we characterize science, we must not try to established sharp boundaries between science and all nonscientific enterprises. For instance, between applied science and engineering science on the one hand and product development on the other, there are fluid boundaries with smaller or larger transition areas. Whatever the concrete meaning of "epistemic connectedness" in a given context is, it comes in degrees and thereby allows for the sort of transition area that is indeed needed.

3.6.4 Revisiting the Examples

I shall now revisit the examples from subsection 3.6.1, above. The purpose is twofold. First, I will provide contexts in which the abstract notion of epistemic connectedness can be made more concrete. This is a necessary part of the explication of a concept that works like the concepts of systematicity or refinement. Second, I will demonstrate the higher degree of epistemic connectedness—and hence the higher degree of this kind of systematicity—of scientific knowledge in comparison to the nonscientific counterparts represented in the examples. The latter part is, of course, my continual holy duty in this chapter in order to argue my main thesis.

Example 1: Knowledge gained in scientific ways by firm A about a market, though more systematically discovered than similar knowledge gained by firm B, is nevertheless not a part of science. The reason is that knowledge about a particular market is, in isolation, scientifically uninteresting; it concerns an isolated fact. This fact may become scientifically interesting if it were integrated into, say, a comparative study of these markets in different countries, or of these markets at different times, or of these markets in comparison to other markets. In other words, isolated knowledge of a particular market has too few epistemic connections to count as scientific knowledge.

Example 2: Knowledge gained in product development, for instance in improving the fuel efficiency of a particular engine, although done in a very systematic fashion, does not count as scientific knowledge. Contrast this with an exemplary project related to the same goal, increasing fuel efficiency, but carried out in the engineering department of a research university. Engineering departments of many universities participate in the "Shell Eco-marathon," where the goal is to build a car that drives as far as possible with the least amount of energy. In 2007, one of the participant groups set a new world record for fuel efficiency at an amazing 5,385 kilometers with hydrogen equivalent to one liter of gasoline. According to the project director, the motivation of the project was "to integrate and test the latest developments in materials, aerodynamics, structures and systems, and many other disciplines into one system." After the race, the main objectives were "the publication of articles and reports that explain the technical details of the system and the transfer of the know-how to all interested groups." Regarding the practical application of the project, the project director declared that he was "convinced that some of the ideas that have been generated in this project will eventually show up on the road and, following our main mission, contribute to saving fuel." All of these statements indicate that the knowledge gained belongs to engineering science.

Now the difference between knowledge in the engineering sciences and product development becomes apparent. Although the knowledge gained in the Shell Ecomarathon contest is explicitly related to practical goals, it is still fairly remote from the development of products that can be sold on the market. It may lay the foundations for a host of future product development by supplying useful knowledge that is more general than knowledge typically gained in the development of one particular product. Due to its higher degree of generality and its remoteness from the design of some marketable car, it has many more epistemic connections to other areas of engineering science and know-how than the knowledge gained in the optimization of the fuel efficiency of a particular engine. It thus has a higher degree of epistemic connectedness and thus a higher degree of this particular kind if systematicity. This is the reason it is a part of engineering science.

Many examples similar to the one just given are available from other areas of the engineering sciences where we have the same contrast between the knowledge gained during product development and the generation of more far-reaching—i.e., more strongly epistemically connected—scientific knowledge. In section 1.2, I have already mentioned the case of earthquake engineering. On the one hand, we have the science that is concerned with a more general task like the experimental study of the seismic behavior of certain types of assemblages. On the other hand, there is the more practical work of designing a building according to such design principles. Also in section 1.2, I have mentioned the case of research in chocolate science and its continuous transition to product development. There is a similar transition area between science and nonscientific domains, namely, where science is applied for nonscientific purposes (compare section 1.2). Take for example the case of meteorological models that are used for daily weather forecasts. The development of such models is clearly a scientific task, whereas their routine application to generate a forecast is not. Again, there is a blatant difference in epistemic connectedness in the two cases. Whereas the concrete weather forecast will have very few manifest connections to other pieces of scientific knowledge apart from being generated by the model, the model itself has countless manifest connections. There are plenty of scientific theories, submodels, and assumptions built into it; it has been tested against much data and been modified as a result; and so on. Thus, clearly any application of scientific knowledge for nonscientific purposes is epistemically less connected than the scientific knowledge itself.

Example 3: Why is chess theory not a part of science? Clearly, chess theory is a highly systematic enterprise apparently fulfilling the former dimensions of systematicity presented in section 3.1 through section 3.5. There is even a mathematical discipline that appears to be its natural home: mathematical game theory. Mathematical game theory can be applied to the game of chess regarding certain specific questions. For instance, in the terminology of game theory, chess can be classified as a specific type of game, namely, a nonrandom, two players, zero-sum game with perfect information. But chess theory itself is not a subdiscipline of game theory. The reason seems to be that chess is so specific a game that the content of chess theory has no connections to other mathematical areas. In other words, chess theory is epistemically isolated from other areas of science such that it is not a part of the (loose) system of science.

Example 4: Compare an article on the latest developments in a crisis region of the Earth in a high-quality newspaper with a scientific paper on the same subject in a learned journal. Suppose that both pieces use more or less the same resources of information. The main difference between the two articles will concern the number of manifest connections to other pieces of scientific knowledge. In contrast to the newspaper article, the scientific paper will contain many footnotes in which the connections to various other pieces of scientific knowledge will be made explicit. The scientist is obliged to declare which data or background theories are used, to explicitly cite the sources, to consider whether the views expressed fit or contradict other current views or theories, and so forth. It is one of the hallmarks of scientific work that presumably new pieces of knowledge have to be fitted into the existing aggregate of scientific knowledge by making explicit the epistemic connections between the new and the old. This sort of obligation does not exist for the journalist, and this is revealed in the different forms of the two products despite possibly nearly identical content.

Here is a concrete example. The example differs slightly from the scenario as described above because the scientist and the journalist are one and the same person, political scientist Danyel Reiche. In an article published in the scientific journal, *Third World Quarterly*, Reiche describes "the politics of sport in Lebanon as a unique case in comparative politics." Whereas in most countries, sport has the potential to unite fragmented societies, the opposite is the case in Lebanon. In this country, sport further divides people. At the same time, Reiche published an article in a German quality newspaper, *Frankfurter Allgemeine Zeitung*, covering the same topic. What is the main difference between these publications? First, the newspaper article is much shorter: only roughly one-fourth of a large newspaper page in comparison to seventeen pages of the journal article. Second, and most important in our context, the article in the academic journal features fifty-five endnotes, covering almost two pages in small print. The notes establish connections to all sorts of other pieces of knowledge, for instance to books about the general political situation in

Lebanon, to articles about the situation of sports in other countries, to other articles in comparative politics, to websites containing relevant data, to interviews that the author conducted, and so on. By contrast, the newspaper article contains only one single note at the end, telling the reader that the author is a professor of comparative politics at the American University of Beirut. Within the newspaper article, none of the central statements are backed up by and therefore connected to any evidence. Whereas the journal article embeds the subject matter in a larger context discussed in comparative political science and explicitly defends its knowledge claims, all of that is missing in the newspaper article. Thus, the higher epistemic connectedness of the scientific article in comparison to the parallel newspaper article is evident, despite their common subject matter.

Example 5: the family tree of the Millers. This example is similar to example 1 in having to do with epistemically fairly isolated facts. Without any further epistemic connections, this family tree is uninteresting for the professional historian. Compare this with the family tree of the Habsburg family. It figures, in one form or another, in countless historical works because it has innumerable epistemic connections to other historical facts and developments. It is a key element in the web of knowledge about European history between the thirteenth and the early twentieth century. Note that the difference between the Miller genealogy and the Habsburg genealogy is not at all a difference in generality. Both cases are basically a set of singular events. But the historians know of many interesting connections of the Habsburg genealogy with other historical events and processes. By contrast, the bare genealogy of the Millers is epistemically isolated from the rest of history and as such cannot be a part of historical science.

All of the examples show that in order to be scientific, it is necessary to be epistemically well connected. What that exactly means depends, of course, on the given situation. But clearly, epistemic connectedness is what sometimes makes the difference between the scientific and the nonscientific. Of course, also in our everyday knowledge, there are epistemic connections between different bits and pieces of knowledge. But they are not what is characteristic of everyday knowledge. Being useful for our normal business is what really counts here. The existence of epistemic connections in scientific knowledge, by contrast, fits it into a larger knowledge web that is also a system, though in a somewhat weak sense. But this provides another sense in which scientific knowledge is more systematic than common knowledge.

3.7 THE IDEAL OF COMPLETENESS

3.7.1 *Some Preliminaries*

One of the most astonishing facts about modern natural science is its remarkable growth. This growth concerns many aspects of science: its scope, its precision, the number of specialties, the number of scientists, the number of doctorates awarded, the number of scientific journals, the number of individual publications, the amount of financial investments, the influence upon other domains of society including everyday life, and many more. At least since the nineteenth century, the humanities and the social sciences have also taken part in this breathtaking dynamics. It is not easy to characterize this growth in quantitative terms because first and foremost, it depends on the aspect one is focusing on. Furthermore, even quantifying a particular aspect of scientific growth poses many problems of detail like data accuracy, counting methods, and so on. With all of this acknowledged, it is nevertheless possible to give a very rough estimate of this growth (getting the "order of magnitude" right, as physicists often put it). From the seventeenth century through the twentieth century, science grew roughly exponentially with a doubling time of roughly fifteen to twenty years. For our purposes, we do not have to care about the details. For us it is enough to note that we are confronted with a probably unique cultural phenomenon: the continual rapid growth of scientific knowledge over several centuries. This feature probably distinguishes science from all other knowledge systems, past and present, European and non-European. It is in need of an explanation.

Of course, the question of why science grows at such an enormous and continual rate is a complex question. Many heterogeneous factors play important roles, including very general social conditions—like the existence of cities, the existence of a general social structure sufficiently supportive of the social structure of science, the existence of sufficient resources and the willingness of those who control them to spend them for science, and many more. I shall here concentrate on one particular cognitive factor that is intrinsic to science, namely, that science itself strives for completeness of its knowledge. In other words, I claim that science has an ideal of completeness.

This is a somewhat unclear claim. First of all, who is the acting subject pronouncing this claim, i.e., who is striving for the completion of science? The first step in answering this question somehow satisfactorily is to break down the object of intended completion—science—into disciplines or subdisciplines. The acting subject would then be the pertinent scientific community that intends to complete knowledge about the subject matter of the discipline. However, this claim also is not overly clear. Is a scientific community really an acting subject with intentions? Intentions of collective agents are a difficult and recently much-discussed subject, and I better stay clear of it. Probably the best way to construe the claim is to say that completion of knowledge is a value in the scientific community. In the last decades, talk of cognitive values that hold in scientific communities has gotten wide currency. So it is quite common to speak about theory decisions by scientific communities that are influenced by values like accuracy, fruitfulness, consistency, and others. I take it that the "ideal of completeness" means that completed knowledge or the completion of knowledge is positively valued in scientific communities.

Second, after some clarification of the conceptual question, the next question must be: how can one investigate whether a certain community is committed to a certain value, in this case the completion of knowledge? The answer must be that in certain decisions by the community, this value must become operative by influencing the decisions. In addition, explicit endorsement of the value by scientists, and no other scientists contradicting, is a further indicator. However, I shall not pursue this line of analysis any further because what is a tricky question in theory is comparatively easy in application, as we shall see in the examples.

Third, in which sense is an ideal of completeness an aspect of science's systematicity? Here I can come back to what I stated in section 3.6.3 about the relationship of epistemic connectedness and systematicity. Scientific knowledge is epistemically connected and forms a system of sorts: not a strict axiomatic system, of course, but rather a loose assembly. The ideal of completeness evaluates positively those contributions to this system that move it into the direction of completion. Again, abstractly speaking, this is fairly vague, but I hope that the examples will remedy this apparent defect.

Before moving to the examples, I should note at this point that science is not only committed to this ideal of completeness, but also transforms this ideal into practice in systematic ways. In other words, the generation of new knowledge is pursued in a systematic way, much more systematic than we are used to in our everyday business. This topic, however, must wait until the next section.

3.7.2 Examples

Roughly speaking, the ideal of completeness manifests itself in the fact that science is never satisfied with some scattered facts about a certain domain. Ideally, any discipline wants to know "everything" about its subject matter, given its particular focus. However, I should add immediately that I am not investigating the question whether a completion of some scientific discipline or even of science in general, is indeed possible. This question is not within the scope of this book; I will leave it to those who enjoy speculating and will patiently wait for answers supported by good arguments. Instead, I will first discuss certain features that can be found in many sciences and are indicators for the existence of an ideal of completeness. Then I will discuss single disciplines or discipline groups in order to show in which ways they specifically embody an ideal of completeness.

Let us look at descriptions first. As I outlined in section 3.1.3, classifications and taxonomies are important tools to increase the order and thus the systematicity of descriptions. These tools to improve descriptions are used in many sciences across the board. Classifications and taxonomies also embody an ideal of completeness, because a classification or taxonomy should, as a matter of course, classify *all* the elements of the domain in question. Just look at two of the examples presented in section 3.1.3 that may stand for many more: biological species and languages. Clearly, the idea here, among other things, is to have a *complete* overview over *all* species or over *all* languages. A similar idea holds for the temporal analog to classification and taxonomy, periodization (see section 3.1.4). The periodization must cover the whole of the process in case of a unique process as in the historical sciences, or must apply to all instances of the process over the whole time interval in cases of classes of processes.

In order to increase the systematicity of their explanations and in some cases also of their predictions, many disciplines articulate and apply theories (see sections 3.2.3, 3.2.5, 3.2.8, and 3.3.4). Theories come in very different guises, depending on the discipline in question. They seem to have in common that the broader their scope, the higher they are valued. This property somehow points toward the value of completeness: as a theory should be as general as possible, in the limit the theory would be all-encompassing in a certain sense. How far disciplines push in this direction differs greatly. As we shall see, physics indeed pushes this line to the extreme by envisaging a "Theory of Everything." Independently of how far it goes, clearly the quest for higher generality of theories is an indicator for an ideal of completeness that is operative in the background.

Let us now move on to specific disciplines and their different articulations of the ideal of completeness. I begin with mathematics in its traditional form. In this kind of mathematics, axiom systems are sought that are complete (or as complete as possible) regarding the domain in question. The axiomatization "should allow for a derivation of *all* the known theorems of the discipline in question." It is an obvious requirement flowing directly from the idea of a strict system of knowledge in which everything relevant is based on and ultimately contained in the axioms. The existence of the ideal of completeness is evident as well as its connection to systematicity, the latter even understood in its most rigid form.

I want to present another example from mathematics in somewhat more detail. It comes under the title of classification. The meaning of "classification" in mathematics is stricter than in other sciences because here, successful classifications always have to come with a *proof* for completeness. Thus, a mathematical classification is a theorem, namely a statement of completeness that has to be backed up by a proof. In the mathematical context, completeness means that all objects of a given sort, i.e., all objects fulfilling a given definition, can be exhaustively listed in detail. Thus, having a classification means to have a detailed overview of all objects fulfilling the definition

definition together with the certainty that this overview is complete. A fairly simple example already discovered in antiquity are the regular polyhedra. What are convex regular polyhedra? A cube is an example of a convex regular polyhedron; it is a regular composition of six squares. A tetrahedron is another example of a convex regular polyhedron, a regular composition of four equilateral triangles. The question may arise of how many different convex regular polyhedra exist. Precisely the two just mentioned? Or three? Or twelve? Or infinitely many? The answer is that there are exactly five different convex regular polyhedra: the tetrahedron, the cube, the octahedron, the dodecahedron, and the icosahedron (don't worry if you have never heard these names). These five make the exhaustive and detailed list of the so-called Platonic solids, and there is a proof that there are exactly five of them. So if you want to know everything about convex regular polyhedra, you do so if you know these five and you know that there are no others. More generally, if you want to know everything about a certain class of mathematical objects, you want to have a theorem that delivers the classification in the sense given above. "This is exactly the sort of theorem that researchers in many areas of mathematics would absolutely love to prove," as one mathematician put it. This is a fitting expression of an ideal of completeness that is operative in certain areas of mathematics.

However, mathematicians not only express their affection for completeness proofs, they also put their money where their mouth is—even at considerable cost. A truly extreme example is the classification of finite simple groups. Never mind what sort of mathematical objects finite simple groups are if you don't know already. At least to the layperson, they sound like uncomplicated objects, being "finite" and "simple." However, if one wants to get a detailed overview of them together with a completeness proof, in other words, if one wants a classification of finite simple groups, one has to enter a world of absolutely bewildering complexity. The classification of finite simple groups consists of eighteen different families plus twenty-six individual groups called the sporadic groups. This looks a little ugly to a mathematical mind, because these twenty-six erratic individuals do not follow any discernible pattern. Nevertheless, the whole thing does not seem to be exceedingly complex. When looked at more closely, however, the mere list of the classification takes several pages. Furthermore, one should know that this classification has been pursued since 1892, and it came to a close only in 2004. The proof consists in something like five hundred journal papers, authored by more than one hundred mathematicians and published on more than ten thousand pages! It is admitted by mathematicians "that no-one in the world today completely understands the whole proof"—small wonder, given its complexity. One exemplar of the groups deserves special mention. It is one of the sporadic groups and is called the "monster group" or, more softly, the "friendly giant group." It is the largest of all sporadic groups and has exactly 808,017,424,794,512,875,886,459,904,961,710,757,005,754,368,000,000,000 elements (the zeros at the end are precise numbers). Nice, isn't it? The classification of finite simple groups is "undoubtedly one of the most extraordinary theorems that pure mathematics has ever seen." Doesn't it embody an ideal of completeness as it lives in mathematics in a wonderful way? And, I dare to say, it is more systematic in the sense of striving for completeness than anything I know of in everyday thinking.

In the natural sciences, manifestations of the drive for completeness abound, and I will have to remain rather sketchy. Physicists constantly push the limits of measuring instruments in order to explore known phenomena more thoroughly and to discover new phenomena accessible only at larger and larger or smaller and smaller scales of some variable. For instance, cosmology tries to cover literally the whole time from the big bang, the assumed beginning of the world, to the present and further—to the end of the world, if it exists. Physical theories make statements about the state of the universe as close as 10^{-35} seconds after the big bang—and of course, one hopes to get even closer. Or, fundamental physics tries to describe all types of fundamental interactions of particles—at the moment, four. Any indication of some new type of interaction would be followed up immediately, of course. It is not just an ironic title that physicists give to the ultimate theory that they search for: the "theory of everything" (T.O.E.).

Since antiquity, chemists have searched for "elements": the ultimate constituents of matter out of which all physical bodies are composed. This enterprise made sense only, of course, if one could come up with a complete list of them. In antiquity, Empedocles and Aristotle seemed to be successful with this enterprise; their four elements they postulated were "the fundamental basis of theoretical chemistry until the eighteenth century." During the Chemical Revolution at the end of the eighteenth century, these elements were abandoned for not being elementary. Instead, other elements were discovered, and a new system of elements began to take shape—the periodic system of elements. In 1869, it was published by chemist Dmitri Mendeleev and enabled chemists often, but not always successfully, to predict the existence of yet undiscovered elements by deriving their properties from their position in the periodic table. The task of empirically finding all the stable elements was only finished in the 1940s. The ideal of completeness was an absolutely self-evident part of the periodic system—as the term "system" already implicates.

Biology displays the ideal of completeness first and foremost in taxonomy, and I have nothing additional to say in view of what I have already discussed in section 3.1.3. However, it is not only the taxonomy of species where biology's drive to completeness becomes visible. It can also easily be found in the molecule-oriented disciplines of biology. The first of my examples—because it is well known to a general audience—is the human genome project. It was a thirteen-year project that was completed in 2003. Its main goals were "to *identify* all the approximately 20,000–25,000 genes in human DNA," to "*determine* the sequences of the 3 billion chemical base pairs that make up human DNA," and some other rather subsidiary goals. Note the clear orientation on the value of completeness. In this case, completeness is not an abstract guiding idea, somewhere in the background of the project, but rather it is a manifest and articulated goal whose accomplishment is essential for the project to count as successful. A much less known but similar case is the project to generate the complete proteome set for *Homo sapiens* and *Mus musculus*, and later also for other species. What *Homo sapiens* is should be well known; *Mus musculus* is the house mouse. The term "proteome" was invented in analogy to the term "genome"; "a complete proteome is the set of protein sequences that can be derived by translation of all protein coding genes of a completely sequenced genome." Again, the idea of completeness is manifestly directly at work: the goal is to build a database of just all human proteins that derive from the genome and similarly for the cute house mouse.

Another example from biology can illustrate how a kind of completeness may not be a goal in itself but a means in order to promote another dimension of systematicity. Although the modern synthetic theory of evolution is practically universally accepted among biologists, there are many details to be filled in. Especially, it is interesting to study evolutionary processes in detail when they occur in front of our eyes. One such project set out to study the evolution of a particular population, namely, the population of finches on one of the Galápagos Islands. It aspired to completeness in the literal sense that some parameters of *all* individual finches on that island were measured—not a single finch was left out. These measurements were repeated over many generations of the finches. They were then correlated with a number of variables that were assumed to be relevant for different selection pressures on the different species of finches and on different individuals within a given species. The knowledge of those parameters of all individual finches on that island was not a scientific goal in itself. But it offered the opportunity to record in detail the effects of different environmental conditions on the finch population and to articulate more fully and test certain claims of the Darwinian theory of evolution. Thus, the completeness of the knowledge about certain parameters of the finch population primarily served other dimensions of systematicity of knowledge, namely the defense of knowledge claims.

Let us now have a look at the Earth sciences. The currently accepted theory of the Earth's lithosphere—that is, the outer solid part of the Earth—is the theory of plate tectonics. This theory has an ideal of completeness built-in at three locations at least. First, describing the lithosphere in terms of different plates presupposes that the lithosphere exhaustively decomposes into plates. Although there seems to be some disagreement about the precise definitions of some of the plates because one can disagree about precise boundaries, the essentials are undisputed: the lithosphere can be completely described in terms of tectonic plates. Second, there are different types of plate boundaries, where different kinds of plate interactions give rise to different phenomena. A complete classification of these boundaries is needed and is delivered by plate tectonics. Plates may move away from each other, they may collide, they may slide horizontally past each other, and there may be broad belts in which boundaries are not well defined. There are also subgroups of these larger classes. Finally, as the theory of plate tectonics is a dynamic theory describing plates in motion, it should have explanations of the plates' movements that are as complete as possible. However, this is an exceedingly difficult task because detailed information about the region below the plates, the mantel, and even farther within the Earth's interior, are very difficult to get. "The fact that the tectonic plates have moved in the past and are still moving today is beyond dispute, but the details why and how they move will continue to challenge scientists far into the future." This statement shows clearly the desire for a complete understanding of plate dynamics.

In the historical sciences, the professionals are well aware of the fact that unrestricted completeness is not an attainable ideal. Any historical process, be it natural or cultural, has innumerably many aspects, many of which are unimportant and boring; they should certainly not be part of the final narrative. Completeness in a naïve sense is thus out of the question for the historical disciplines. However, there are restricted, or focused, forms of completeness operative in the historical sciences. In the section on descriptions, I have already discussed some peculiarities of historical descriptions that take on the form of narratives (see section 3.1.7). I mentioned there that historians guide the necessary selection from the historical material by several principles, among then the principles of factual relevance and of narrative relevance. The principle of factual relevance demands that all material that is necessary because of the particular subject matter of the story must be included in it. Thus, there is an idea of completeness in the background demanding that the historian include all the necessary aspects with respect to the given subject—whatever these aspects are and however they are determined. Similarly, the principle of narrative relevance demands that the historian include everything that is relevant in order to produce a continuous narrative. Again, there is an idea of completeness operative demanding that in order to avoid any jumps in the story, everything generating the necessary transitions must be told.

These were very general principles of historical work that embody ideals of completeness. Besides them, there are many examples where historical work follows a very concrete ideal of completeness. Here is a randomly picked example from the history of science, namely, the history of astronomy. From everything we know about the history of astronomy, in all cultures from which we have written records, eclipses were events that for obvious reasons found highest attention by the astronomers. Now, one historian of astronomy tried to collect "all known accounts of timed eclipse observations and predictions made by early astronomers, and to give detailed description of the sources in which they are found." The author collected the pertinent records of all cultures around the globe dating roughly from the beginning of written records until the sixteenth century (the invention of the telescope). This is an impressive work displaying an ideal of completeness in most explicit terms.

The motive of having a complete record of some sources is pervasive in all text-based disciplines, thus covering most of the humanities. Important authors of all sorts get complete editions of their work. In a similar manner, "critical editions" of some particular work try to collect all different versions of it, thus striving for completeness regarding variants. They may also try to reconstruct some original version of a text that has been changed for whatever reason.

Sometimes, great efforts are invested in a thorough stocktaking of all knowledge in encyclopedias. The history of encyclopedias seems to have begun in antiquity with Pliny the Elder, who in the first century AD collected all known facts about the natural world in his *Historia Naturalis*. An additional milestone in the history of encyclopedias was the *French Encyclopédie* by Denis Diderot (1713–1784) and Jean le Rond d'Alembert (1717–1787). Its thirty-two volumes were published between 1751 and 1777; more than 140 authors contributed altogether more than seventy thousand articles covering not only the whole universe of scientific knowledge but also the arts and the crafts. Today, encyclopedias typically have a restricted range, trying to cover all knowledge of a given discipline. For instance, during the last fifty years, two important English language encyclopedias covering all of philosophy have appeared in print and similarly in other languages. The latest development is to produce scientific encyclopedias on the Internet and not as printed books, for which in philosophy the *Stanford Encyclopedia of Philosophy* is an example.

What we have discussed in this section is a meaning of "systematic" in the sense of "thorough and complete." In all of the sciences, one or another variant of this form of systematicity is present. The contrast to our everyday knowledge is as tremendous as it is obvious.

3.8 THE GENERATION OF NEW KNOWLEDGE

3.8.1 Some Preliminaries

As we have seen in the last section, one of science's most important driving forces is its ideal of completed knowledge. Clearly, this is also an aspect of systematicity as this ideal requires systematically identifying missing pieces, which in turn presupposes some idea of a complete (loose) system of knowledge about the domain in question. We shall see in the next section how this supposed systematicity of scientific knowledge about some domain is reflected in the representation of knowledge. In this section, however, we ask about the ways the sciences go about realizing their goal of science completion, or, more modestly, about the generation of new scientific knowledge in general. As it will turn out, also the procedures of knowledge generation exhibit a high degree of systematicity, and again we will encounter a strong dependence of the relevant notions of systematicity on the scientific field in question. We will see what a Herculean task we are confronting, because the extreme diversity of scientific fields is also reflected in the vast differences of their approaches to knowledge generation. In relation to this immense variety, only a small number of examples must suffice.

For some readers, the supposition of systematicity in scientific knowledge generation may sound odd. Isn't it the case that in the most fundamental theoretical advances in, say, physics, chemistry, or biology chaos rather than order is the rule? Aren't there countless stories about scientific discoveries around in which the acting heroes rather half-consciously stumbled upon their discoveries instead of systematically directing their investigations to the novelty? Without attempting to critically assess these sometimes worshipping stories, it is true that the creative process of coming up with new and even revolutionary ideas is often not a very orderly procedure. This, however, is not a contradiction to our main thesis. As should always be remembered, our main thesis is comparative: it only states that scientific knowledge is *more* systematic than comparative knowledge from other domains, especially everyday knowledge. Thus, the existence of somewhat chaotic processes in the generation of novel scientific knowledge does not contradict the thesis. First, one may doubt whether in common knowledge there are processes of the generation of novel knowledge that can appropriately be compared to the generation of novel scientific knowledge. Second, if there are indeed processes of generation of novel everyday knowledge, it is a plausible guess that this novel knowledge comes about in equally chaotic processes as in science. If we had in our daily life somehow systematic procedures to generate novel knowledge, the sciences would immediately take up these procedures and systematize them further.

So we shall not worry about the messy elements in scientific creativity. Instead, we shall concentrate on the vast area of systematic knowledge generation and compare it with everyday procedures, and shall probably not be surprised if the former turns out to be more systematic than the latter. I shall start my discussion in the next subsection with the vast and diverse array of data collection. Subsection 3.8.3 will take up science's systematic exploitation of knowledge from other domains. In the final subsection, I will characterize science's generation of new knowledge as an autocatalytic process (subsection 3.8.4).

3.8.2 Data Collection

In the course of my discussion of the defense of knowledge claims, I have stressed the preeminent role played by data in all empirical sciences with regard to the defense of knowledge claims. It was important to note that here, one has to take the term "empirical sciences" in its wide sense, covering the natural sciences as well as the social sciences and the humanities (see section 3.4.1). However, the defense of knowledge claims as described in section 3.4.1 is by far not the only role of data in science. Clearly, empirical descriptions (3.1) have to be based on data as well as explanations (3.2) and predictions (3.3). Thus, data are a sort of vital principal of empirical science, and it is therefore not surprising that most empirical sciences are constantly on the move to systematically improve existing data and to gain new ones. The ways different fields of research go about this process are extremely different, and in the following, I shall discuss a few of them.

Let us start with the simplest case where data can be collected by unaided visual observation, without experimental interventions. This case is not very relevant for today's science, but in the not too distant past, it was relevant in a number of fields. Take early zoology or botany as an example. In the Western tradition, these fields emerged in the works of Aristotle (384–322 BC) and his pupil Theophrastus (371–287 BC). The description and classification of different species of animals and plants was based on their traits that were open to inspection (dissection of specimens also played a role early on). This descriptive enterprise was by no means restricted to this very early period of science. To present just one later example, Charles Darwin's voyage with the Beagle from 1831 to 1836 was largely devoted to the description of plants and animals and also included the description of geological formations. In the eighteenth and nineteenth centuries, there were many expeditions of the same or similar sort; the main means of gathering data was unaided observation. The central aim of such expeditions was to systematically collect data in order to proceed with the systematic descriptive tasks of botany, zoology, and geology.

However, science did not settle for unaided observation, and the history of astronomy is a particularly good case in point. Already before the invention of the telescope, various instruments were in use, mainly aiding the measurement of the position of celestial objects. With the invention of the optical telescope and its use in science from 1609 on by Galileo Galilei, however, a new era of collecting data about astronomical objects had begun. Immediately from its invention on, the optical telescope was continually and systematically improved, and already Galileo had started this development. Many new designs of the optical telescopes have been invented and continue to be invented. Ever since 1609, the optical telescope has been permanently used in astronomy. In the beginning, its light-gathering power was something like ten times that of the unaided human eye, corresponding to an objective lens diameter of roughly two centimeters. Four hundred years later, the most powerful optical telescopes feature mirrors equivalent to an aperture of almost twelve meters, corresponding to a light-gathering power three million times more than the human eye. There are even plans to build gigantic optical telescopes exceeding the power of today's largest telescopes by a factor of up to one hundred. In addition to the optical telescope operating with visible light, other sorts of telescopes were invented, systematically using all other media like X-rays, infrared radiation, or radio waves to gather information about astronomical objects. In addition, space-based telescopes of various sorts systematically gather data of extreme importance to cosmology. For instance, the Wilkinson Microwave Anisotropy Probe (WMAP) is a NASA explorer mission dedicated to measuring the cosmic microwave background radiation, especially its anisotropy (the radiation's uneven distribution over the sky). In 2001, the measuring instrument was put on a spacecraft and put in an orbit at a distance of 1.5 million kilometers from the Earth. Until 2010, it measured data of unprecedented accuracy and highest relevance to cosmology. For instance, from its data, the age of the universe was determined to be 13.73 billion years within 1 percent.

Not only were astronomical instruments systematically improved, but also their use is systematic and has become even more systematic. Just one topical example is the Sloan Digital Sky Survey. It uses an Earth-based 2.5-meter telescope and is one of the most ambitious and influential surveys in the history of astronomy. It started in 2000, and within its first eight years of operation, it systematically covered one quarter of the entire sky. It created three-dimensional multicolor images that contain more than 930,000 galaxies and 120,000 quasars. Furthermore, our own galaxy was investigated, and the spectra of 240,000 of its stars were recorded. Finally, systematic surveys of variable objects on the sky discovered some five hundred supernovae of a particular type that are important in determining details of the cosmic expansion over the last four billion years. The data produced by the Sloan Digital Sky Survey have been used in a wide range of astronomical investigations around the world, and, as of September 2008, they have contributed to more than two thousand articles published in scientific journals.

Astronomy is highly dependent on purely observational data. This means that the objects, events, or processes that are observed came into being without any human intervention. By contrast, laboratory sciences like physics or chemistry often produce the phenomena they are interested in themselves. The obvious advantage is that they can get exactly the kind of data they want at the time they want them by producing the pertinent phenomena. Systematic experiments became an essential part of science only in modern times, and the standard story is that it was Galileo who started systematic experiments with falling bodies and on the inclined plane. Whatever the historical details are, it is uncontroversial that since the seventeenth century, experiments gained more and more space in the sciences. Very schematically and simplified, in experiments there are two types of variables involved. The first type of variables comprises those that can be manipulated: they can be given some predetermined values by the experimenter; they are called independent variables. Variables of the other type are somehow dependent on the independent variables; their values can be observed or measured. The experiment consists of experimentally giving the independent variables some values and then measuring the values of the dependent variables. To illustrate with an exceedingly simple example that I have already used earlier (see section 3.4.3): if one wants to know what the relation of pressure and volume in a particular gas sample is, an experiment can deliver the desired information. The volume of the gas sample can easily be manipulated, so it is the independent variable. The resulting pressure can then be measured—the dependent variable. In fact, this experiment was performed in 1662 by Robert Boyle, and he found what is today known as Boyle's law, namely, the ideal gas equation $pV = const.$

Thus, experiments produce data, namely, the values of the dependent variables as functions of the values of the independent variables. These data can be put to various uses: finding a functional dependence, testing a functional dependence, applying a functional dependence for technological purposes, and so on. In the context of the present section, however, we are only interested in the systematic production of data by experiments and not in their further use. A topical example for systematic data production in experiments is elementary particle physics. In order to find out what properties particles have, one has to let the particles collide. To do so, one has to accelerate the particles first, and this is done in so-called particle accelerators. There are various types of particle accelerators, and the two principal types differ in the geometry of the particle paths: in linear accelerators, the particles travel a straight line, and in the other type, particles travel in spirals or circles, depending on the particular subtype of accelerator. The first accelerators were developed in the late 1920s and 1930s and were comparatively small machines. For instance, the first accelerator in which particles traveled a spiral path, called a cyclotron, was put to use in 1932 by Ernest Lawrence of the University of California at Berkeley, and its diameter was 5 inches (12.7 cm). In rapid succession, Lawrence and his team developed machines with larger diameters: 11 inches (30 cm), 27 inches (68 cm); in 1936: 37 inches (94 cm); in 1939: 60 inches (1.52 m); and in 1946: 184 inches (4.67 m). Impressive as this development is, these are still miniature machines in comparison to today's largest accelerator, the LHC. "LHC" stands for Large Hadron Collider, and this machine is quite large—8.6 kilometers in diameter (27 kilometers in circumference). To get an idea how incredibly complex it is to prepare and run an experiment on this machine, one should have a look at a paper in which a certain aspect of proton collisions was investigated: it lists no less than 3,172 authors (the so-called ATLAS collaboration) affiliated with roughly two hundred institutions.

What is the reason for building ever larger accelerators? Roughly speaking, the larger the accelerator, the larger the available collision energy is. And what is so attractive about higher collision energies? At higher collision energies, other processes take place than at lower energies. So in order to learn as much as possible about particles, their collisions should be observed in an energy range as large as possible, especially as high as possible. In other words, systematically increasing the energy of accelerators means systematically increasing the range of available data about particles. Obviously, the continual increase of collision energies for particles over some eighty years is an extremely systematic endeavor to generate experimental data.

Also in the historical natural sciences, there are many extremely systematic investigations underway whose purpose it is to produce new data. Think of paleoclimatology, the study of climate prior to the widespread availability of records of temperature, precipitation, and other instrumental data, roughly before 1850. This is a very important field in a time when human civilization presumably strongly influences world climate, and one therefore tries to understand natural and anthropogenic climate change. There are various sources of data of past climate. One particularly systematic way of generating new data about past climate is the investigation of ice cores. Ice cores are taken from continental glaciers by hollow drills. Several ice cores of more than 3,000 meters in depth have been taken, both from the Northern and the Southern hemispheres. As ice cores are formed by precipitation, mainly snow, they contain a record of the past atmosphere regarding temperature, precipitation, gas content, chemical composition, as well as remnants from volcanic eruptions and the like. The oldest, deepest layers of the ice in ice cores are roughly two hundred thousand years old, and a yearly resolution is possible for the last forty thousand years. The systematic aspect of investigations of ice cores does not only concern the individual cores, but also the possibility to compare different cores and extract information from their similarities and their differences.

It probably does not come as a surprise that the social sciences and humanities also generate data in very systematic ways. Here is a particularly intriguing example at the interface of psychology, psychiatry, and medicine. It is the longest longitudinal study of adult life ever conducted. On the study's web page, it is described as follows: "For 68 years, two groups of men have been studied from adolescence into late life to identify the predictors of healthy aging. This study has allowed us to examine the psychological traits, social factors, and biological processes that characterize adolescents and forty-year-olds who evolve into vigorous and engaged octogenarians. The study has created an unprecedented database of life histories with which to view the dynamic character of the aging process." There are two fairly different groups of people, so-called cohorts, that make up the study. They consist of several hundred men from the Boston area, selected for the study in the early 1940s. Every two years, the members of the study completed questionnaires about various aspects of their lives, and every five years, health information was collected from the members themselves and their physicians. Of course, gaining data about human development and aging in this way is incredibly more systematic than our everyday practice of, say, seeing our grandparents growing older and hearing them telling stories of their lives.

Let us now turn to the humanities. One of their extremely systematic enterprises is the edition of collected works and letters of some author. Of course, such an edition is an invaluable and authoritative source of data relevant to the study of the respective author. Take the Leibniz Academy Edition as an example. Leibniz lived from 1646 to 1716 and was one of the most prolific writers of his times, covering practically all areas of learning. In 1901, the project of editing all of his works, manuscripts, and letters was initiated by the European Association of Academies of Humanities and Sciences, and this project is still underway. More than eighty thousand manuscripts had and have to be identified, analyzed, transcribed, and annotated, among them about fifteen thousand letters from and to 1,100 correspondents. So far, most of the documents have never been printed. It is estimated that the project will continue for some thirty years to come, and then, it will have taken 140 years!

One particularly interesting aspect of all of these extremely systematic enterprises aimed at data generation in extremely heterogeneous fields is that they can be described as systematically forcing chance. The history of science of full of stories about chance discoveries where a lucky discoverer stumbled upon something he or she was not looking for at all, but which turned out to be extremely consequential. The discovery of penicillin by Alexander Fleming is a famous case in point. The story begins in 1928 when a culture plate containing certain bacterial colonies (staphylococci) was inadvertently contaminated by spores of a species of the mold *Penicillium*. The mold spores developed into a large colony which, as Fleming writes, "in itself did not call for comment but what was very surprising was that the staphylococcal colonies in the neighborhood of the mould, which had been well developed, were observed now to be showing signs of dissolution.... This was an extraordinary and unexpected appearance and seemed to demand investigation." As experimental scientists are not especially eager to wait for the lucky moment when an unexpected discovery is made, one way of systematically beating chance is by so-called brute force approaches in which every possibility within a certain range is examined. Some of the examples described above fit this description very well, for instance the systematic survey of a certain region of the sky or the measurement of the effects of particle collisions in a particular energy range. Another example of this kind is an important strategy of contemporary drug development where some chemical compound is systematically varied and its possible pharmacological effects are investigated.

It should be obvious by now that the sciences may be extremely systematic in their efforts to collect data. There can be no doubt that the degree of systematicity exhibited is by far higher than in any comparable everyday activities.

3.8.3 *The Exploitation of Knowledge from Other Domains*

There is another way in which the sciences systematically improve their ability to generate new knowledge: they systematically exploit knowledge from other domains that may serve their own purposes. This concerns mainly knowledge of neighboring and auxiliary disciplines as well as technological knowledge of all sorts. Let us look at some examples.

The emergence of modern natural science in the seventeenth century is characterized by several features, among them by what has been called the mathematization of nature: nature became the object of descriptions and of theorizing in mathematical terms. This move was enormously consequential, both for the natural sciences, initially only for physics, and for mathematics. Physics became a quantitative science, which it was not before, and later other sciences followed this path. Using a quantitative language made many mathematical means available that proved of utmost importance for the further development. Mathematics could then systematically be exploited to develop physics. On the other hand, mathematics itself gained enormous impulses for its own development, and ever since the seventeenth century mathematics' development has been strongly influenced by the challenges posed by natural science.

In the 1940s, a similar development began with the emergence of computers and software technology. Computers and software technology have been invented and developed largely in the context of scientific applications, especially in physics. However, as is well known, information technology has long left this narrow confinement and pervades scientific life in general and also nonscientific life. Today, literally all disciplines use information technology in one way or another, and some disciplines have even been fundamentally revolutionized in the process. Again, the computing opportunities are being systematically exploited in various ways by the sciences by making use of the available computing power for the purpose of the respective discipline. Even new hybrid disciplines have been created like bioinformatics, the field of science in which biology, computer science, and information technology merge to form a single discipline, or business informatics, which covers the interface between information science and business studies.

On a much more general scale, technology is systematically exploited and sometimes further developed by all disciplines that use instruments for observation and experimentation. Scanning the examples given in the previous section 3.8.2 on the data collection techniques used in various sciences makes this obvious. Also in many humanities, especially the historical sciences, cutting edge technologies are put to various purposes, e.g., for dating and other analyses of materials. For instance, radiocarbon dating (also called the C_{14} method) is the most important chronometric technique in archaeology today (it is also important in historical natural sciences). The principle is that plants are in constant exchange with the atmosphere that contains a small proportion of a radioactive isotope of carbon, namely C_{14}. The plants' content of carbon thus reflects the proportion of carbon isotopes in the atmosphere, and animals eating plants ingest the same proportion of carbon isotopes. Once the organism is dead, the direct or indirect exchange with the atmosphere terminates, and the decaying carbon isotope C_{14} is not replenished anymore. Thus, the dead organism's C_{14} content slowly decreases in time, corresponding to the half life of C_{14} of approximately 5,730 years. In organic materials, the proportion of C_{14} to the stable isotopes is thus a measure for the time since the organism's death. In this way, organic remains of all sorts can be fairly accurately dated, up to a maximum of fifty to sixty thousand years of age.

It is not only technological knowledge that scientific disciplines make use if in various ways, but also contributions of other scientific disciplines are systematically imported whenever needed, sometimes resulting in new hybrid disciplines. Examples abound. Biology imported chemical knowledge in order to understand biological phenomena. Several new disciplines emerged, like biochemistry, protein chemistry, or molecular biology. Zoologists studying animal flight imported knowledge from physics; a new subdiscipline developed, flight biophysics. With the help of neuroscience and other disciplines, economists study in detail people's economic decision making; a new subdiscipline arose, neuroeconomics. And so on. Whenever scientific knowledge from other domains can be productively used, it will be systematically imported and utilized.

Of course, also in daily life, we turn to other knowledge resources. Many of us use computers for e-mail. Many of us turn to the Internet in order to get pertinent medical information. And so on. However, in everyday practice, the use of other knowledge resources is typically spotty and selective—just the opposite of the systematic exploitation of other knowledge resources as it is done by the sciences.

3.8.4 *The Generation of New Knowledge as an Autocatalytic Process*

Finally, I would like to characterize science's knowledge generation on a more abstract and general level. The level of abstraction that I mean can easily be apprehended by giving the traditional view of science's knowledge generation at that level, namely, that scientific knowledge is produced by the application of the scientific method (or scientific methods). In section 1.1, I have already hinted at the insight that this view cannot be upheld. An alternative positive view was developed by the historian and philosopher of science, Thomas S. Kuhn (1924–1996), and I think that his view is correct. In scientifically nonrevolutionary times, the basic orientation of scientific work does not come from any abstract rules but from the scientific work already done. More specifically, the scientific problems already solved in some area of research provide the main guideposts for future work. They do so in various respects at the same time. First, they display by example what a typical solvable research problem in the respective area looks like. Second, they present the legitimate means by which a research problem at that time in that area is to be solved. Third, they show what can be expected from an adequate solution to a research problem in terms of accuracy, explanatory power, and the like.

In other words, in the sciences, the stock of already existing knowledge is systematically used in order to create new knowledge. This is true across all disciplines. Every piece of newly gained knowledge provides additional resources for potentially further increasing knowledge. Thus, science is a self-amplifying or, in the language of chemistry, an autocatalytic process. As the potential growth of science is proportional to the already existing knowledge, science is capable of exponential growth, given sufficient resources and no other constraints. Over several centuries, this has indeed been observed for modern natural science, when scientific growth was measured by fairly different indicators.

Again, it is plausible also for everyday knowledge that we expand it on the basis of existing knowledge. However, we seldom do that systematically. We simply add what our situation demands and what is in our reach.

3.9 THE REPRESENTATION OF KNOWLEDGE

3.9.1 *Some Preliminaries*

The topic to be discussed in this section will somewhat overlap with other sections in this chapter, especially with section 3.1 on descriptions. There, I discussed the systematicity-enhancing devices of classification and taxonomy, and they typically come with a special nomenclature. For instance, the Linnaean classificatory scheme for plants and animals was introduced together with the binominal nomenclature in which a particular species is identified by its genus and by a specific trait of that species within the genus. A particular nomenclature is one of the topics of this section, in being a systematic device of knowledge representation. So the systematicity of descriptions may be somehow interwoven with the systematicity of representation. Still, it is worthwhile to devote an entire section to the representation of knowledge as we will find it independent enough from the other dimensions to constitute an autonomous dimension of systematicity. Special techniques of representation in science are by no means restricted to nomenclature introduced in the context of classification. However, as we shall see, systematicity of representation mostly contributes to other dimensions of systematicity, further interlocking different dimensions of systematicity.

As in many other sections of this chapter, the quantity of the pertinent material is overwhelming. I shall have to pick almost randomly a few examples that illustrate my thesis. Unsatisfactory as this is, especially for a book on systematicity that should live up to its own standards, I have no other choice.

3.9.2 *Examples*

Let us begin, as usually, with mathematics. Mathematics is characterized by, among other things, its tendency toward abstraction. Mathematical objects are thus abstract objects, and our everyday language, or more generally, our everyday means of representation, usually do not provide for referring to such mathematical objects. Thus, mathematics is forced to invent new means of representation of its specific objects. Going back to the beginnings of mathematics in antiquity, this coercion may not be immediately visible for geometry. Geometry deals with circles, squares, and other figures. The abstract nature of these figures when conceived as *mathematical* objects is often not immediately evident, because they are represented in concrete drawings. However, when turning to logic, the character of mathematical objects as something special, something different from the objects we usually refer to by ordinary means, becomes visible. When Aristotle invented the concept of logical form as a key element of his codification of syllogistics, he also had to introduce means of presentation of logical forms. To us today, they look very familiar because we were confronted with them in algebra, a discipline much later invented and developed by Indian and Arab scholars. So for Aristotle, the logical form of "All humans are mortal" can be represented as "All A are B," and the other logical forms he was concerned with in a similar way. This is a precise representation of the logical form Aristotle was interested in; it represents exactly the logical form—not more, not less. Furthermore, it perfectly suits the systematic exposition of syllogistics.

Roughly 2,300 years after Aristotle, Gottlob Frege set out to revolutionize logics by putting it in a thoroughly mathematical framework. In 1879, he published his *Begriffsschrift* (translated as "concept notation" or "concept writing"), a book that has been widely seen as the most important contribution to logic since Aristotle. I do not have to go into the details here because we are only interested in the representational aspect of this important event. Together with the proposed conceptual change, Frege had to invent a new formalism in order to represent what he now took to be the essential elements of the logical form. Already in its subtitle, Frege draws attention to the representational aspect of his work: he calls the Begriffsschrift "a formula language, modeled on that of arithmetic, of pure thought." Frege's formula language is as systematic as it is cumbersome because it is two-dimensional. Fortunately enough, not much later, much simpler notations were invented in which the systematicity of this part of logics is transparently represented.

Another representational device extremely widespread in mathematics is the graph by which the members of a certain class of mathematical functions can be easily visualized. Roughly speaking, a mathematical function unambiguously assigns to the values of a given variable ("independent variable") the values of another variable ("dependent variable"). For instance, calling the independent variable x and the dependent variable y, the function $y = 3x$ assigns to every value of x of the independent variable the value $3x$ to the dependent variable. This kind of assignment is fairly abstract, and it can be much more complicated. However, it can easily be translated into a graphical representation in which we get, in its most typical form, for $y = 3x$ a straight ascending line. These graphs are quite familiar in life, for instance the representations of the time change of the stock market or of the exchange rate of foreign currencies—they are to be found in almost every issue of almost any newspaper. However, originally, these were scientific means to visually represent functional dependencies. They have a high degree of built-in systematicity as they are themselves well-defined mathematical mappings from the functional dependence to the graphical representation, or, roughly put, the graph depends on the represented function in a law-like manner.

Before getting lost in mathematics, let us turn to chemistry. The representation of the systematic order of the chemical elements in the periodic system has received almost iconic character: it is found in many classrooms around the world as a wall decoration, and no introductory chemistry textbook can do without it. There is much information contained in the specific display of the elements because it graphically represents certain recurrent features of the elements—thus its name, "*periodic* system." Furthermore, certain similarities between the elements are displayed by their two-dimensional ordering in the system: elements belonging to the same column and elements belonging to the same row share certain properties. These features can be inferred from the similarities in their electron configurations. As it happens, there are quite a few variants of the standard representation of the periodic system, but none of them equal the popularity of the standard form in the least.

Staying with chemistry now, another remarkable representational feature of chemistry is its naming systems. Chemistry has the problem of an abundance of chemical items: as of December 2012, chemical knowledge covers more than sixty-nine million unique organic and inorganic chemical substances, such as alloys, coordination compounds, minerals, mixtures, polymers, and salts, and more than sixty-four million sequences. In addition, there are more than forty-six million known single- and multi-step chemical reactions. Of course, very few of these items have traditional names like water, salt, or caffeine, but all of them have to be named in order to refer to them in scientific discourse. One brutal but on the other hand unique naming procedure consists in assigning numbers to all of the substances and sequences, the so-called CAS Registry Numbers. Up to December 2012, more than 133 million such numbers have been assigned, and, on average, fifteen thousand new numbers designating new substances are added *each day*. The CAS Registry number is a unique numeric identifier that designates only one substance, but has no chemical significance. However, it is a link to a wealth of information about the substance contained in the pertinent database. Needless to say, the CAS system aims at completeness. It thus interlocks the systematicity of representation with the systematicity dimension of completeness that I discussed in section 3.7.

However, this is by far not the only naming system in chemistry. For chemical compounds, there are additional ways of naming them, among them by empirical formulae, structural formulae, systematic names, and trivial names. They are used in different contexts. In routine contexts, the trivial names like caffeine or salt are used, but they are uninformative about the composition or the structure of the substance: for instance, "caffeine" only indicates that the substance is somehow related to coffee. The empirical formula displays the information about what kinds of atoms are contained in the compound and in which proportion. For instance, the empirical formula for table salt (sodium chloride) is $NaCl$. The formula contains the information that table salt is composed of sodium (Na) and chlorine (Cl) in equal proportions. This way of denoting compounds is useful when describing those aspects of chemical reactions in which the preservation of the atomic constituents of the reaction partners is in focus. However, the empirical formula is silent about the specific spatial configuration of the atoms in the compound. This is what the structural formulae are designed to convey, together with the type of chemical bonding within the compound. Structural formulae come in very different forms. A representation of the molecular configuration of a compound is extremely useful for a visualization of chemical syntheses and reactions.

Finally, there is a system of nomenclature for compounds maintained by the International Union of Pure and Applied Chemistry, IUPAC. The idea is to have a name that uniquely refers to a compound and that, at the same time, systematically encodes the compound's composition and structure. The latest version of the rules for naming inorganic compounds is contained in a book of 377 pages, the so-called "Red Book." The Red Book describes the function of chemical nomenclature as giving "systematic information about a substance." This goal can be achieved at four different levels, each level containing more information about the compound. For instance, the second level is described as follows: "When a name itself allows the inference of the stoichiometric formula [i.e., the empirical formula, P.H.] of a compound according to general rules, it becomes truly systematic." Thus, the motif of systematicity is clearly expressed regarding the aims of the nomenclature. Organic chemistry has its own nomenclature; its rules are contained in the so-called "Blue Book." Of course, the more information that is contained in the nomenclature, the longer the names get. For instance, whereas the empirical formula for caffeine is the comparatively short $C_8H_{10}N_4O_2$, its IUPAC name is 1,3,7-trimethylpurine-2-,6-dione (and there are even longer versions).

Here is a brief excursion into biology. I want to discuss only two examples out of a host of others in which biological knowledge is systematically represented. Both examples were extremely consequential. The first example concerns the only diagram contained in Charles Darwin's groundbreaking book *On the Origin of Species*, published in1859, between pages 116 and 117. The diagram depicts something quite familiar today, namely the increase of the diversity of life of a (hypothetical) genus in the course of evolution, accompanied also by the extinction of several species. This is the first depiction of the evolutionary tree of life. Darwin spends roughly ten pages to explain the content of the diagram, but once one has understood what the message is, it is all contained in the diagram. This is a fitting example for the saying that a picture is worth a thousand words.

My second example from biology concerns the physical model of the structure of DNA that James Watson and Francis Crick devised in order to understand DNA's basic functioning; DNA is the genetic material of most organisms. As Watson and Crick set out to detect the structure of DNA, they assembled possible DNA structures out of its known molecular building blocks. They used physical models of these molecules that consistently represented their magnitude and, even more important, their chemical structure. The possible structures for DNA were additionally constrained by X-ray diffraction data gained from crystallized DNA in another laboratory. In the end, Watson and Crick experimentally, i.e., by playing around with different possibilities, found a constellation of the building blocks satisfying all constraints. Furthermore, this constellation immediately suggested "a possible copying mechanism for the genetic material" which was a central element of the great mystery of life: how are organisms capable of reproduction? Once it was completed and further confirmed, Watson and Crick's model clearly *represented* biologically relevant knowledge of a certain molecular structure. However, the representation of the molecular components of DNA by toy models also provided the opportunity to generate this knowledge, at least in hypothetical form. The generation of new knowledge was the topic of the previous section in this chapter (section 3.8). What we see again here is that some means enhancing the systematicity in one dimension are also instrumental in increasing the systematicity in another dimension. The different dimensions of systematicity are sometimes mutually reinforcing each other.

Let us now turn to geography that has a long tradition of the representation of knowledge by maps. Of course, not all maps can count as representations of scientific knowledge, and there is a large grey transition area between the scientific and the nonscientific. The ancient Greeks had already realized that the Earth is a sphere, and having a technique of projecting the surface of a sphere onto a plane, they were able to draw world maps. A particularly well-known example is the world map of Ptolemy, originated c. 160. However, scientific mapmaking seems to have emerged only in the course of the seventeenth century. These maps were based on land surveying techniques and more precise time measurements that were often based on observations of Jupiter's moons. Geometry is, of course, a vital ingredient to mapmaking because it delivers different projection techniques in order to translate the surface of a sphere into a flat map. Mapmaking has produced hundreds of different sorts of maps. In addition, map-like visual representations of nongeographical subjects have found virtually unlimited applications in science and nonscience alike. Especially since the advent of powerful computers and plotters, data visualization of all sorts permeates the sciences, in particular the natural and engineering sciences. But also the social sciences use countless graphical means like charts, maps, and diagrams in order to visually display their results. For instance, the "landscape" of the sciences is the subject of many different maps contained in the *Atlas of Science*. Various relationships among the sciences concerning paradigms, history, themes, research communities, cross-citations, and so on can be graphically depicted in such maps facilitating to get the "big picture" quickly.

Clearly, the idea behind all of these representations of knowledge is to have a representation of some body of knowledge that, due to its specific visual quality, can be grasped quickly and accurately. It is obvious that the sciences were forced to develop means of representation vastly more systematic than our everyday measures because of the vastly larger amount of information generated in the sciences. This is, of course, a consequence of the ideal of completeness operative in science, together with the partly highly systematic procedures of knowledge generation; I have discussed these two features in the two preceding sections 3.7 and 3.8. Again, we have an interplay of different aspects of systematicity: more systematicity in one dimension may reinforce an increase of systematicity in another one.

4

Comparison with Other Positions

IN THIS CHAPTER, I want to compare, in a somewhat schematic way, the position of systematicity theory with older answers to the question "What is science?" that were given in the history of philosophy science. My interest in this chapter, however, is not "antiquarian" in the sense of just collecting these answers and contrasting them with systematicity theory in order to embed the latter in the historical course (or just to show off with my historical literacy). On the contrary, the purpose of this chapter is mainly argumentative: we shall get, from a different angle, additional arguments for the persuasiveness of systematicity theory. First, as a sort of prelude, by contrasting systematicity theory with earlier positions, the specifics of the new theory will become clearer. Second and more important, by comparing the older positions with systematicity theory, it will become obvious that in many cases, systematicity theory is a generalization of the former theories. It will turn out that some of the older positions are not just wrong, but they are somehow one-sided by overemphasizing, or even pushing to the absolute, one or the other aspect of systematicity. Furthermore, the specific meaning of systematicity they employ implicitly in one or the other dimension of systematicity may be restricted. In other words, what I intend to show in this chapter is that some of the older positions are special cases of systematicity theory. In outline, this may already be obvious with regard to the first two historical phases of the answers to the question "What is science?" that I sketched in section 1.1. In the first phase, where the specificity of scientific knowledge was seen in its absolute certainty, the third aspect of systematicity, the defense of knowledge claims, was pushed to the extreme by asking for demonstrative proof for scientific statements (the axioms exempted, of course). In the second phase, where the scientific method was stressed, the aspect of an orderly procedure, which may be an aspect of systematicity, was overemphasized. Quite generally, "systematicity" is a wider concept than "methodicity": everything that is methodical is also systematical, but systematicity covers more than methodological rules of generation or justification of knowledge.

Why should this mode of historical comparison supply additional argumentative support for systematicity theory? Why is it a positive argument for systematicity theory if it turns out to be a generalization of some of the former theories? Regarding the older answers to the question "What is science?," there are two principal possibilities. Either these answers were just plain wrong at their own time, or they were at least, in some sense, approximately right. If the first possibility obtains, we may simply dismiss the past and count ourselves happy to live now and not then, and concentrate serenely on our present business. However, given that many of these answers showed some historical stability—which means that they survived some critical discussion over a longer period of time—the assumption that they were just plain wrong is not very plausible. It is more likely that they had some degree of persuasiveness, possibly due to getting something about the contemporary science right. That we today find these positions unacceptable may have two heterogeneous reasons. First, there may be flaws in these positions not clearly visible to the contemporaries. To us, however, after prolonged discussion and perhaps some disillusionment about the force of reason, these positions may thus appear untenable. For instance, the insight that inductive arguments may be quite problematic was gained only during the eighteenth century, so we may be more hesitant to trust them than people in the seventeenth and early eighteenth century. Second and more important, the subject matter of these earlier positions was their contemporary science, but science has evolved and changed in the meantime. So, what may have been right about earlier science may be wrong, or rather too narrow or one-sided, with respect to today's science. The standards of legitimate science may have changed, to wit both in the direction of more or less strictness. What may have been accepted as a proof in mathematics two hundred years ago may be unacceptable as a proof today (for example, the older proof may have involved so-called infinitesimals, that is, infinitely small quantities that have later been seen as illegitimate). What may have been a procedure too hypothetical or speculative in natural science three hundred years ago may be acceptable today because of a relaxed view of the role and legitimacy of hypotheses (for example, the existence of some sort of acceptance of string theory despite its lack of the slightest direct empirical confirmation). In other words, if the subject matter of a theory of science is a moving target, which it is, also that theory has to get moving. A theory that is more general than its predecessors, thus containing them as special cases, stands the chance of being able to explain the success of the former and, at the same time, of gaining plausibility for itself. This will be the heuristic strategy in this chapter: we will try to understand the earlier answers to the question "what is science?" as special cases of our answer—in other words, as restricted versions of the idea of systematicity in nine dimensions.

Let us now discuss the positions of various authors or schools that played an eminent role in the history of philosophy of science. For every author or school, my intent is to sketch their view of the nature of scientific knowledge and to show how that view relates to the view developed in this book. This holds for the first six positions discussed in this chapter. However, I have to warn the reader: it really will only be a sketch of the respective positions—and a one-sided view on top of this. This is due to the aim of the exercise: not a historically balanced total view of the respective positions, but a much focused view on those aspects that can serve as points of the desired comparison with systematicity theory. Furthermore, I have to admit there is an element of arbitrariness present, this time in my choice of the authors discussed. There are many more authors who could be considered. For example, of the classical authors involved in reflections about systems and systematicity, Leibniz, Wolff, Lambert, and Hegel are conspicuously absent, to name but a few. However, I think that my choice represents a variety of views about the nature of science that is broad enough to fulfill its purpose: to show that many of the valuable insights developed in the history of philosophy about the nature of science are indeed incorporated in the position presented here.

The discussion of the last two positions in this chapter is somewhat different (sections 4.7 and 4.8). These authors present challenges to systematicity theory or a related though markedly different enterprise also concerned with systematicity. The challenges must be met, and the differences to the other enterprise must be delineated.

4.1 ARISTOTLE

4.1.1 The Position

Aristotle (384–324 BC) is a good starting point for our discussion because he can be considered the first philosopher of science in the Western tradition. The emergence of a philosophy of science presupposes the existence of some science and a practice of philosophical reflection that can then be directed at that science. It was in ancient Greece that these presuppositions were indeed fulfilled for the first time in Western history. Aristotle did research in most of the existent subfields of what we call today natural sciences, formal sciences, social sciences, and humanities. He wrote the very first treatise fully devoted to the philosophy of science, *Posterior Analytics* (Aristotle's *Prior Analytics* is devoted to formal logic). Aristotle proposed an ideal for scientific knowledge that was, in his times, realized in mathematics. We are mostly familiar with it through the *Elements* of Euclid, a work that appeared shortly after Aristotle's times, and that was probably influenced by *Posterior Analytics*. Aristotle's ideas about scientific knowledge proved enormously consequential: their influence lasted for roughly two millennia.

Aristotle's ideal for scientific knowledge can be denominated as being *categorical-deductive*—a science that realizes it is a demonstrative science. The term "categorical" denotes a concept that is opposed to "hypothetical," thus indicating the quality of knowledge of being absolutely certain, i.e., that its truth is beyond doubt. "Deductive" concerns the architecture of scientific knowledge, i.e., that the fundamental relation between different pieces of knowledge is logical deduction. More to the point, there are "principles" that serve as a basis for deduction of other sentences that are proven in this way. As the principles are conceived of as absolutely certain, so are all the sentences that are deduced from them. The resulting body of scientific knowledge therefore consists of principles and proven sentences, and that is what demonstrative science is all about. In addition to their epistemic status as proven, the content of sentences is explained by their deduction from first principles. Thus, by deduction, we get not only a guarantee for their truth but also an explanation for what they are about. In this way, a demonstrative science supplies two unsurpassable epistemic goods: truth of its claims and full explanation of their content.

The absolute certainty of principles, or their necessary truth, can, of course, not be established by proofs. Nevertheless, according to Aristotle, humans have the capacity to grasp the necessary truth of these principles without proof, by a sort of direct insight into our experience. Of course, this latter position is problematic, but for centuries it was a widely accepted position. Fortunately, given the purpose of our discussion of Aristotle, we do not have to delve into these problems.

4.1.2 *Comparison with Systematicity Theory*

There are several aspects of this ideal of scientific knowledge that can directly be compared with what I contended about scientific knowledge in the last chapter. These aspects concern the defense of knowledge claims, explanations, the ideal of completeness, and the representation of knowledge.

With respect to the defense of knowledge claims, Aristotle posits the absolute maximum, namely, proof for every claim that is derivative and the sort of immediate justification that is appropriate for principles. As I have discussed in section 3.4, this is the most rigorous form of systematicity applicable to the defense of knowledge claims. The same holds with respect to explanation, where Aristotle posits deduction from principles. Again, this is a special and particularly rigorous form of systematicity of explanations. An ideal of completeness is only implicit in Aristotle's vision of scientific knowledge. Basically, it consists in the complete characterization of the subject matter of the respective science by the principles. Finally, a science in the Aristotelian sense is highly structured and, correspondingly, very orderly represented. It has the form of what we call today an axiomatic system.

We can conclude that Aristotle's position is a special case of the systematicity position. For knowledge to count as scientific, Aristotle stresses in four dimensions the requirement of systematicity in particularly rigorous and therefore narrow forms. The history of science has shown that this rigor—desirable as it may be—is, in fact, beyond human reach. With respect to principles, we have become much more modest. Neither in mathematics nor in the natural sciences, let alone the other areas of learning, do we believe any longer in the attainability of any sort of immediate certainty. With respect to a rigorous axiomatic structuring of a science, it is only in mathematics that this idea is still fully alive. In other areas of learning, it is only in a few highly mathematicized areas where strict axiomatization is aspired to. In a less rigorous form, the idea of axiomatization, namely, to concentrate knowledge in a few general propositions and draw consequences from them, is of course still important for the organization of scientific knowledge, at least in some disciplines.

4.2 RENÉ DESCARTES

4.2.1 *The Position*

Descartes (1596–1650) is usually and rightly seen as the champion of method as constitutive of science. The work most relevant in the context is his first published book, *Discours de la Méthode*, as it is called in abbreviation. The full and nicely descriptive title translated in English is *Discourse on the Method for Properly Conducting Reason and Searching for Truth in the Sciences, as well as the Dioptrics, the Meteors, and the Geometry, which are Essays in this Method*. Usually, in philosophy, only the philosophical introduction of this book is read, and in many editions, the scientific essays on optics, meteorology, and geometry are not even included, although they are extremely important parts of Descartes' scientific work. For our purposes, the second part of the philosophical introduction is most important because it contains Descartes' famous four rules on method. The context of his articulation of these rules is his frustration with the sciences of his times, which he takes to be unreliable, with the exception of mathematics. But, being aware of their shortcomings, Descartes believes

> that it has been my singular good fortune to have very early in life fallen in with certain tracks which have conducted me to considerations and maxims, of

which I have formed a method that gives me the means, as I think, of gradually augmenting my knowledge, and of raising it by little and little to the highest point which the mediocrity of my talents and the brief duration of my life will permit me to reach.

What does this "method" consist of by which knowledge can be gradually augmented? Descartes' idea was to seek a method that would comprise the advantages of logic, geometrical analysis, and algebra and be exempt from their defects. In his view, their basic defects were that they were, for various reasons, not really conducive for the augmentation of knowledge. One of these reasons was that there were too many precepts, including injurious or superfluous ones. So Descartes had to reduce their number and focus them on his only goal, the augmentation of knowledge. And he thought he was successful: "I believed that the four following [precepts] would prove perfectly sufficient for me, provided I took the firm and unwavering resolution never in a single instance to fail in observing them." Thus, it is constitutive of these precepts that they are not just rules of thumb that could be modified and assimilated to the situation at hand, but that they were to be followed strictly. Here are the four precepts that Descartes commits himself to:

First precept: "never to accept anything for true which I did not clearly know to be such; that is to say, carefully to avoid precipitancy and prejudice, and to comprise nothing more in my judgment than what was presented to my mind so clearly and distinctly as to exclude all ground of doubt."

This first rule tries to prevent the intrusion of false statements into science. Descartes applied a special criterion for the truth of statements: they present themselves as "clear and distinct." This is part of his well-known rationalism. Descartes supposed that the mind had the power, by itself, to judge the truth of statements: their truth was evident by the statement's being clear and distinct such that all possible grounds for doubt disappear.

Second precept: "to divide each of the difficulties under examination into as many parts as possible, and as might be necessary for its adequate solution."

This rule articulates a rather abstract heuristic strategy that is supposed to make complex problems manageable by considering their component parts.

Third precept: "to conduct my thoughts in such order that, by commencing with objects the simplest and easiest to know, I might ascend by little and little, and, as it were, step by step, to the knowledge of the more complex; assigning in thought a certain order even to those objects which in their own nature do not stand in a relation of antecedence and sequence."

This rule is, again, rather abstract. It postulates an orderly procedure. One must begin with the simplest in the sense that is the simplest with respect to the knowing subject, not necessarily the simplest by its nature. From there on, one has to proceed in a stepwise manner to the more complex. This methodological ordering is conceived of as overruling possibly different or nonexisting ordering relations that are based in the nature of the objects themselves.

Fourth precept: "in every case to make enumerations so complete, and reviews so general, that I might be assured that nothing was omitted."

Again, this is a rather abstract and rather vague rule: be careful that, if you have a number of subproblems, you don't forget anything in the end.

It is obvious that these rules present many problems, both with respect to content and to practical application, given their highly abstract level. But we will not pause to delve into questions of interpretation that have kept Descartes scholars busy for centuries. Instead, we will consider immediately the relation of Descartes' position to the position developed in this book.

4.2.2 Comparison with Systematicity Theory

On an abstract level, it is immediately visible that Descartes' four precepts can be appropriately characterized as rules for a systematic procedure in the sciences. This is no surprise, because Descartes is justly seen as one of the main champions of the scientific method, and methodicity is a special case of systematicity: every procedure that is methodical, i.e., that is committed to rules, is also systematic. The converse, however, does not hold because procedures may be fully systematic in some sense without a firm commitment to rules. But let us look into more detail about how Descartes' position relates to the systematicity position.

The first precept is a special way to defend knowledge claims: it states that only true statements are acceptable, and it formulates a criterion for truth. Whether that criterion—clarity and distinctness—is still acceptable today does not matter in our context.

From today's point of view, the second and the third precepts are probably best seen as part of a rational heuristics, i.e., as rules on how to tackle complex problems in a rational way. However, from today's perspective, it is not really important whether a scientific problem was solved by being attacked in a rational, transparent, and stepwise way as it is prescribed by the two precepts, or in a more chaotic manner. What ultimately counts is whether it can be justified that a proposed solution to some scientific problem really is a solution. Thus, the second and third precepts are certainly not binding for today's science. They may, in some field, be part of a heuristic toolbox. As such, they may be seen as a special case of our eighth dimension of systematicity, the generation of new knowledge. Descartes' second and third precepts represent a very specific, systematic way of knowledge generation.

The fourth precept is clearly covered by what I have called the ideal of completeness.

We can summarize this evaluation of Descartes' four precepts as follows. The first precept concerns a special case of the defense of knowledge claims (dimension 4). The second and the third precepts concern a specific way of knowledge generation, a rational heuristics. They are therefore a special case of our dimension 8. Finally, the fourth precept emphasizes an aspect of completeness and is therefore a special case of our dimension 7. Again, we find that Descartes' position is a special and partial case of the systematicity position developed in this book.

4.3 IMMANUEL KANT

4.3.1 *The Position*

Kant (1724–1804) is a well-known champion of systematicity as the defining characteristic of science. In his *Critique of Pure Reason*, he is very explicit about the role of systematicity with respect to a prospective science: "[S]ystematic unity is what first raises ordinary knowledge to the rank of science, that is, makes a system out of a mere aggregate of knowledge." In his *Metaphysical Foundations of Natural Science*, he is equally clear: "Every discipline if it be a system—that is, a cognitive whole ordered according to principles—is called a science." One commentator goes so far as to say that the "concept of *system* is perhaps the most central idea of Kant's theory of knowledge." It thus appears that, apart from our formulation of the systematicity position in comparative terms, Kant is pushing exactly the same idea as we are: it is systematicity that indicates the difference between "ordinary knowledge" and scientific knowledge. However, we must take a closer look into what Kant exactly means by "system" and how it applies to the sciences.

What is a system for Kant? He explains as follows: "In accordance with reason's legislative prescriptions, our diverse modes of knowledge must not be permitted to be a mere rhapsody, but must form a system.… By a system I understand the unity of the manifold modes of knowledge under one idea." This passage requires some explanation. First, it must be noted that Kant thinks that the drive toward systems is a "legislative prescription" by reason. It just belongs to our rational nature that we want to organize knowledge into the form of a system. Second, the expression "diverse modes of knowledge" may be misleading. As can be seen from the German original, Kant means the multitude of different pieces of knowledge. Third, these different pieces of knowledge are not permitted to be "a mere rhapsody," or, as he put it in the quote cited above, "a mere aggregate." By these expressions, Kant means a multitude of things that are put together without any order, "as the result of a haphazard search," as he puts it elsewhere. Fourth, by contrast, a system exhibits order among its parts by unification of these parts "under one idea." We must not misunderstand the term "idea" here because for Kant, it is a technical term. Kant explains it in the sentence that follows immediately after the quoted passage: "This idea is the concept provided by reason—of the form of a whole—in so far as the concept determines *a priori* not only the scope of its manifold content, but also the positions which the parts occupy relatively to one another." Thus, the "legislative prescriptions" by reason, mentioned above, consist of an idea—we would today say a normative conception—of what the body of knowledge should look like. And this normative conception is the conception of systematically ordered knowledge: what belongs to that system of knowledge and what does not, and what relative position different parts should have.

This is the abstract idea of a system that is, for Kant, constitutive of science. Kant applies this idea to three different subject areas, according to his classification of what we call the natural sciences in three main groups: historical doctrine of nature, improperly so-called natural science and properly so-called natural science. The historical doctrine of nature "contains nothing but systematically ordered facts about natural things." Kant thinks here of "*natural description*, as a system of classification for natural things in accordance with their similarities, and *natural history*, as a systematic presentation of natural things at various times and places." Conspicuously absent from the historical doctrine of nature are laws of any kind. This is the distinctive feature of both forms of what Kant calls natural *science*. It is important to note here that Kant's use of the word "science" is stricter than ours. Properly so-called natural science in his sense contains principles or laws that hold a priori, i.e., principles that are justified by exclusively nonempirical means. As a consequence, *proper* science's statements hold apodictically, or with necessity. By contrast, "improperly so-called science" contains principles or that are valid on empirical grounds, i.e., that are gained by induction. As a consequence, statements of improperly so-called science in Kant's sense do not hold apodictically, which means that they do not hold with necessity. The reason is that induction can never establish apodictic certainty of some general statement because not all possible cases to which it applies can be overviewed. As an example of such an improperly called science, Kant uses chemistry. Its grounds and principles are only empirical "and thus the whole of cognition does not deserve the name of a science in the strict sense; chemistry should therefore be called a systematic art rather than a science."

What Kant calls "historical doctrine of nature" and "improperly so-called natural science" is familiar to the modern reader, despite their fairly unfamiliar names: it is simply what we call empirical science. Stranger to us may be "properly so-called natural science" because of its containing principles that hold a priori and thus with necessity. This is a kind of science that Kant thought was possible basically due to his "transcendental philosophy": he thought that certain core elements of Newtonian mechanics could be justified on a priori grounds, i.e., without the help of any empirical investigations. I am not going to make plausible why Kant thought this form of mathematics-like science to be possible but rather turn to the different conceptions of systematicity that are involved in the three kinds of disciplines.

Kant's "historical doctrine of nature" contains, in his own words, "nothing but *systematically* ordered facts about natural things" (my italics). There are two different sorts of systematicity hidden under the title of "systematically ordered facts." First, synchronically, there is "a system of classification for natural things in accordance with their similarities." Second, diachronically, there is "natural history, as a *systematic* presentation of natural things at various times and places" (my italics). What Kant calls the "historical doctrine of nature" does not reach the level of science (in his sense) because it lacks laws of any kind.

Once laws or, as Kant often puts it, principles are involved in organizing a whole of cognition, we have a system that can be called a science. If the system is built up such that its parts are related as "ground" and "consequence," then the science can even be called a *rational* science. The somewhat unfamiliar words "ground" and "consequence" are important here because they are intended to cover two relations that are thought of as one piece by Kant: the logical relation of premise and conclusion and the causal relation of cause and effect. For Kant, "ground" and "consequence" stand thus in a logical and a causal relation at the same time. Due to their ordering by "grounds and consequences," rational sciences are deductive systems that have explanatory power: the consequence is causally explained by the ground.

According to Kant, the systematic order of scientific knowledge serves important functions for science: "The unity of the end to which all the parts relate and in the idea of which they all stand in relation to one another, makes it possible for us to determine from our knowledge of the other parts whether any part be missing, and to prevent any arbitrary addition, or in respect of its completeness any indeterminateness that does not conform to the limits." Thus, if we only have an aggregate of knowledge, i.e., some pieces of knowledge without apparent order, we cannot judge whether something is missing or whether a potential addition really belongs to this body of knowledge, or whether we have reached completeness. By contrast, a system of knowledge licenses judgments of this sort, thus exhibiting its epistemic superiority by giving directions for further research. In addition, "the systematic connection which reason can give to the empirical employment of the understanding not only furthers its extension, but also guarantees its correctness." Systematicity in Kant's sense is thus not only relevant with respect to the completeness of scientific knowledge but also with respect to its correctness.

4.3.2 Comparison with Systematicity Theory

The notion of systematicity that I have developed in chapter 2 is fairly flexible and open-ended in the sense that it can be adapted to various dimensions and to various disciplines. Kant's notion of systematicity is also surprisingly flexible. It is also meant to distinguish everyday knowledge from scientific knowledge. It turns out that Kant covers quite a few of the nine dimensions of systematicity that are developed in systematicity theory, even if partly in more specialized form such that systematicity theory is a generalization of Kant's approach. I am discussing now all of the areas where Kant applies a notion of systematicity to knowledge, even if he denies the resulting body of knowledge the title of science. In other words, I am going to discuss also what Kant calls the "historical doctrine of nature" and "improperly so-called natural science," in addition to what he calls "properly so-called natural science." From our point of view, also the "historical doctrine of nature" covering, for instance, botany and Earth history, and the "improperly so-called natural science" covering, for instance, chemistry, are sciences. As Kant ascribes systematicity to these areas as well, the comparison with systematicity theory suggests itself.

With respect to properly and improperly so-called science, Kant's notion of systematicity refers to systems of knowledge that are based on and ordered by certain principles. The highest principles of proper and improper science must be justified, strictly by a priori means in the former case and empirically in the latter case. Clearly, such a system of knowledge covers most of the dimensions of systematicity theory: it leads to systematic descriptions (dimension 1); to explanations in the case of rational science, probably the normal case in Kant's understanding (dimension 2); to at least the possibility of predictions (dimension 3); to the defense of knowledge claims due to derivation from the principles (dimension 4); to epistemic connectedness because of the relatedness in the system (dimension 6); to an ideal of completeness because the principles involved should cover the whole field (dimension 7); to at least the spotting of knowledge gaps due to the system character of the body of knowledge and therefore to a precondition for the generation of new knowledge (dimension 8); and clearly to a systematic presentation of knowledge (dimension 9). However, it must be noted that Kant's concept of systematicity for the proper and improper sciences is very specific and quite restricted: it is systematicity in the sense of a system of knowledge that is based on and ordered by principles.

With respect to what Kant calls the historical doctrine of nature, he uses a much more relaxed form of systematicity. Having, for example, botany and Earth history in mind, he thinks of the systematic ordering of the facts belonging to these disciplines. This certainly belongs to the systematicity of description (dimension 1), where we also dealt with classification, periodization, and similar topics that bring order to a variety of data. In addition, some epistemic connectedness may be established in this way (dimension 6), together with a systematic presentation of the pertinent knowledge (dimension 9).

We may conclude that Kant covered a lot of ground regarding the conceptualization of the specifically scientific in what we see as scientific knowledge. To him, as well, the concept of systematicity played a key role. Given the much less developed state of the sciences in his days, and given his adherence to the ideal of scientificity as strict demonstrability, it is not surprising that his notion of systematicity was more narrow and less developed than the one we are using here. However, there can be no doubt that Kant's enterprise to capture what scientificity is stands in general agreement with the spirit of this book.

4.4 LOGICAL EMPIRICISM

4.4.1 *The Position*

Logical empiricism is not, contrary to what the subheading suggests, a position. Instead, it is a family of positions that are in constant historical change as recent scholarship in history of philosophy has shown. Given the aim of this chapter and of this section in particular, however, we can use a fairly coarse image of logical empiricism. As our goal is not a subtle historical reconstruction of positions that are somehow related to systematicity theory, an image that captures some of the main themes and some of the main stances will suffice. As I indicated earlier, the heuristic hypothesis is that earlier positions in the history of the philosophy of science usually embodied certain aspects of systematicity theory, typically in a less sophisticated and one-sided form. To this purpose, I shall look at the pertinent topics that logical empiricism dealt with regarding philosophy of science. Here are some of the main topics that can be correlated with the main topics of this book.

- The protocol sentence debate: roughly speaking, the question was in what language observation reports should be articulated in order to serve their purpose for empirical science.

- Classification and taxonomy.
- The confirmation of hypotheses by empirical evidence.
- The confirmation of theories by empirical evidence.
- Scientific prediction.
- Scientific explanation.
- Reductive relations between laws, theories, and disciplines.
- The unity of science.

The general strategy of logical empiricism was to tackle these questions by logical analysis.

4.4.2 Comparison with Systematicity Theory

In the treatment of all these topics by logical empiricists, motives of systematicity play an important role.

The protocol sentence debate, the confirmation of hypotheses, and the confirmation of theories by empirical evidence all deal with the systematic defense of knowledge claims (dimension 4). Regarding protocol sentences, a form of them was sought that established the empirical basis for the empirical sciences. Regarding the confirmation of hypotheses and theories by empirical evidence, the problem of induction was systematically tackled in order to show how the knowledge claims of general scientific statements could be defended on the basis of singular observation sentences.

Regarding classification and taxonomy, Carl G. Hempel begins a paper, one of the central publications on this topic, with the following words: "This paper attempts to provide a *systematic* background for a discussion of the taxonomy of mental disorders. To this end, it analyzes the basic logical and methodological aspects of the classificatory procedures use in various branches of empirical science . . . ." (my italics). Thus, a particular topic of the systematicity of descriptions (dimension 1) is treated.

Regarding scientific prediction and scientific explanations (dimensions 2 and 3), I have already noted in section 2.1.1 that logical empiricists used systematicity terminology in order to describe the role of scientific predictions and explanations. To quote Hempel: "all scientific explanation ... seeks to provide a *systematic* understanding of empirical phenomena by showing that they fit into a nomic nexus." Similarly Nagel: "It is the desire for explanations which are at once *systematic* and controllable that generates science." Hempel discusses explanations, predictions, and postdictions (the latter more commonly called "retrodictions") under the rubric of "scientific systematization."

The treatment of reductive relations between laws, theories, and even disciplines is relevant with respect to scientific explanations (dimension 4) and epistemic connectedness (dimension 6). Logical empiricists favored reductionist explanations and the reduction of disciplines of higher aggregation levels to those of lower aggregations levels. Pushed to the most extreme level, all disciplines were supposed to be ultimately reducible to physics. As we have seen, reductionist explanations have a highly systematic character. Reducing one discipline to another increases the epistemic connectedness; reductive relations among all disciplines result in an especially strong form of unity of science and therefore in an especially strong form of epistemic connectedness.

We may summarize that quite a few of the topics we treated in this book are already present in logical empiricism, although not under the title of systematicity. Systematicity theory may thus be seen as a sort of unification and generalization of many aspects of logical empiricism, in accordance with the heuristic strategy in this chapter. This view accords with the view expressed in the introduction to Nagel's magisterial *The Structure of Science* of 1961, entitled "Science and Common Sense." Nagel contrasts common sense with science, and he diagnoses the source of their difference as follows: "A number of ... differences between common sense and scientific knowledge are almost direct consequences of the systematic character of the latter."

4.5 KARL R. POPPER

Although Karl Popper (1902–1994) lived in Vienna in the 1920s and 1930s and was interested in the philosophical analysis of science like the Vienna circle philosophers, he was not a member of the Vienna circle and was not a logical empiricist. He was interested even in the same central topics as the logical empiricists, namely scientific explanation, scientific prediction, and the defense of knowledge claims, and worked with roughly the same methods in philosophy, namely, logical analysis. Nevertheless, among the many disagreements between Popper and the logical empiricists, two rather deep disagreements stand out, together with one major difference of interest.

4.5.1 The Position

The first deep disagreement between Popper and the logical empiricists concerned the legitimate procedures relevant for the defense of general statements like scientific hypotheses, laws, theories, and the like. Regarding this topic, logical empiricists continued the inductivist tradition that goes back at least to the beginning of modern natural science in the seventeenth century. This tradition believes, in some variant or other, that there are procedures that justify the generalization of empirical data to general hypotheses; the core of these procedures is a "principle of induction." Popper, however, thought that such a principle does not and cannot exist—it is a logical impossibility. He thought that there is no really positive justification of the general statements of science. Instead, there are only tests of such statements, by deductively deriving consequences from them that can be confronted with experience.

The second deep disagreement between Popper and the logical empiricists concerned the so-called verifiability criterion of meaning, which, according to logical empiricists, supplied a criterion that allowed the judging of whether a given statement was "cognitively meaningful" or not. It is not necessary to go into all the details of it here; what is relevant in the present context is that the criterion was meant to eliminate metaphysics from the realm of cognitively significant topics. The argument put forward by means of the criterion was not that metaphysics was wrong here and there, but that metaphysical statements were typically void of any cognitive meaning whatsoever. In other words, metaphysical statements are, by their very nature, not even candidates for being true or false. Popper, however, believed that this analysis was deeply flawed; judged from the viewpoint of the verifiability criterion of meaning, all general scientific statements like laws or theories would also be rendered meaningless—clearly an unacceptable consequence. He therefore thought that the problem should be given a certain twist, namely, how should empirical science be in order to be different from metaphysics and pseudoscience? The well-known answer to this question, called the demarcation criterion, states that scientific statements must be falsifiable in principle when confronted with empirical data.

The difference of interest concerns the dynamics of science. Logical empiricists showed little interest in the question of the development of science. Is there some recurrent pattern in this process? How must scientific progress be characterized? Can scientific development be described as an approach to truth? It seems that in the twentieth century, Popper was the first one who seriously contemplated these questions.

4.5.2 Comparison with Systematicity Theory

Undoubtedly, the two disagreements between Popper and the logical empiricists regarding their philosophical positions are important, and they have been duly discussed in the philosophical literature. From the special point of view of the current section, however, these differences are minor. The first disagreement concerns variants in the defense of knowledge claims—whether general scientific statements should be defended positively, by induction, or negatively, by survival of tests. Both procedures stress the importance of dimension 4, the defense of knowledge claims, and both procedures are equally systematic, at least roughly. The second disagreement concerns variants in the results of the analysis of contrast between empirical science and metaphysics. This is a topic we have not touched upon so far because it did not come up in our approach in which we contrast scientific knowledge primarily with everyday knowledge. I will discuss the demarcation problem later in chapter 5 where I investigate consequences of systematicity theory. For the very specific aim of the present chapter, the difference between Popper and the logical empiricists regarding the relationship between science and metaphysics is not relevant. This is because systematicity theory does not claim to articulate a generalization of either of these positions. According to current wisdom, both positions appear to be untenable. I will articulate an alternative to them later but can leave the issue aside at this point. Finally, Popper's interest in the dynamics of science finds, at this point, no counterpart in systematicity theory. Later, we shall see that Popper's view of the dynamics of theories can be seen as a possible special case of systematicity theory's view.

The result of this section is this: Popper worked on roughly the same problems as logical empiricism. His answers regarding the defense of knowledge claims differ significantly from those of logical empiricism, whereas his views of scientific explanation and prediction are remarkably similar. Viewed from the perspective of whether systematicity theory is a generalization of earlier philosophical positions, both positions have the same status, in spite of their differences. Both positions can be seen as elaborating philosophies of science that fit well in the more general picture of systematicity theory.

4.6 THOMAS S. KUHN

Arguably, Thomas Kuhn (1922–1996) and Karl Popper were the most influential philosophers of science of the second half of the twentieth century. Thomas Kuhn very vigorously took up the question of the dynamics of science that Popper had tackled and was much less interested in many of the questions that were at center stage in logical empiricism. With respect to the dimensions of systematicity so far discussed, there is little overlap with Kuhn's philosophy of science. It is predominantly the generation of new knowledge (dimension 8) where Kuhn's position can be compared fruitfully with systematicity theory. Even there, however, the comparison can only be partial.

4.6.1 *The Position*

As is well known, Kuhn has claimed that in the process of scientific development, different phases can be distinguished that follow some pattern. This pattern can be described as a phase model, which represents the typical development of a discipline or research field in the basic disciplines of the natural sciences. What is often overlooked, however, is that Kuhn was cautious enough not to claim that this phase model is always strictly followed; it is only *likely* that this pattern obtains because it is the *usual* developmental pattern, there are *minor variants* to it, and, as I already stated, its intended range of application is only the basic disciplines of the natural sciences. The phase model is cyclical: after an initial phase of "pre-normal science," a phase of "normal science" is entered. When normal science encounters serious anomalies, the development switches to a phase of "revolutionary science." Revolutionary science is successful if it leads to a new phase of normal science, and then the cycle continues. I am not explaining this phase model any further, nor will I discuss the question of its range of validity or the possible principal objections against it. The reason for introducing it here is its featuring the phase of normal science that is especially suited for a comparison with systematicity theory.

Normal science is a tradition-bound practice in which the members of the scientific community agree about the fundamentals of the field. Based on this consensus, a particular form of scientific practice emerges. According to Kuhn, the core of the consensus in a scientific community consists of paradigmatic solutions to exemplary research problems. They are the point of departure for further research: problems are selected in analogy to the exemplary problems, and solutions are modeled on the paradigmatic solutions. As long as this practices works, it is irrefutably progressive, at least for the members of the community, because they unanimously endorse the exemplary problems and their solution in their function of guiding research in this way. Kuhn classifies the typical problems tackled in the phase of normal science in three groups; in all three groups there are more theoretical problems and more observational or experimental problems. First, there is the determination of facts that are relevant for the concrete application of theory. Second, there is the improvement of correspondence between theory and observation or experiment. This involves both the quantitative improvement of existing comparisons between theory and empirical results and the procurement of new opportunities for comparison. Third, there is the further articulation of the theory or the theories that form an integral part of normal science, be it explicitly in areas where theory dominates or more implicitly in areas where, at least at first sight, experimental inquiry appears to operate almost independently from theory.

4.6.2 *Comparison with Systematicity Theory*

Normal science, as described above, can be fruitfully compared with systematicity theory, mainly with respect to dimension 8, the generation of new knowledge.

Normal science is a practice of science in which the potential of paradigmatic problems and solutions for further research is systematically exploited. Both paradigmatic problem and paradigmatic solutions are vitally important. Paradigmatic problems are just those problems that point the way to other problems that are likely to be solvable. To be sure, there is always a plethora of interesting problems in any science. However, the important point is to identify those problems that are likely to be solvable with the resources at hand. This is the function of the paradigmatic problems: new problems that display analogies to them are also likely to be solvable. Similarly with the paradigmatic solutions, solutions to the new problems should display analogies to them. This is both heuristically important as well as relevant for assessing the acceptability of any proposed solution: solve a problem in analogy to the paradigmatic solution, and accept a proposed solution only if it is sufficiently similar to the paradigmatic solution. Thus, the paradigmatic problems and solutions provide some kind of systematical guidance for normal science. Although this guidance is not as strict as the guidance by rules to be followed slavishly, it is still systematic in asking for an indispensable orientation toward the paradigmatic problems and solutions. They provide the unity of the respective research tradition of normal science. By implication, the systematic generation of new knowledge also increases the systematicity in other dimensions, i.e., of descriptions, explanations, or predictions, depending on the details of the case.

It is much less obvious that the phases of pre-normal and of revolutionary science as described by Kuhn also exhibit systematicity in the generation of new knowledge. They are certainly less constrained and often more chaotic than normal science. As such, the element of order and thus of systematicity in them is less conspicuous. I shall come back to this topic in section 5.1.3, where I shall ask whether Kuhn's characterization of revolutionary science, although possibly not supplying direct evidence for systematicity theory, is at least compatible with it.

4.7 PAUL K. FEYERABEND

Paul Feyerabend is a special case in this section. His position will not, like the other positions, be embraced and then devoured as a special case of systematicity theory. His role in this book is different. As I already mentioned in the preface, Feyerabend is both the starting point and the antithesis of systematicity theory. He claims that nothing is special about science in comparison with other forms of knowledge; I claim that it is systematicity that sets science apart from everyday knowledge. In order to locate the clash between Feyerabend's position and systematicity theory, let us first investigate what Feyerabend exactly claims and how he argues for what he claims.

4.7.1 The Position

Unfortunately, the reception of Feyerabend's position is riddled with many stereotypes. I am not going to rehearse them here but will simply present his position. Feyerabend's position seems to be somehow summarized in his famous slogan "anything goes," best known from his book *Against Method*, originally published in 1975. However, this slogans means different things to different people, and unfortunately, only to few of them does it mean what Feyerabend wanted to express by it, admittedly quite provocatively. In order to truly understand its meaning, "anything goes" must be put into the context in which Feyerabend used it.

The target of Feyerabend's attack in *Against Method* was a specific epistemological (self-)understanding of the sciences—one that reduces the special quality of scientific knowledge to the strict application of rules for practicing science. This understanding of science had accompanied modern natural science from the very beginning and, in its essentials, can be traced back to the Greeks in antiquity. Strict rules to achieve a certain target are called "methods." The rules of practicing science are respectively called "scientific methods" or, summarily, "the scientific method." In his book, Feyerabend questioned the existence of such strictly binding scientific methods—thus the title *Against Method* and its subtitle, *Outline of an Anarchistic Theory of Knowledge*: anarchism as antithesis to the unconditional reign of one or more methods.

The main thesis of *Against Method* claims that science neither is an endeavor which *is* special because of strictly binding methodological instruction, nor that it *could be*, and consequently, *should not be* such an endeavor. This thesis in no way claims that science is an endeavor in which one can do whatever one pleases. Rather, it only claims that it is not an endeavor that can be characterized by following absolutely binding rules, like, for example Descartes' precepts in his *Discours de la Méthode*, which was discussed in section 4.2.1. The existence of methodological instructions in science and also its (limited) success is not denied in any sense. Feyerabend only claims that such rules in science are not de facto slavishly followed all the time and should not be so followed. There are always situations in which a rule—that until now has been fruitful—must be broken if one wants to prevent hindering the progress of science. Soberly formulated, Feyerabend only claims, as one of his papers is entitled, "the limited validity of methodological rules." But how is this quite moderate view compatible with "anything goes" that Feyerabend asserts for science? First of all, one must consider the rhetorical or, more precisely, ironic component of the slogan. "Anything goes" is an ironic answer to those who insist that there must be absolutely binding rules in the practice of science. Yes, if you insist, Feyerabend says, then I'll give you such a rule; namely, "anything goes!" With this, Feyerabend in no way provides incorrect information because, indeed, one can state this as an absolute rule in the practice of science or any other practice for that matter, as it cannot be broken *because it is entirely empty*. The strict validity of the rule, independent of the concrete circumstances to which it can be applied, is thus bought with its absolute emptiness. Furthermore, when Feyerabend first published the statement "anything goes," it came with an ironic footnote about his surprise that people had not noticed that he was joking.

How does Feyerabend justify the limited validity of all methodological prescriptions in the sciences? Rather casually, one finds an abstract justification. The justification is that each methodological rule for increasing knowledge (or for testing or confirming knowledge) is only reasonable relative to certain substantive assumptions, about reality and its interaction with the understanding subject. These assumptions are by no means indisputable, but they can change during research and in fact have changed often enough. The strict adherence to methodological rules thus implies a dogmatization of their underlying substantive assumptions, which of course may hinder research and could, in an extreme case, even bring it to stagnation.

Feyerabend puts more weight on the historical support of his main thesis, especially in the chapters about Galileo. The idea of the argument is to find, for any suggested methodological rule, an episode in the history of science containing what is generally accepted as an incidence of crucial scientific progress that was only possible by breaking the rule in question. Feyerabend works through several candidates of methodological rules that seem, prima facie, plausible. To give some examples: he considers the methodological rules that one should not introduce ad hoc hypotheses, that new hypotheses should not be in contradiction with established data or other established theories, that new hypotheses in comparison with those that they replace should not have less content, and so on. He always presents historical examples in which breaking the particular methodological rule in question was essential to the progress of knowledge, thus showing that the proposed rule cannot be universally valid. Feyerabend is completely aware of the limited scope of his argumentative strategy: "In this book [*Against Method*] I try to support the thesis by historical examples. Such support does not *establish* it; it makes it *plausible*."

Feyerabend even generalizes his thesis that science is not ruled by a universally valid method to the thesis that there are no elements whatsoever that are both necessary and sufficient for science. This is his thesis of *Against Method* as he summarizes it in the Introduction to the Chinese edition of the book: "[T]*he events, procedures and results that constitute the sciences have no common structure*; there are no elements that occur in every scientific investigation but are missing elsewhere." So there are not only no universal methods in science but also no other elements that are universally present in science and only there. In other words, there are no necessary and sufficient conditions by which science could be generally defined. There is another thesis by Feyerabend whose relation to the former thesis is not immediately obvious: "[Science] is a collage, not a system." Feyerabend argues for this thesis by stating that "[s]cience itself has conflicting parts with different strategies, results, metaphysical embroideries." From this quote it becomes apparent that Feyerabend thinks of a system as something that is united by some elements that are common to all parts of the system. As he had already argued that there are no such common elements in the sciences, they do not therefore form a system in that sense.

Feyerabend goes even further. He thinks that the argumentative strategy he employs "indicates how future statements about 'the nature of science' may be undermined: given any rule, or any general statement about the sciences, there always exist developments which are praised by those who support the rule but which show that the rule does more damage than good." In other words, Feyerabend claims that for any general statement about the sciences there will be episodes in the history of science that were undoubtedly progressive ("praised by those who support the rule") but would have been forbidden by a strict application of the general statement or rule to the case in question. Again, this is of course a generalization of his argumentative strategy against the universal validity of certain methodological norms. This generalization indicates a deep conviction of Feyerabend: historical processes in the sciences are, on the whole, so diverse and multifaceted that there simply are no substantial generalizations under which they can be subsumed, and they cannot be understood as the result of strict rule application.

4.7.2 Comparison with Systematicity Theory

Feyerabend thus confronts systematicity theory with two challenges that must be met. The first is that science "is a collage, not a system," which may be seen as expressing some tension with systematicity theory. The second challenge is that according to Feyerabend, there are neither valid generalizations about science nor strictly valid rules of procedure governing science. There will always be episodes in the history of science in which such generalizations or rules will be violated, to the benefit of scientific progress. By contrast, systematicity theory states something general about the relationship between scientific knowledge and everyday knowledge. Fortunately enough, I do not have to adjudicate on the truth of Feyerabend's extremely general claims. For the sake of argument, I take them for granted and will show that they do not contradict systematicity theory. The ultimate truth of Feyerabend's claims may thus be left undecided.

Let us recall what Feyerabend means when saying that science is a collage, not a system. As I explained in the previous subsection, Feyerabend has a concept of system in mind in which the parts of the system all share something that unifies them and thereby makes them parts of this system. Systematicity theory also claims that the sciences form some sort of unity. However, there is nothing concrete that all of the sciences have in common. It is only the *abstract* notion of systematicity that provides some tenuous sort of unity, as I have already discussed in section 2.2. The abstract notion of systematicity is extremely thin; its concretizations, covarying with the nine dimensions, differ among each other in different sciences and subfields, and they vary in time. What Feyerabend opposes to is the claim that there are some concrete properties that all sciences share, or some general rule by which scientific knowledge must be produced. By contrast, systematicity theory does not claim anything of this sort. It claims a sort of unity that is subtler than the rather straightforward and brute form of unity that Feyerabend believes does not exist.

Feyerabend's second challenge encourages opponents of systematicity theory to search for specific historical examples of science in which the main thesis of systematicity theory is violated, but the scientific quality of the specific example is undoubted. Let me note a difficulty to which I will return in much more detail in section 5.1.1. When comparing two areas with respect to their degrees of systematicity, we are referring to an overall systematicity aggregating all nine dimensions. The aggregation of the systematicities belonging to the nine dimensions to an overall systematicity can be done in countless ways. This leaves enormous leeway as to what a statement of a higher degree of systematicity of one area, in comparison to another one, concretely means. In other words, the statement of a difference in the degree of systematicity between two areas is extremely abstract and therefore "thin" or, to put it in more unfriendly terms, very vague. Not much is put on the table by such a statement, and correspondingly, it will be difficult to refute it. However, this property of systematicity theory is not a vice, but a virtue. Necessarily, all general statements of systematicity theory are abstract and cannot have much content because of the enormous diversity they are designed to cover. They are an attempt not to shut up in confrontation with the diversity of countless disciplines, subfields, and their historical development, but to state something that is general and abstract enough to deal with this diversity, without becoming entirely empty. Therefore, a refutation of the general theses of systematicity theory by empirical examples is in principle possible, in spite of the theory's enormous flexibility. However, if it is the normal course of the development of a discipline or field to increase in systematicity, as I will argue in section 5.1, then the existence of counterexamples to systematicity theory will become quite unlikely. So I will patiently wait until someone executes Feyerabend's envisaged strategy of "how future statements about 'the nature of science' may be undermined."

4.8 NICHOLAS RESCHER

Nicholas Rescher is, as far as I know, the only philosopher in the twentieth century who has more than casually dealt with systematicity in a sense relevant to our concerns. In particular, the book that he published in 1979 already signalizes in its title, *Cognitive Systematization: A Systems-Theoretic Approach to a Coherentist Theory of Knowledge*, Rescher's interest in the relation between knowledge and systematization. Early in the book he explains that he "seeks to examine the systematic aspects of our knowledge and to show why—and how—this represents one of its crucial features." He rightly deplores "that no work published in the [twentieth] century affords any substantial treatment of these matters." Rescher's goal is "to take some small steps towards remedying this large deficiency." It is evident that Rescher's work is highly relevant regarding my enterprise. I will first present his position and then analyze how it relates to systematicity theory as it is understood and developed in this book.

4.8.1 The Position

The context of Rescher's investigation is a philosophical position called "coherentism." Rescher explains coherentism and its connection to systematization as follows: "The guiding thought of [coherentism] is the idea that systematization is not merely a way of *organizing* our knowledge, but—more fundamentally—a criterial *standard* for determining what it is that we indeed know." In other words, to the coherentist, systematization is not only interesting from a bookkeeping point of view, which focuses on bringing order to what we know. Rather, it is also interesting—and more fundamentally so—from an epistemological point of view, where the justification and understanding of what we believe to know plays a preeminent role. Roughly speaking, the position of coherentism contends that any (successful) justification of a knowledge claim has to refer necessarily to a specific relation of this claim to other knowledge claims, namely, the relation of "coherence." The best way to realize this relation of coherence for a given knowledge claim is to embed it into a *system* of knowledge claims. The paradigmatic example of an extremely successful systematization of this kind has always been Euclid's system of geometry. Why is this kind of systematization so important? Because "it has been insisted … that men do not genuinely *know* something unless this knowledge is actually *systematic*." According to Rescher, this idea of systematization has been endorsed throughout the history of

Western philosophy, at least until Hegel, and it obviously contains a fundamentally coherentist conception of (genuine) knowledge.

This idea of systematization is even more important when it comes to science. "[T]here can be no science without system. Systematicity is the very hallmark of a science: a 'science' is—virtually by definition—a branch of knowledge that systematizes our information in some domain of empirical fact," claims Rescher. However, this idea of systematization of information should not only be applied to individual scientific disciplines, but also to science as a whole: "The systematic idea in the context of science embraces not only the more modest view that the several branches of empirical inquiry exhibit a systematic structure severally and separately, but also the more ambitious doctrine that the *whole* of natural science forms a single vast and all-comprehending system." In other words, scientific knowledge exhibits systematic order, be it as individual disciplines or as a whole. However, this statement should not only be understood descriptively but also normatively: "Systematicity is thus not only a prominent (if partial) aspect of the structure of our knowledge, but is a normatively *desirable* aspect of it—indeed a requisite for genuinely scientific knowledge. It is, accordingly, correlative with the regulative ideal presented by the injunction: develop your knowledge so as to endow it with a systematic structure." In other words, in order to make your knowledge scientific, bring it into the form of a system.

What I have presented so far is the background and starting point of the position that Rescher develops. He has explicitly put himself in the tradition of Leibniz, Wolff, Lambert, Kant, and Hegel in the sense that also for them, "the prospect of organizing a body of claims systematically is crucial to its claims to be a science." Of course, Rescher cannot simply follow the fully rationalistic and/or aprioristic programs of these authors because also in philosophical respects, times have changed, and such programs do not seem to be realizable anymore. Instead, Rescher's general philosophical orientation is pragmatism, in our context a position that he calls "methodological pragmatism." In order to understand what this position consists of and why systematicity plays the role in it that it does, we have to understand the context in which methodological pragmatism is introduced and developed.

The context we are dealing with is a theory of knowledge. From the very beginning, Rescher restricts this wide area to factual knowledge, thus excluding the realm of purely formal knowledge like mathematics or logic. Knowledge comes with a claim to truth. In the traditional understanding at least since Aristotle, truth is the correspondence of thought with thought-external reality. There is a deep problem with this idea of correspondence: how can I actually judge the presence (or absence) of that correspondence without bringing putative thought-*external* reality *into* my thought, such that I am not dealing with thought-external reality any longer? In other words, it seems impossible to strictly apply the idea of truth as correspondence to any actual situation. There are several ways out of this predicament. The first is to keep the idea of truth as correspondence and conclude that truth is in principle beyond human reach. Period. The second is to give up the idea of truth as correspondence and replace it by some more manageable idea. Typically, the nonepistemic idea of correspondence (the correspondence does or does not obtain for a certain thought, completely independent of what we believe) is replaced by some epistemic notion, i.e., a notion that is somehow tied to our beliefs or knowledge. For instance, truth could be seen as the result of a rational investigation of the relevant matter by an ideal scientific community. It is almost needless to say that both ways are highly controversial.

However, there is a third way, and this is the route taken by Rescher; he calls it methodological pragmatism. For this position, a distinction between a *definition* of truth and *criteria* for truth is fundamental. The traditional definition of truth as correspondence is kept because it appears to be indispensible. However, this definition is not practically applicable because of the predicament discussed above. Instead, the position seeks to develop criteria that allow estimating whether certain claims to truth are fulfilled or not. Of course, these estimates can never be 100 percent certain, because that would be equivalent to a truth judgment on the basis of the correspondence definition of truth, which is impossible. Thus, we cannot expect to find criteria that are logically sufficient for truth, but only criteria that are reasonable indicators for truth. These criteria can be called "pragmatic" because they do not refer to the meaning of truth, but to tangible indicators of truths (like predictive success or effective applicative control).

This sketch of the program of methodological pragmatism should suffice to see why Rescher is fundamentally interested in the systematization of knowledge and how his idea of systematization is continuous with the classical idea of systematization. For Rescher, systematization belongs to the large domain of truth criteria in the sense discussed above. If a problematic truth claim can be harmoniously fitted into a nexus of other truth claims that have already been somehow validated (by whatever truth criteria), then this fact is an indicator for the truth of the problematic truth claim. Under these circumstances, it becomes intelligible why for Rescher systematization is not simply an orderly bookkeeping of previously established truths. On the contrary, the established coherence of a problematic truth claim with other truth claims is a criterion for truth in the sense discussed above: it is far from being a sufficient criterion, and it is only one among other criteria, though a very important one. Generally speaking, the main purpose of cognitive systematization is quality control of knowledge claims. Rescher's investigations of systematicity are directed toward an elucidation and legitimation of its precise role as one of the truth criteria. In short and in his own words, his answer is: systematization is "a testing-process for acceptability."

The continuity with the older systematicity conceptions consists in the idea that the embedding of a truth claim into an appropriate system is relevant for the very quality of the respective truth claim. The discontinuity with the older systematicity conceptions may consist in the fact that in methodological pragmatism, the embedding of a truth claim into an appropriate system is neither intended to nor capable of definitively establishing its truth, at least in the nonformal sciences.

4.8.2 Comparison with Systematicity Theory

In order to determine the differences between Rescher's project and systematicity theory as conceived in this book, two interrelated points must be emphasized: the difference in kind of the two projects and the difference between the two notions of systematicity employed in these projects.

Rescher's project belongs to the theory of knowledge. As such, it is deeply concerned with the question of truth: what does truth mean and how can it possibly be attained. By contrast, systematicity theory belongs to the philosophy of science. As such, it is concerned with the question of how the sciences function. There are various ways in which this question can be made more concrete. The concretization chosen by systematicity theory is the question of how the functioning of the sciences differs from the knowledge-gaining practices of everyday thought (and other prescientific practices). Its main goal is the characterization of the specific difference between scientific knowledge and everyday knowledge. Truth does come into play, but only implicitly in dimension 4, the defense of knowledge claims. It does not enter the discussion in the form of an evaluative question, for instance: Is science more successful than everyday thinking in defending its knowledge claims *and therefore* in reaching the truth? Systematicity theory compares the ways everyday thinking and scientific thinking go about when defending knowledge claims, but the ultimate aim of this comparison is not evaluative. This is continuous with the discussion of the other dimensions: the main goal is to make the contrast between scientific knowledge and other forms of knowledge as explicit as possible, but this is not taken as a mere prelude to a comparative evaluation between the enterprises. Of course, there are other areas of philosophy of science much closer to Rescher's enterprise than systematicity theory. For instance, the realism-antirealism debate in the philosophy of science is also a question about truth, namely, whether the assertions of science about theoretical entities are at least approximately true or not. However, this is simply not the focus of systematicity theory.

The second main difference between Rescher's project and mine is somewhat subtle. It concerns the notions of systematicity employed in both projects. As we will see in a moment, the two notions are markedly different. However, this difference is totally masked in statements that one can read in Rescher's pertinent publications. Take, for example, his statements "that the proper, the *scientific* development of our knowledge should proceed systematically," or "Systematicity is the very hallmark of a science." Clearly, sentences like these are extremely close to sentences that one can find in the context of systematicity theory. However, they mean very different things. The reason is that in Rescher's context, "systematicity" and "systematically" do not mean the same as in the context of this book's systematicity theory.

As we have seen above, Rescher's project is focused on a *system* of knowledge. He is concerned with the organization of knowledge such that it forms a (coherent) system, in contrast to a set of independent pieces of knowledge or an unordered aggregate of knowledge with undefined or unspecified relations among each other. Rescher therefore often refers to the *systematization* of knowledge, which of course means bringing knowledge into the form of a system. He also uses *systematical* and the correlated *systematicity*, which denote the defining property of any system, namely—trivially—being in the form of a system. "Systematicity" is thus, for Rescher, the system-endowing or system-bearing character of certain bodies of knowledge. In Rescher's context, for example, "striving for systematicity" means "trying to bring into the form of a system."

By contrast, in the context of the present book, the concept of systematicity is not derived from the noun "system" as in Rescher's case. I use "systematicity" as derived from the adjective "systematic" (Rescher seems to prefer, for good reasons, the adjective "systematical," which seems to be more strongly tied to the noun "system.") This implies a subtle but essential difference of my concept of systematicity to Rescher's. The adjective "systematic" covers more than just "being the essential property of a system." Whereas in "system"—an ordered whole possessing interrelated parts is in view—the adjective "systematic" as it is commonly used lacks this strong connotation of an ordered whole. It is rather just a contrast term to "unorderly" or "unstructured" but leaves open whether the implied order can be translated into a full-blown "system" of the parts or elements involved. There is some conceptual space between "being (completely) unordered" and "being ordered in the form of a system." It is this conceptual space that is also occupied in this book, for instance in its main thesis that scientific knowledge differs from other kinds of knowledge by being more systematic. Of course, it is not *excluded* that this difference of the degree of systematicity is sometimes due to some parts of science indeed having the form of a system, but this is certainly not a *mandatory* requirement. In other words, "systematicity" as derived from the adjective "systematic" (my use) is a wider concept than "systematicity" as derived from the noun "system" (Rescher's use). Everything that is systematic in Rescher's sense is also systematic in my sense, but not vice versa. It is obvious that the difference in the two notions of systematicity employed in Rescher's and in my project is an immediate consequence of the difference of the two projects themselves: Rescher is fundamentally interested in *systems* of knowledge, whereas for me, this is only a special case. Clearly, neither Rescher nor I are just "right" or "wrong"—we are simply pursuing different interests in our two projects.

Despite all differences, it should not be overlooked that Rescher's project and the present project may be mutually enlightening. Rescher's investigation of the epistemic consequences of putting a given truth claim into a system of other truth claims can productively inform systematicity theory's discussion of the defense of knowledge claims. It may also make us aware of a not undisclosed connection between the defense of knowledge claims (dimension 4) and epistemic connectedness (dimension 6). More generally, viewing systematicity theory from Rescher's point of view may lead to a decoding of other dimensions' role for the defense of knowledge claims. On the other hand, systematicity theory may make Rescher's enterprise aware of truth criteria (in his sense) that are systematic in the weaker sense, i.e., located underneath the threshold of being part of or contributing to a system. Although not immediately useful for the part of Rescher's project that deals with (his) systematicity, these weaker criteria may still conduce to the larger project of identifying stronger truth criteria in Rescher's sense.

5

Consequences for Scientific Knowledge

IN THIS CHAPTER, we shall discuss some consequences of systematicity theory for scientific knowledge. It should be obvious that a particular understanding of the specificity of scientific knowledge in comparison to other forms of knowledge also influences and shapes features of science that have not been in direct focus so far. More specifically, systematicity theory immediately suggests a certain way that new scientific disciplines are born from nonscientific roots and how an established science further develops (section 5.1). We then proceed to a discussion of the relationship between common sense and science in section 5.2. As I have repeatedly indicated, systematicity theory is primarily a descriptive theory. However, it is an interesting question whether normative consequences can be drawn from such a theory without inviting the objection of committing the naturalistic fallacy (section 5.3). This section will also be preparatory for the final section 5.4 of this chapter, which deals with the question of the demarcation of science from pseudoscience. This "demarcation problem" has fueled much of the debate about the nature of science in the twentieth century. Unfortunately, this debate has produced few tangible results, and it has not been vigorously treated during the last decades. We shall investigate whether systematicity theory allows a new take on this question. Due to developments on the pseudoscience front during the last decades, especially regarding so-called creation science, this problem has also stirred up some political and even legal interest.

5.1 THE GENESIS AND DYNAMICS OF SCIENCE

Our characterization of scientific knowledge as being more systematic than other comparative forms of knowledge immediately suggests hypotheses both about the primordial genesis of scientific knowledge out of prescientific knowledge and about a pattern of the further development of some scientific discipline or field. It is important to note that the hypothesis about the genesis of scientific knowledge only concerns those cases in which a science developed from nonscientific knowledge, be it from everyday knowledge or from knowledge and know-how of artisans or of technicians. Clearly, a scientific discipline can also emerge from sources that are already scientific themselves. For instance, a new scientific discipline or field may arise through a partial fusion of hitherto separated disciplines or fields, like physical chemistry out of physics and chemistry, or molecular biology out of scientific knowledge regarding molecules and biology. These cases are a little more complicated to describe in terms of systematicity theory. Here, I am only concerned with something like the primordial events leading to a science from nonscientific origins, i.e. from human knowledge practices that do not count as scientific. For these cases, the hypothesis will be that it is an increase in its systematicity that sets the newborn science apart from the earlier, nonscientific knowledge practices.

The hypothesis about the further development of a scientific discipline or field states that in this development, the degree of systematicity of this discipline or field will increase. The first hypothesis concerns what may be called the *initial* development or the genesis of a scientific discipline or field; the second hypothesis concerns the *further* development of a scientific discipline or field. Both hypotheses may be condensed into the single hypothesis that states that the development of a scientific discipline or field is characterized by an increase in systematicity. However, before articulating these two hypotheses any further, some conceptual clarifications are in order, because it is not clear what "an increase in systematicity" really means. Afterwards, the two hypotheses will be discussed in more detail in the two subsequent subsections.

5.1.1 Conceptual Clarifications

What does "an increase in systematicity" mean? Clearly, this expression is extremely vague. We have to confront this vagueness and discuss which aspects of it are unavoidable at this point and in which respects the existing vagueness can and should be reduced. We must consider several different points here.

First, why should at least some part of the vagueness of the expression "an increase in systematicity" be unavoidable? When speaking about the genesis of scientific knowledge or about its development *in general terms*, we are situating ourselves at an extremely abstract level. This is because we are referring to an immense variety of different concrete historical processes that took place at different times and are still taking place in the present. They belong to a large number of disciplines and scientific fields. Remember that regarding disciplines, we are talking about several hundred, and regarding scientific fields, we are talking about several thousand. If we can find any communality among developmental processes in the vast and extremely heterogeneous landscape of the sciences at all, it must be very abstract indeed and hence cannot be very substantial. Thus, a large part of the apparent vagueness of the expression "increase of systematicity" is necessary and is due to its abstractness. This abstractness, however, is intended because when talking in general terms, we want to cover the whole variety of the sciences (in the wide sense, of course, also covering the social sciences and the humanities).

Second, as we have especially seen in chapter 3, there are many different concrete concepts of systematicity that are pertinent in the characterization of the sciences. The differences among them are due to the dependence of any specific concept of systematicity on the respective dimension of the systematicity of science, on the specific discipline, and even on the specific scientific field in question, and on the specific historic time we are interested in. As I explained in chapter 2, all of these different concepts of systematicity are connected by family resemblance relations only and not by a set of necessary and sufficient conditions. The same also holds, of course, for the different concrete concepts of "an increase in systematicity." As a matter of course, all of these different similarities and dissimilarities between the concrete concepts of "an increase in systematicity" disappear when moving to the abstract level of "an increase in systematicity in general." Thus, it is not only the abstractness of the envisaged level of discourse that contributes to the apparent vagueness of "an increase in systematicity" but also the nature of the things from which it is abstracted. Nothing of their particularities survives the ascent to the abstract level because they have almost no general properties in common that would also show up at the abstract level.

Third, when speaking about an increase in systematicity in the development of some scientific discipline or field, one necessarily speaks about an aggregate effect. The "degree of systematicity" of some discipline or field (however badly the "degree of systematicity" may be defined) can be nothing but an aggregate effect of the "degrees of systematicity" in any of the nine dimensions. The same is true for an "increase in systematicity," which must be, in any case, an aggregate effect of an increase of systematicity in the nine dimensions. This immediately shows the very substantial intrinsic vagueness of the notion "overall increase in systematicity" fed by two independent sources. First, it is not clear how the aggregation of the different dimensions of systematicity should be accomplished. For instance, should we speak of an overall increase in systematicity only if there is an increase in systematicity in every single dimension? Further, should we allow for compensation between different dimensions of systematicity such that a decrease in systematicity in, say, one dimension may be compensated by a strong increase in the other dimensions, resulting in an increase in total systematicity? Second, it is not clear how the different dimensions of systematicity should be weighed. Do they all have equal weight? This is not very likely. For instance, dimension 4 regarding the defense of knowledge claims plays a central role in science. If this dimension is not existent or only insufficiently developed, an increase in the other dimensions will hardly suffice to improve the respective discipline. If the overall systematicity is somehow a measure for the scientific quality of a discipline or scientific field, too low a level of systematicity regarding the defense of knowledge claims will in general not be really compensated by an increase in the systematicity regarding other dimensions. If, however, the different dimensions of systematicity have unequal weights in the overall aggregation, then one can expect that the weights are discipline or even field dependent. Different dimensions of systematicity are more or less important in any given discipline, at least to some degree. For instance, increases in the systematicity of predictions (dimension 3) may be extremely important in some disciplines, whereas it may be legitimately dismissed as irrelevant in others that do not in general aim for predictions.

It may seem that due to the different sources of its vagueness, the notion of an "overall increase in systematicity" may be absolutely hopeless. This, however, is not the case. It cannot be expected that on a very abstract level, the notion of an "increase in systematicity" will have much substantial content. This does not exclude, of course, that in quite a few concrete cases of application, it may be obvious whether we have an overall increase in systematicity or not. Imagine the case that in some scientific discipline or field, a new measuring technique is introduced that increases the relative accuracy of the measurement of a central quantity by a factor of 10. This may immediately affect descriptions that may become more accurate and also the defense of knowledge claims that may be carried to greater accuracy than before. If, for the sake of argument, the other dimensions of systematicity are either unaffected or also positively affected, then it is obvious that an increase in overall systematicity can be stated—irrespective of all complicated principal questions of how to weigh and aggregate dimensions. Of course, it is equally simple to construct cases in which these questions present insurmountable problems for a unique assessment of whether an increase in systematicity has taken place. What I shall assume in the following is that in a large enough number of concrete cases, an assessment of an increase in systematicity is indeed possible such that the proposed hypotheses at least make sense.

Given the admitted abstractness and therefore vagueness of the notion of an overall increase in systematicity, it is clear that at a concrete level it can be realized in countless ways. We have nine fairly independent dimensions possibly having very different weights that may contribute to an identifiable overall increase in systematicity. Therefore, the following two hypotheses that state an overall increase in systematicity assert something on an abstract level that can, on the concrete level of a particular discipline or a scientific field, be realized in countless different ways. It is therefore true that these general hypotheses do not have much empirical content. However, they are not empirically empty, although in many imagined cases, they may be indeterminate because of a necessary weighing and aggregating process whose details are indeterminate. With these words of caution uttered, let us now move on to the articulation of the hypotheses.

5.1.2 *The Genesis of a Science*

Particular sciences or scientific fields can originate from very different conditions. For instance, a new field may emerge that bridges two existing disciplines, like physical chemistry or chemical physics that both connect physics and chemistry, though in different ways. Or a new field or even discipline may emerge in reaction to some recalcitrant problem of an established discipline. For instance, virology emerged from the problems that the theory of infectious diseases had with certain diseases that were long known, like influenza, in which bacteria, however, could be excluded as infectious agents. These are cases that are not relevant in the present section. Here, I want to discuss the cases in which a science arises from nonscientific origins. Of course, ultimately every science has nonscientific origins if one goes back far enough in time, and it is this genesis of a brand-new science in this sense that I am focusing on in this section.

The hypothesis that I am advancing with respect to a new science states that the newborn science is more systematic than the knowledge practice from which it emerged. Clearly, this thesis is immediately suggested by our general thesis that science is more systematic than the corresponding everyday knowledge. The predecessor knowledge practice and the knowledge that the newborn science produces have to be compared with respect to their systematicity. The typical case may be that the new science emerges from an everyday or professional knowledge practice first by reflection on that practice and later by further development of that knowledge. The first step of reflection may be an attempt to bring some order into the variety of pertinent phenomena and to gain insight into the similarities within certain groups of them.

As an illustration, take the case of mathematics. This case is not completely straightforward, because it is not unambiguously clear since when a *science* of mathematics has existed. It is uncontroversial that in Egypt and Mesopotamia, mathematics began to develop that later informed the Greeks and triggered the emergence of their mathematics. However, whether Egyptian and Mesopotamian mathematics should be called science is strongly dependent on the question of what standard for scientificity one applies. It is, however, fairly evident that in some of the remaining sources, a theoretical interest in mathematical questions becomes visible that goes somewhat beyond immediate practical interests. There are sources in which problems are treated that undoubtedly have no direct practical origin. There are also sources in which procedures are described on how to achieve solutions to specific classes of problems. Even near-equivalents to what are today variables can be found. There seems to be some implicit knowledge of certain algorithms and theorems (like the Pythagorean theorem), which is contained in lists of exemplary problem solutions, probably for teaching purposes, in which the same procedure is applied to somewhat differing problems. However, what is missing relative to today's understanding of mathematics as a science are explicit steps of justification of problem solutions or of general theorems: the idea of a mathematical proof was not yet developed. So there are rudimentary forms of mathematics that are somewhat independent of direct application. As such, the Egyptian and Mesopotamian forms of mathematics may qualify as embryonic sciences.

Judged from the viewpoint of systematicity theory, we have indeed the beginning of a science. On the one hand, there is a practice of calculations of various sorts in various contexts of everyday life. On the other hand, there is a sort of mathematical knowledge that transcends the purely practical purposes a little. In this form of mathematical knowledge, there is some ordering of the pertinent "phenomena," namely, in lists of similar problems together with their solutions. Furthermore, there are first glimpses at general rules. Comparing now the knowledge present in practical calculations with this embryonic mathematical knowledge, it is obvious that the latter is more systematic than the former. This holds in at least two dimensions. Regarding the description of pertinent mathematical "phenomena," there is some systematic order in the grouping of similar problems. Regarding epistemic connectedness, the knowledge of general rules is clearly more systematic than knowledge of individual problem solutions because general rules connect all of those cases to which they apply. With respect to the other dimensions, probably nothing has changed significantly. In such a case, the aggregation of the dimensions of systematicity is easy: the embryonic mathematics is more systematic than the everyday practice of calculations.

However, one may be somewhat reluctant to bestow the title of a full-blown science on the embryonic forms of Mesopotamian and Egyptian mathematics, in spite of the somewhat higher systematicity that they feature in comparison to the calculational practice of the day. The main reason may be that with respect to the defense of knowledge claims (dimension 4 in the list of systematicity dimensions), there is nothing explicit to be found. As I stated in section 3.4, the defense of knowledge claims is, of course, an indispensable element of scientific knowledge. Because of the conspicuous absence of this element, it may appear quite appropriate to deny Mesopotamian and Egyptian mathematics the title of a science. This is especially true in view of the later development of mathematics in which the specifically mathematical defense of knowledge claims became exemplary: a proof that deduces the desired statement, then called a "theorem," from explicitly given premises, thus making the theorem certain (relative to the premises).

Proofs have been introduced by Greek mathematics in the fifth century BC, possibly by Thales of Miletus (c. 624–546 BC), and have in subsequent times been fully developed into the axiomatic structure of Euclid's *Elements*, written toward the end of the fourth century BC (I shall come back to this development in the following subsection 5.1.3). The main differences in the earlier Mesopotamian and Egyptian mathematics consist in the formulation of explicitly *general* mathematical statements (mostly geometrical), accompanied by *proofs* that establish their truth. There is general agreement that this is the birth of a kind of mathematics that undoubtedly deserves the honorable title of a science. Let us now see how our hypothesis about newborn sciences fares with respect to this example. Comparing the degree of systematicity of the older calculational techniques of Mesopotamian and Egyptian mathematics with this new kind of mathematics, massive increases in systematicity in at least two dimensions are conspicuous. First, the explicit generality of (proven) statements increases epistemic connectedness because the general statements can be applied to infinitely many concrete cases. Second, a proof is an explicit recognition of the necessity of a defense of knowledge claim and, at the same time, an utterly persuasive answer to that necessity. There is certainly no decrease of systematicity in the other dimensions. Thus, the aggregation of the changes of systematicity in different dimensions to an overall systematicity assessment is absolutely unproblematic: clearly, the overall systematicity also increased. Thus, the transition to Greek mathematics superbly exemplifies our hypothesis that a newborn science is more systematic than the knowledge practice from which it emerged.

Of course, one convincing example for a general hypothesis cannot really establish it. However, it is quite plausible that in the very beginning of any science, distinguishing relevant phenomena from irrelevant ones and bringing some order to the former may be an important first step. This is an act of reflection about the known phenomena. Further steps must follow, and their concrete direction can certainly not generally be determined because there are many different paths. Additional phenomena may be sought; the known phenomena may be more accurately described; more or less general explanations for the phenomena may be sought; connections to other sorts of phenomena may be established; knowledge claims about the phenomena may be critically scrutinized; communities of critical discourse may form; new forms for the representation of the achieved knowledge may be invented; and so on. It is obvious that all of these steps increase the overall systematicity of the new form of knowledge that may be a new science after it has become sufficiently distinct from its predecessor.

5.1.3 The Dynamics of Science

Let us now turn to the second hypothesis of this section, which states that in the further development of a scientific discipline or field, its overall systematicity will increase. If this hypothesis holds at all, it only holds on the condition that we are dealing with progressive developments. This condition shall first of all exclude extreme cases such as the development of biology in Soviet Russia from the 1940s to the 1950s, connected with the name Lysenko. During this time, the development of Soviet biology was strongly influenced by outdated views of evolution and genetics, politically enforced by the government. This imposed development was certainly not progressive in any scientific sense, and our hypothesis is not designed to cover it. Furthermore, there may be episodes of stagnation of some scientific discipline or field caused by whatever factors, and clearly our hypothesis is not supposed to apply to such cases. Before turning to more general considerations in support of our hypothesis, let us take a look at one example that illustrates our hypothesis in a particularly impressive way.

As I mentioned in the last subsection, there is a highly interesting development in Greek mathematics after the introduction of proof that led, in somewhat less than three centuries, to Euclid's *Elements*, written roughly at that end of the fourth century BC. This book is the most successful scientific book of all times and cultures, used as a textbook for more than two millennia, comprising almost all of the mathematical knowledge of its times. The book features a strict axiomatic deductive structure. It begins with a number of definitions, followed by five postulates and five axioms. The definitions deal with concepts like point, line, parallel lines, and so forth; the postulates articulate geometrical truths that appear evident (including Euclid's famous fifth postulate, the parallel postulate); and the axioms formulate more general truths needed in mathematics, such as things being equal to the same thing are equal to each other. The principal procedure in *Elements* is the formulation of some mathematical proposition that is followed by a proof. The proof shows that the proposition is a necessary consequence of the definitions, postulates, and axioms. *Elements* covers plane geometry, solid geometry, arithmetical topics, and incommensurable magnitudes.

Comparing now the development of Greek mathematics from its beginning as a science, when it featured a few proofs for some theorems, to the Euclidean edifice, the increase in systematicity with respect to the most important dimensions is overwhelming. Due to the explicit definitions, the systematicity of descriptions has increased (dimension 1). The defense of knowledge claims by proofs whose premises are transparent and whose conclusions are compelling set a standard that lasted for more than two millennia, reaching far beyond the borders of mathematics (dimension 4). All mathematical knowledge was internally strongly epistemically connected as all theorems could ultimately be traced back to the postulates and axioms. Furthermore, due to the generality of the theorems, they could be and were applied to countless concrete cases within and outside of pure mathematics (dimension 6). Clearly, an ideal of completeness was embodied in the choice of definitions, postulates, and axioms, because all true mathematical propositions should be derivable from them (dimension 7). Finally, the representation of mathematical knowledge in *Elements* was so outstanding regarding its organization and clarity that it served as a model for scientific work in general for something like 1,500 years to come (dimension 9). It seems entirely unproblematic to state now that the aggregated, overall systematicity of Greek mathematics as evidenced by Euclid is much higher than it was at the beginning of the Greek science of mathematics; aggregation problems just do not arise.

After this (successful) illustration of our hypothesis about an increase in overall systematicity in the course of the development of a discipline or scientific field, let us now turn to more general considerations. As I stated earlier, we shall only deal with progressive developments of a science. At this point, we may, for the sake of argument, introduce a distinction similar to one that has been made famous by historian and philosopher of science Thomas S. Kuhn. Later in this section, I will drop this distinction again, but for present purposes of illustration, it is quite useful. In the basic disciplines of the natural sciences, Kuhn distinguished between progressive developments during normal science and progressive developments due to scientific revolutions. Although there may be episodes of science that may be difficult to assign to one or the other type, for many situations it is fairly clear of which type they are. In normal science, the general framework in which the research is conducted is fixed; it is also called "paradigm-bound" research. Also, in what Kuhn called pre-normal science, a sort of paradigm-bound research may be found, namely, on a more local level within specific schools. Within a given school, certain assumptions are shared that provide—at least temporarily—the unchallenged basis of the school's work. By contrast, revolutionary science tries to replace a given framework (shared by a school or by the whole scientific community) with a new one, because the given framework ceases to be a useful guide to research. An example for paradigm-bound research is research in atomic physics between 1915 and 1922, which was firmly based on Bohr's so-called planetary model of the atom. Research in atomic physics during this time took the fundamentals of Bohr's model for granted; researchers wanted to apply the model to hitherto not well-understood or even utterly mysterious atomic phenomena. The situation began to change in about 1922 when more and more problems in atomic physics turned out to be unsolvable on the basis of Bohr's model despite contrary expectations. Researchers started to look for modifications or even wholesale replacements of the fundamentals of atomic physics, which finally led to the revolutionary invention of quantum mechanics in 1925.

Although the distinction was originally designed by Kuhn to cover the basic natural sciences only, he himself applied it also to historiography of science and to philosophy. It is quite obvious that the distinction can indeed be applied in scientific areas different from the basic natural sciences, so long as we can identify certain traditions of research that are based on some framework. This holds also for research in which no general consensus across the whole scientific community about the fundamentals of the field is achieved, as is typical of the social sciences and the humanities. Roughly speaking, research may be tradition-bound or tradition-shattering, with quite a few shades in-between. The modes of progress differ for these two possibilities; this is why I draw the distinction here. Generalizing Kuhn's account, I shall speak in this section of "normal science" and of "revolutionary science" as ideal types that are more or less precisely realized in actual scientific practice in all sciences (in the wide sense), across the board. Their difference concerns the fundamentals of a discipline or scientific field: either they are taken for granted and constitute the basis of a given research tradition, or research is directed at a replacement of such hitherto accepted fundamentals.

Consider first a progressive normal science tradition. I have already pointed out in section 4.6.2 that normal science exhibits a systematic way of knowledge generation and that, by implication, other dimensions of systematicity profit from this new knowledge. In other words, a progressive tradition of normal science will increase the overall systematicity of the respective discipline or field. With regard to the development in so-called extraordinary science that may lead to a scientific revolution, the case is a little more difficult. Still following Kuhn, a scientific revolution will be achieved when the new paradigm is superior to its predecessor. This judgment is based on the ensemble of scientific values pertinent to the respective scientific community. The scientific values in question include accuracy, consistency, scope, simplicity, and fruitfulness. The value of problem-solving capacity functions as a sort of super-value summarizing the aforementioned values: by better realizing the aforementioned values, the capacity to solve scientific problems also increases. It is plausible that a successor paradigm outperforming its predecessor paradigm with respect to these values also has a higher overall degree of systematicity than its predecessor. Higher accuracy means better descriptions, explanations, or predictions; higher consistency means higher epistemic connectedness, which also holds for broader scope; higher simplicity may translate into different dimensions of systematicity, for instance, more systematic means of the defense of knowledge claims; the same holds for greater fruitfulness, which may translate into higher epistemic connectedness and more efficient generation of new knowledge.

It has to be admitted that these considerations are fairly vague. This is due to the high level of abstraction on which we are situated. We are covering a great variety of very different historical developments of very different fields of learning when describing them either in terms of Kuhn's scientific values or in terms of the dimensions of systematicity. Certainly, we will not be able to reach more than some plausibility when summarily asserting that these developments will be in the direction of a fuller realization of Kuhn's cognitive values or, alternatively, in the direction of a higher degree of overall systematicity.

We may even abstract from the potential difference between normal and revolutionary science and still make plausible that a progressive development of some research field implies an increase in overall systematicity. What does it mean to say that a research field displays a progressive development? It means that the field is somehow improving. In which ways can a scientific field improve? There are, of course, many ways, but the dimensions of systematicity are certainly among them: producing better descriptions, explanations, predictions, defenses of knowledge claims, and such surely will make the field move forward. In other words, an increase in systematicity is certainly *sufficient* for a progressive development of a scientific field. Is it also necessary, i.e., does any progressive development imply an increase in overall systematicity? This is at least plausible because the dimensions of systematicity have been consciously chosen in order to represent the differences that delineate scientific knowledge from everyday knowledge. Thus, the specificity of scientific knowledge is presumably captured by the nine dimensions, and therefore, any improvement of that knowledge should be somehow mapped onto some of these dimensions by increasing the pertinent systematicities. To be sure, this is only a plausibility argument that seeks to establish a correlation between an improvement of some scientific field and its overall degree of systematicity.

Whether the correlation between scientific progress and systematicity is conceptual or empirical in nature is a question that can probably not be answered, possibly even in principle. First of all, it seems clear that the correlation cannot be straightforwardly conceptual in a simple way. The reason is that there are cases in which an apparently unambiguous increase in systematicity leads to scientifically less acceptable results. Take the situation of decision making under uncertainty as an example. The uncertainty mentioned concerns the lack of information that would be relevant for the decision. In recent years, the surprising discovery in cognitive psychology was that in various cases like this, "simple heuristics were more accurate than standard statistical methods that have the same or more information." More systematic ways of prediction thus do not necessarily provide better results, as quasi-empirical comparisons of the different procedures have shown. Less systematic appears to be scientifically preferable in these cases. Therefore, there is no straightforward conceptual connection between an increase in systematicity and scientific progress. If a positive correlation between systematicity and scientific progress exists, it is at least partly empirical.

However, it may be possible in such cases and similar cases that recourse to other dimensions of systematicity, possibly even the introduction of new dimensions of systematicity, could restore an increase of overall systematicity. In this way, the idea of a conceptual correlation between systematicity and scientific progress could be upheld. The crucial question is whether this move was credible or whether it smacked of an immunization strategy on behalf of the defense of the conceptual connection. In the abstract, it seems impossible to decide this question. In the final chapter, I shall come back to the question of the nature of the relation between science and systematicity.

5.2 SCIENCE AND COMMON SENSE

Systematicity theory offers a particularly well-suited platform in order to investigate the relationship of the sciences and common sense. This is due to the dynamical version of its main thesis discussed in the last section: science develops out of common sense of the respective historical time or out of a nonscientific knowledge practice due to an increase in systematicity. The further development of a science is also characterized by increasing systematicity. Thus, we can determine the relationship between science and common sense by investigating what the effects of this increase in systematicity are, first upon common sense itself and later during the ensuing scientific development. I shall first discuss elements of common sense that are not affected by the emergence of a science and its further development. The existence of such unaffected elements of common sense should be plausible because the incipient increase in systematicity does not transform just every aspect of the pristine common sense knowledge. I will then turn to various sorts of breaks with common sense that result from the incumbent increase in systematicity.

5.2.1 *The Preservation of Common Sense*

When discussing the continuity of common sense with an emerging and then further developing science, two aspects should be distinguished. First, there are the objects of scientific investigation that have also been objects of common sense. Second, and interwoven with the first aspect, there are fundamental ontological and epistemological common sense presuppositions that are furthermore used in science. I shall discuss these two aspects in turn.

In the beginning of any science, the objects of investigations must already be known to some degree, and they are known from common sense or the nonscientific knowledge practice out of which the respective sciences emerge. Take, for example, biology, astronomy, mathematics, or dramatic theory. When these sciences emerged in the Western tradition in ancient Greece, animals and plants, fixed stars and planets, numbers and geometrical figures, and dramas were already known. These common sense objects became objects of scientific investigations, and, at least in the beginning, their assumed nature and their principal properties were not affected by their transfer into the scientific realm. However, only the objects of biology and dramas survived the further development of the pertinent sciences more or less unscathed, whereas the ideas about celestial and mathematical objects underwent serious change in the seventeenth and nineteenth centuries, respectively. I shall discuss such changes in the following subsection in more detail.

A second and most important aspect of conservation in the transition to a science concerns commonsensical fundamental conceptions of physical objects and of our epistemic access to them. Since antiquity, for Western common sense, the physical world has consisted of well-defined physical objects that have certain properties, some of which are stable and some of which are changeable in time. In common sense, it has been unquestionably assumed that we have epistemic access to this world by perceptual and conceptual means, i.e., we can see, touch, and otherwise sense these objects, and we have a language to name and classify them. In other words, in common sense we have always been realists about the objects of the world, and we have always believed that we can gain objective knowledge about them. Clearly, in most natural sciences and many social sciences and humanities, this is the dominant ontological and epistemological attitude as well, in straight continuity with common sense. Note that in some scientific fields in which the common sense conception has been given up with regard to their *objects* of inquiry, this conception has not been given up with regard to the physical *means* of inquiry. For instance, in the quantum realm, objects are conceptualized differently from ordinary objects. However, even the subtlest experimental setup designed to disclose mind-boggling aspects of the quantum world is treated as an assembly of ordinary physical objects. Thus, the common sense conception of physical objects has survived even there.

In those sciences that postulate so-called "theoretical entities," i.e., objects to which we do not have direct observational access, these entities are usually treated in the same way as comparable observable entities; philosopher Arthur Fine has called this custom the "natural ontological attitude." We usually accept confirmed scientific theories, including their theoretical posits, in the same way as we accept the evidence of our senses. Of course, one must be more careful with assertions about such theoretical entities than about observable entities (compare section 3.2.3). However, with respect to their ontological status, theoretical entities are mostly treated like observable objects once they are established. This becomes especially obvious when the borderline between the observable and the unobservable has shifted, and previously unobservable entities have become observable or even manipulable. On these occasions, scientists usually react gratified because of the confirmation of their previously more theoretically based views; there are no signs of shock because of a potential ontological upheaval. This is nicely illustrated by the first manipulation of individual atoms by means of a scanning tunneling microscope. In 1990, Don Eigler and Erhard Schweizer shifted thirty-five individual xenon atoms on a nickel surface such that the logo "IBM" resulted. This image made it to the front page of *Nature* and further around the world. As it was later described, this event "changed the nanoworld," but only in the sense that individual atoms were now practically manipulable, almost like ordinary objects. It added nothing to our fundamental understanding of atoms but only reinforced the view that they can be similarly treated as the larger physical things we are familiar with in everyday life. Thus, even theoretical entities as they are postulated in the sciences do not usually break the mold of the common sense conception of physical objects.

It is not only the conception of what the objects of scientific inquiry are that have been taken over from common sense and that have mostly stayed in science, but also the associated ideas of objectivity and truth. For common sense, the notions of objectivity and truth are quite uncomplicated in themselves and very often completely uncontroversial in their practical applications. A report, for instance, is objective if it represents the relevant facts as they are, without any distortions or additions by the observing and reporting subject. A statement is true if it is adequate to the state of affairs. Mary's statement that she went to the cinema yesterday is true if she really went to the cinema yesterday—for common sense, usually no problem arises from this understanding of truth. It is the same "naïve-realist" understanding of objectivity and truth that is dominant in many sciences. Of course, there may be complications, for instance, about the notion of objectivity in history, but these complications have often not eroded the allegiance to the concept. And of course, there are also serious deviations from these common sense conceptions of objectivity and truth, for example, in all of those social sciences and humanities that understand themselves as "constructivist." I shall come to such deviations in the next section.

It is characteristic for such elements immediately inherited from common sense that they are mostly not explicitly defended in science, neither at the beginning of a science nor during its further development (unless they are challenged; see further below). Their validity is simply taken for granted, which means that the question of their defense does not even come up, as their validity was taken for granted in common sense. To a critical philosopher, this may be a dangerous omission, even a confirmation of Martin Heidegger's infamous dictum: "Science does not think." However, one should reflect upon what science is up to. Scientists want to explore the world, and in the beginning of a science, they want to surpass common sense (or the respective prescientific knowledge practices) by being more systematic. Given this goal, it is reasonable just to take over those elements of common sense that are useful and apparently unproblematic and not to scrutinize them. Otherwise, they might turn into skeptical philosophers who question states of affairs that for the practice of science need not be questioned. Furthermore, it is important to realize that in any enterprise concerned with knowledge claims, not literally every knowledge claim can be explicitly defended; at least, not all knowledge claims can be defended at the same time. In any defense of a particular knowledge claim, other knowledge claims have to be taken for granted, at least for the time being. For a nascent science, only claims that are new, relative to, or deviant from common sense must be defended. The basis for this defense is given by the uncontroversial heritage of common sense. However, this unchallenged status of certain areas of common sense does not necessarily hold forever; what was taken for granted at one time may be completely revolutionized at a later time.

5.2.2 *The Deviations from Common Sense*

Let us now turn to those elements of a science that were not inherited by common sense or a former knowledge practice, but were created in the course of a nascent science or during its further development. As variously stressed, both processes are characterized by an increase in overall systematicity. The deviations from common sense come in three different shades. First, there is new knowledge that results from a specification of common sense knowledge. Second, there is new knowledge that is unrelated to common sense knowledge. Third, there is new knowledge that breaks with common sense. The boundaries between these areas are not entirely sharp; there are transitions, and I will return to one of them at the beginning of the next subsection.

Let us first discuss common sense knowledge that is adopted by a nascent science but specified. This is probably fairly typical for the initial transformation of common sense knowledge to scientific knowledge, which clearly increases its systematicity. For example, before anything like a science of astronomy existed, people were already aware of the existence of stars, of the movements of the planets, of the existence of extraordinary events like eclipses, of the change of the length of daylight over a year, and so on. The beginning of the science of astronomy is marked by systematic observation of celestial phenomena that were known beforehand, and their systematic descriptions. For instance, early astronomy produced catalogs of stars, descriptions of striking astronomical processes, the discovery of periodicities in the latter, and the like. In this way, common sense knowledge was made more precise, in the form of catalogs or even with the help of mathematics. This increase in systematicity of astronomical knowledge was, of course, based on other prescientific convictions that were taken for granted. For instance, there was no doubt that one could truly see celestial bodies and many of their properties. This common sense precondition to scientific knowledge was probably not at all called into question, just as little as the religious, mythological, and astrological connotations of celestial objects and configurations. These things were taken over from the prescientific understanding of world, and there was no need to defend them. What had to be defended were knowledge claims deviant from what was accepted by common sense. This includes, for example, a claimed periodicity in the appearance of lunar eclipses, or a mathematical description of the change of the length of daylight over a year.

More generally speaking, it is probably typical of a nascent science that the common sense descriptions and explanations of the familiar objects falling in the domain of the respective science will be specified. The same is true for the defense of knowledge claims: common sense procedures will be used more consciously and more carefully, thus more systematically.

In the further course of a new science, sooner or later, knowledge will be generated that stands in no relation to common sense knowledge: neither is it just an advancement of common sense knowledge, nor does it contradict common sense. This sort of new knowledge may be due to the discovery of new phenomena, or it may be due to new theoretical developments. For instance, in the observational and experimental sciences, new instruments may open up or produce phenomena that are not accessible to the unaided senses and are thus unknown to common sense. Similarly, theoretical developments may lead to postulates of entities or techniques of investigation that are foreign to common sense. Think of the postulates of elementary particles in physics, or of rigorous proof as the main instrument to defend knowledge claims in mathematics. Though the generation of such knowledge distances the respective field from common sense, it does not contradict it. This is different for the third kind of new knowledge to which I now turn.

Many sciences also produce a kind of knowledge that directly contradicts common sense knowledge. The history of modern physics provides a plethora of examples of knowledge claims strongly contradicting common sense, from its beginnings until today. In its beginnings, modern physics developed in opposition mainly to an older

Aristotelian tradition of physics. The latter is largely a very successful specification of common sense; it survived for something like eighteen centuries. Here are three examples of how the modern physical tradition deviated from and contradicted the Aristotelian world picture. First, the observed motions of the celestial bodies in the sky, i.e., of stars, planets, the sun, and the moon, appear to us as the true motions of these bodies themselves. Also in the Aristotelian tradition, they were conceived as such. However, in the first half of the sixteenth century, the Copernican theory urged to reconceptualize these motions as *apparent* motions, resulting from the true motions of the bodies themselves and the motion of the observer. This resulted in a reconceptualization of observations in blatant contradiction to common sense and its implied interpretation of these motions. Second, common sense tells us that the motion of a body naturally comes to rest after some time. However, in the course of the seventeenth century, several laws of inertia were articulated, starting with Galileo and terminating with Newton. A law of inertia claims that without external influence, the motion of a body would continue forever. From a commonsensical point of view, this is not plausible at all because all actual observations contradict it. However, physics very successfully adopted the (Newtonian) law of inertia. Third, in everyday life we are entirely sure that material bodies have colors and that they may have characteristic smells. However, in the course of seventeenth century science, these assumptions have been challenged, distinguishing these properties as "secondary qualities" (in some variant) from "primary qualities." Only the latter, conceived of as inherent properties of the objects, were thought to be possible subjects of science, whereas secondary qualities, being in one way or another related to a perceiving subject, were not in the main focus of science.

Today, these views of modern physics have largely been incorporated into common sense, at least to some degree. However, in the twentieth century, physicists claimed things that were so provocative to common sense that even today there are vociferous groups claiming the falsehood of physics, especially of Albert Einstein's relativity theory. Take, for example, the notion of simultaneity. It is a very successful everyday notion stating that two events, wherever they are located, are either simultaneous or not. This notion of (absolute) simultaneity is also a part of Newtonian physics. The background is that time is universal and that therefore, two events either occupy the same time point on the scale of universal time or not. As far as I know, simultaneity and the associated idea of universal time were never seriously challenged in science until the twentieth century. Until the end of the nineteenth century, most scientists were not even aware of the fact that absolute simultaneity is an assumption that is not necessarily true, i.e., an assumption that could be challenged. It was only in 1905 when Einstein challenged this assumption in his special theory of relativity. In order to obtain maximal coherence of all of the relevant data, Einstein postulated that simultaneity was not an absolute notion but a relative one, relative to the respective frame of reference in which the events were described. Thus, the challenging of the common sense notion of simultaneity was not the result of a philosophical reflection bringing to the fore that this notion was more fragile than was previously assumed. Instead, it was the realization that a relativized notion of simultaneity would increase the coherence and thus the systematicity of explanations for some class of phenomena. There are quite a few other claims in the physics of the twentieth century that are slaps in the face of common sense, like space being bent or some entities being both particles and waves. It is not very audacious to predict that also in the twenty-first century, provocations of this sort will continue to be produced by fundamental physics.

Of course, it is not only physics that has produced scientific beliefs that are in square opposition with common sense. Two more examples will have to suffice. Take first biology that has been, as far as individual animals or plants are concerned, fairly continuous with common sense. However, when it comes to classification, deviations from common sense occur. For instance, as (almost) everyone knows today, whales are not fish, but mammals. Furthermore, the common sense concept of biological species has turned out to be much more complicated when looked upon in different biological contexts. Finally, here is an example from the humanities. In literary theory, there are the concepts of "effective history" and "reception history." In certain quarters of literary theory, these concepts are seen to be very important (although they are highly controversial). There exist several variants of them, but I shall not go into details. The area in which these concepts play a decisive role is often described as "reader-response criticism." The basic idea here is that the meaning of a (literary) text is not something that has been fixed, once and for all, by the author of the text. On the contrary, the meaning of the text is something that is influenced (or even constituted) by the later reception of the text. Certainly, this idea is alien to common sense. The common sense idea is probably that the meaning of any utterance, literary or otherwise, is just determined by its author, and it is either understood, or not understood, or misunderstood. However, one of the core arguments in the pertinent literary theory is that the meaning of a literary text "has no effective existence outside of its realization in the mind of a reader." This is a reflection that is certainly quite foreign to common sense (whatever the merit of this argument in the final analysis is).

5.2.3 *Additional Remarks*

There are many more interesting aspects of the relation between science and common sense. I shall focus on four of them.

First, as I noted in the last subsection, the division into specifications of, additions to, and breaks with common sense is not supposed to be entirely sharp. More specifically, there is a blurred area between what is just unknown to common sense and what common sense *implicitly* assumes not to exist. For instance, common sense knows very little or nothing about bacteria living in symbiosis with us. However, people are shocked when learning how many different kinds of bacteria live symbiotically in our organs, for instance in the nose, on the conjunctiva, in the mouth, on the skin, in the stomach. The reaction of shock indicates that implicitly, the assumption was that when a person is healthy, no or at least not so many and so many different kinds of microbes populate the body. Generalizing from this example, it is plausible to assume that with respect to many areas where common sense does in fact not know anything, it implicitly assumes that there is nothing to know about. It is characteristic of common sense that it is often not aware of its knowledge lacunae (which is, of course, one of the main breeding grounds of philosophy).

Second, the specific mixture of continuity and discontinuity with common sense present in any given discipline or scientific field drastically varies with that discipline or field. Thus the distance of some discipline or field's scientific knowledge from common sense is extremely variable. It depends on the duration and the specific quality of the process in which common sense was first modified and of the resulting science's subsequent development. For instance, contemporary mathematics, theoretical physics, chemistry, molecular biology, economic theory, parts of sociology, or literary theory are not accessible to the layperson in unpopularized form. By contrast, many parts of the humanities are accessible to the layperson, especially many branches of history or parts of the disciplines dealing with the arts and literature.

Third, what are the reasons an increase in systematicity leads to breaks with common sense? Why, for instance, have some common sense notions like simultaneity not survived the revolution in physics that took place in the twentieth century? Roughly speaking, these notions could not find a place in the new theories that were advanced in response to difficulties that the predecessor theories were entangled in. These new theories were accepted because they outperformed the older theories in terms of the scientific values that the respective scientific community was committed to. As I have already explained in section 5.1.3, this transition can also be described as an increase in overall systematicity, and its victims sometimes are common sense notions.

Fourth, many of the breaks of scientific knowledge with common sense are of the same kind. This specific kind of break is due to a fundamental stance of common sense to reality that can be characterized as "objectivist." An integral part of this objectivist stance is a specific form of realism that is often characterized as "naïve realism," also called "direct realism" or "common sense realism." This form of realism assumes that in general, the physical things that we perceive are, in themselves, as they appear to us. Thus, these things are in space and time and have the various properties and relations we perceive. Built into this view is a theory of perception, namely, that perception is basically a passive affair: perception just represents its objects, without any additions. Naïve realism's stance toward physical things can therefore be described as "objectivist." However, by the "objectivist stance," I mean something slightly more general than naïve realism that is usually focused on physical things only. The objectivist stance takes *everything* that appears to be "out there" as real and independent of the perceiving subject: not only physical things and their properties but also psychological and social phenomena. There is one area where common sense has, from antiquity until today, been aware that the objectivist stance does not seem to work, namely, in matters of taste, literally and figuratively understood. "There's no accounting for taste" does not only refer to certain flavors of meals and drinks, but also more generally to qualities that at first sight appear to be rooted in the objects themselves, like aesthetic qualities, though they are apparently not. The commonsensical proverb articulates a critical objection against an objectivist understanding of such preferences. However, in all other matters, common sense's stance toward reality has unswervingly been objectivist.

Science, however, has again and again been forced to give up the objectivist stance with respect to certain things, and hence opposed common sense. Many of the examples I gave above are of this kind. In modern science, it all started with the Copernican theory of the planetary system that deprived the celestial motions of their objectivist status. And so it is with smells, colors, and the like, whose nonobjectivist status became terminological in their being called "secondary" qualities; only the "primary" qualities continued to be objectivisticly interpreted. In the twentieth century, "simultaneity" suffered the same fate. In common sense and in classical physics, it was understood objectivisticly, as a relation pertaining (or not pertaining) to pairs of events. However, relativity theory taught that this absolute sense of simultaneity is an illusion because the relation also depends on the pertinent frame of reference. Similarly with wave-particle dualism, an entity is not just a particle *or* a wave as common sense would have it, but this ascription is dependent on the experimental setup in which the entity is observed. Finally, the example from literary theory is of the same kind. In reader-response criticism, the meaning of a literary text is not understood objectivisticly but as something to which the contemporary reader contributes.

However, not all of the examples of serious derivation from common sense are of that kind. The law of inertia contradicts common sense in analytically taking two factors apart that common sense takes as essentially belonging together. For common sense, motions always come to rest because of their very nature. Classical physics, however, distinguishes two aspects: force-free motion on the one hand and effective forces, among them friction, on the other. Coming to rest is thus not essential to motion, but accidental: it is an effect of a frictional force. This is somewhat similar to the case of the motion of celestial bodies: common sense sees only one source of motion (proper motion), whereas Copernican theory sees two (proper motion and the observer's motion). However, the move to the Copernican theory is a move away from the objectivist assumption, whereas the result of the move to the law of inertia is as objectivist as its predecessor. Finally, the reclassification of whales as mammals is still another case. Here, common sense's classification is criticized as insufficient because of its superficiality, for not taking relevant features of whales into account. Again, it is not a criticism based on a refutation of objectivism.

5.3 NORMATIVE CONSEQUENCES

I have emphasized in the beginning that the main thesis of systematicity theory is descriptive (see section 2.1.2, second remark). This status of systematicity theory is important regarding the possible arguments that can be adduced in its support (see section 2.3). The long journey of chapter 3, where we went through all nine dimensions and many disciplines and research fields, was necessary because our thesis is descriptive and concerns the sciences as they really are. However, to many readers, it may now appear that this cannot be the full story. Hasn't it become evident that in all of the dimensions a higher degree of systematicity is a *good* thing? Don't the descriptions of science given by systematicity theory immediately imply that the sciences *should* strive for higher and higher degrees of systematicity? These questions clearly suggest normative consequences of systematicity theory: systematicity theory licenses the evaluation of something as good and recommends a certain course of action.

However, in situations like these, philosophers will immediately see a red blinking warning device, accompanied by a loud alarm sound signaling "the naturalistic fallacy": it is impossible to derive normative statements from descriptive sentences alone. This insight was gained in the eighteenth century by David Hume (1711–1776) and is today a standard item in a philosopher's toolbox. Descriptive statements have, by themselves, no normative content. Given that a logical or analytical inference can only result in what is already (implicitly) contained in a statement, nothing normative can be logically or analytically deduced from a descriptive statement. Therefore, any presumed deduction of a normative statement from a descriptive one is a fallacy. Given this objection, we have to be quite careful about deriving normative statements from systematicity theory. Let us now look at an example in which the derivation of normative consequences suggests itself. We want to find out whether such a step to normative consequences is justified.

The example concerns the defense of knowledge claims in medical studies, more precisely in clinical studies. A clinical study is a medical study that investigates the validity of diagnoses of illnesses, the effectiveness of treatments, and the like, and it involves patients. During the last decade, a classification of clinical studies in terms of their "level of evidence" has been spreading in the medical community. It is basically a five-level classification (with some subclassifications) of clinical studies that ranks them according to the evidence that entered a given study. What the levels concretely mean somewhat depends on the type of study that is evaluated: there are therapeutic studies, prognostic studies, diagnostic studies, differential diagnostic studies, and economic and decision analyses. The currently most widespread versions of the classification of clinical studies in the medical literature are identical with or small variants of the March 2009 version of the *University of Oxford Centre for Evidence Based Medicine*. Since the early 2000s, medical journals have adopted the practice that every new submission of a clinical study has to specify its level-of-evidence rating.

What are the "levels of evidence"? The levels are a measure of the quality of a clinical study with respect to the evidence adduced. For our purposes here, a very rough sketch of the levels will suffice in order to get an idea of the enterprise. Level 5 is just an expert opinion. Level 4 refers to case-series, i.e., to a number of patients that were treated one way with no comparison group of patients treated another way. Levels 3 to 1 refer to studies that involve control groups in one way or another (compare the discussion of treatment-control studies in section 3.4.4). Level 3 simply demands the involvement of a control group. The highest level 1 demands a "systematic review" of several random control studies in which the individual studies are free of worrisome deviations from one another (compare the discussion of "systematic reviews" in section 3.4.4). Level 2 indicates a quality of the evidence between levels 1 and 3.

In terms of systematicity theory, the difference between the levels can be very simply expressed, if only in an abstract fashion. We are clearly dealing with defenses of knowledge claims, i.e., dimension 4. The gradation between levels 5 through 1 corresponds to an increase in the systematic efforts to avoid empirical error. In ascending from level 5 to level 1, more and more possible sources of error are taken into account and systematically eliminated.

This characterization of the different levels of evidence is purely descriptive. For instance, to state that level 4 studies only involving case-series are less systematic in eliminating certain types of error than level 3 studies involving control groups is a purely descriptive statement. However, it seems to follow *immediately*, i.e., without any further premise, that in general, level 3 studies are better than level 4 studies.

This is due to the fact that in our context, namely, science, the concept of error elimination comes with a highly positive value connotation. It appears to be *obvious* that that more effective error elimination is better than less effective error elimination. In spite of this blatancy, it is clear that *stating* something about error avoidance is something else than *evaluating* error avoidance. Just imagine a situation in which being caught in a certain error for a while is more profitable in some sense than immediately finding out the error. Thus, the apparent immediacy of the normative consequences from certain descriptive ones is due to the implicitness and the naturalness of certain norms in the given context. In the context of systematicity theory, all nine dimensions are of this kind. As in dimension 4, it appears that a higher degree of systematicity is always a good thing. For instance, more detailed or more unified, i.e. more systematic, explanations immediately strike as better explanation in science, and the same holds for descriptions, predictions, and the other dimensions. If one assumes that the nine dimensions descriptively cover those properties of science whose higher degree of systematicity sets science apart from other knowledge claims, one will tend to assume that science can be evaluated according to these dimensions.

Also with respect to the dynamics of science, we made a purely descriptive statement: science progresses by increasing its overall systematicity (section 5.1.3). However, given the validity of this description, the statement that science *should* increase its overall systematicity immediately suggests itself. This normative statement is, however, by no means an immediate consequence of the given description. The normative statement involves the additional normative premise that increased systematicity in the nine dimensions is a good thing. The apparent obviousness of this additional normative premise should not conceal what it is: an additional premise.

Given that the inclusion of the normative premises licenses normative statements by systematicity theory, one should still not be overly optimistic about the resulting normative power of systematicity theory. In many cases where scientists (or philosophers of science) disagree about evaluations of scientific matters, systematicity theory will be of little help. For example, consider string theory. String theory is a theory about the fundamental constituents of matter which should, in principle, be *the* foundational theory of all of physics: the "theory of everything." However, there is a problem with string theory that in recent years has increasingly troubled physicists. Although string theory has been around for several decades, it has not produced one single empirically testable consequence, despite earnest efforts. The defenders of string theory do not seem to be terribly worried about these missing empirical checks, ascribing this fact roughly to the nature of such a fundamental theory that requires more patience with respect to empirical predictions. Critics, however, have mockingly characterized string theory as "turning theoretical physics into recreational mathematics." The problem consists in weighing the missing empirical tests and the promising theoretical potential of string theory. Scientists disagree on this matter, and I do not see how the philosophy of science could come to the rescue. When viewed from the perspectives of current philosophies of science, none of the relevant philosophies reaches a unanimous conclusion. The same holds for systematicity theory. The reason is easy to understand. Systematicity theory is a descriptive theory, describing what exists in science. When systematicity theory is brought to produce normative statements in the way described above, neither their descriptive input nor their normative input can go beyond what already exists in science itself. Systematicity theory may bring reflective clarity to scientific issues that are unclear or implicit, but it cannot substantially contribute in matters where the issue is clear but the scientific norms are insufficient to decide the case. The facts and the question regarding string theory are clear: should string theory be pursued in spite of its not having produced empirical predictions for decades? The scientific community is somewhat divided about the issue, and systematicity theory can clearly not contribute additional descriptive or normative aspects, unknown to the scientists, that would decide the issue.

I think that this situation is fairly typical for any philosophy of science that is primarily descriptive. Such theories lack the resources to be normatively helpful in scientific disputes in which the scientists themselves cannot reach a decision. The inability of a scientific community to reach a decision in such cases may be due to descriptive deficiencies (not enough is known about the case) or to normative deficiencies (for instance, different values contradict each other). The philosophy of science has no additional resources to complement the scientist's resources and to bring the case to a close. However, when it comes to normative problems whose origins are conceptual, descriptive philosophy of science in general and systematicity theory in particular may be of some help. Such a problem is the so-called demarcation problem, where pseudoscience is to be demarcated from real science. This is clearly a distinction that is normatively loaded because, as its name indicates, pseudoscience poses as science without being science, and that is surely a bad thing.

5.4 DEMARCATION FROM PSEUDOSCIENCE

As I said, the so-called demarcation problem concerns the question of how proper science can be distinguished from other enterprises that may resemble science, or even pose as science, but are not science. "Science" is to be taken here in the wide sense but excluding the formal sciences. We are thus concerned with all disciplines that are empirical in the widest sense. The attempts to solve the demarcation problem consist in the formulation of a criterion that, applied to a field in question, gives an unambiguous and correct assessment of its status as being a science or not. Such a criterion is therefore called a "demarcation criterion" between science and nonscience. Before discussing what systematicity theory has to offer in this realm, let us have a quick look into the history of the demarcation criterion in the twentieth century.

5.4.1 A Little History

It was Karl Popper (1902–1994) who made the demarcation problem prominent in the twentieth century; he took it to be one of the two fundamental problems in epistemology (the other is the problem of induction). Popper moved away from the logical empiricists' project of a criterion of meaning for empirical sentences and replaced it with the project of the demarcation of science from nonscience (see section 4.5.1). The logical empiricists' project was designed to destroy traditional metaphysics and did not take into account the so-called pseudosciences at all. Thus, Popper's focus on the demarcation problem, including pseudoscience, was a momentous innovation. With respect to demarcation, he thought that metaphysics and the pseudosciences had equal status when compared to real empirical sciences and should therefore be treated in one sweep.

With respect to metaphysics, Popper gave the ongoing discussion a new twist. The subject matter of this discussion was so-called metaphysical sentences. The realm of these sentences was not strictly defined but loosely determined by examples, specifically from texts of the philosophical tradition that is today sometimes called "continental." Take as an example the idea known from early Greek philosophy that the universe is governed by love and hate. Logical positivists denied metaphysical sentences like this any cognitive meaning whatsoever, i.e., in their view, the utterance of such sentences was, contrary to appearance, cognitively undistinguishable from undefined noises. Popper, by contrast, did not go so far in his criticism of metaphysics. He contended that metaphysical sentences often do have meaning but lack something else that would qualify them as scientific (the demarcation criterion was, of course, assigned the job of specifying this "something else"). However, he granted metaphysical sentences the possibility of turning into scientific ones, given some pertinent conceptual and scientific advance. For Popper, scientific sentences were judged as having a higher cognitive status than metaphysical ones, but the latter were at least not cognitively completely empty and useless as the logical positivists had thought.

With respect to pseudosciences, Popper's leading examples were the Marxist theory of history, Adler's individual psychology, and Freud's psychoanalytic theory. Popper thought that these fields were typical pseudosciences, i.e., fields for which a scientific status was claimed but which in fact lacked it. The principal idea was that although for the theories in questions, numerous confirming instances could be found, it was impossible in principle to show on empirical grounds that they were wrong. In Popper's view, these theories were formulated so flexibly that they could account for any phenomenon—and therefore for none. What was missing in these theories was a kind of definiteness such that they could principally collide with empirical findings, and for Popper, true sciences had exactly this property.

According to Popper, the criterion that distinguishes scientific sentences from metaphysical and pseudoscientific sentences is principal falsifiability. In the literature, this criterion has not always been properly understood. Abstractly speaking, the falsifiability of a sentence denotes a logical relationship between this sentence and so-called basic sentences. Basic sentences are sentences whose truth or falsity can *in principle* be decided by observation. The falsifiability of a sentence demands that there are logically possible basic sentences (whose actual truth value can remain undecided) that stand in logical contradiction to the sentence in question. In other words, the falsifiability of a sentence guarantees that the empirical falsification of the sentence is not excluded for logical reasons, i.e., that it is in principle empirically testable. Put simply, the falsifiability of a sentence *logically* allows for the possibility of observations that show that the sentence is false. Of course, it is an entirely different question whether such observations can be made in practice. It may be impossible to make such falsifying observations for purely technical reasons, for instance, because the entities referred to in the basic sentences are too small to be observed with today's technical means. Or it may be impossible to make such falsifying observations because the sentence is simply empirically true. For instance, that a normal clear daytime sky is blue cannot in fact be falsified. But certainly there are *logically possible* basic sentences that falsify the sentence in question, for instance: "This is a normal day with a clear sky, it is daytime, and the sky is green."

It must be noted that the status of a sentence as falsifiable or not does not depend on the sentence alone, i.e., its status as scientific or not is not an intrinsic property of the sentence. This is the reason a sentence can change its status with respect to falsifiability. This possibility arises because the falsifiability of a sentence also depends on the embedding of the terms contained in it into a wider context. For instance, a sentence containing a theoretical term, i.e., a term referring to things not directly observable, may or may not be falsifiable. It depends on whether there are bridge principles sufficiently connecting the theoretical term with observable manifestations of the thing it refers to. Complete lack of such bridge principles makes the sentence unfalsifiable; an appropriate connection to observable phenomena may generate a contradiction to logically possible basic sentences, hence falsifiability. Here we have the reason Popper did not fully condemn metaphysics as the logical positivists did. Metaphysical and hence unfalsifiable sentences may become falsifiable and hence scientific when the entities figuring in them are appropriately connected in principal to observable phenomena. Due to this possibility, metaphysics may play a positive heuristic role in the development of science.

Before assessing the merits of Popper's demarcation criterion, three of its features should be highlighted. First, the scope of the criterion is *global*. This means that the criterion can be applied uniformly to all fields that make empirical claims, independently of their subject matter, and that it is atemporal, i.e., it does not change in time. Thus, it is decided by the very same criterion whether, say, current high-energy physics, post-structural literary theory, nineteenth-century psychology of perception, or ancient dramatic theory are scientific or not. Second, for the application of the criterion, it is enough to have a *static* representation of the field in question. Thus, a flash picture of a field is enough to decide whether it is scientific; especially, the field's history is irrelevant. Third, the criterion takes into consideration only *intrinsic* properties of the field in question; its relations to other fields are irrelevant. Although it is not fully accurate and is potentially misleading, it is convenient to characterize Popper's criterion of demarcation as global, static, and intrinsic.

Popper's criterion of demarcation was severely criticized. First, there is the problem of existential statements (like "there are atoms" or "there is a planet closer to the sun than the Earth"). In order to be scientific, such statements must be falsifiable. But the method of producing empirical counterexamples that works so well with universal statements clearly does not work for existential statements. The falsification of an existential statement is equivalent to the verification of a universal statement that seems to be impossible (in fact, Popper's whole theory is triggered by the empirical unverifiability of universal statements—the problem of induction). Second, Popper's demarcation criterion has the untoward consequence that under its regime, many undoubtedly pseudoscientific theories would count as scientific, namely those pseudoscientific theories whose falsehood can, in principle, be empirically shown. As long as they can formulate a logically possible observation sentence contradicting their theories, these theories count as scientific. And even empirically refuted pseudoscientific theories count as scientific because their empirical refutation proves their status of being empirically refutable and hence of being scientific! All of this is fairly implausible. Third, with respect to Popper's examples, it is far from clear that, for example, astrology or Freud's psychoanalysis are indeed pseudoscientific when judged according to Popper's standards.

After Popper, there have been very few systematic attempts to articulate a demarcation criterion. None of them was so convincing as to gain the support by a substantial group of philosophers; the problems they confronted were apparently too large. It seems that many philosophers got the impression that the problem of demarcation is unsolvable and should therefore be dropped. For many, this impression was probably rooted in the insight that the sciences are too heterogeneous as to be assessed by one single yardstick. As philosopher of science Larry Laudan put it in 1983: "*The evident epistemic heterogeneity of the activities and beliefs customarily regarded as scientific should alert us to the probable futility of seeking an epistemic version of a demarcation criterion.*" We should not, however, become so despondent because we can try to activate the resources of systematicity theory in order to tackle the demarcation problem.

5.4.2 Systematicity Theory's Demarcation Criterion

Let us first determine how we understand the demarcation problem in the following. In its most general form, the demarcation problem is taken to demarcate the scientific from the nonscientific. However, I will treat a restricted version of it with the more modest task of demarcating science only from pseudoscience. In particular, I shall not discuss the case of so-called indigenous or traditional knowledge and its relationship to science. This relationship is interesting but beyond our scope here. I shall not try to really define pseudoscience but will stay content with typical presumable examples, like creation science or astrology. However, I shall not definitively presuppose that these fields are pseudosciences. For the following, it will suffice to treat them as candidates for pseudoscience, but not more.

It is an important observation that the pseudosciences typically compete with established sciences. For example, creation science competes with evolutionary biology, and (certain brands of) astrology competes with psychology. However, in the discussion about Paul Thagard's demarcation criterion (more on this a little further below), it was objected that there were also situations in which a vast majority of scientists reject a field as pseudoscientific, and there were no competitors. I shall postpone this case and first discuss the more common case in which a competitor exists for a given putative pseudoscience. Thus, for any (putative) pseudoscience of this sort, we can define a "reference science" as that science with which the pseudoscience competes.

The demarcation criterion that I am proposing is diametrically opposed to Popper's regarding the three characteristics that I delineated above: Popper's criterion is global, static, and intrinsic, whereas the one discussed here is local, dynamic, and comparative. It is an advancement of a criterion that philosopher Paul Thagard proposed in 1978. This is Thagard's "principle of demarcation":

A theory or discipline that purports to be scientific is *pseudoscientific* if and only if:

1 it has been less progressive than alternative theories over a long period of time and faces many unsolved problems; but

2 the community of practitioners makes little attempt to develop the theory toward solutions of the problems, shows no concerns for attempts to evaluate theory in relations to others, and is selective in considering confirmations and disconfirmations.

The fundamental idea here is that the comparison of a putative pseudoscience with another discipline regarding the progressiveness of its development will display the pseudoscientific character. Thagard explains the progressiveness of a theory as "a matter of success of the theory in adding to its set of facts explained and problems solved." We can generalize this idea of progressiveness in terms of systematicity theory and make more explicit the relation of the putative pseudoscience's development to the "alternative theories." For simplicity, I shall speak of a (putative) pseudo*science* although only a single theory may be meant.

For a test of whether a given field is pseudoscientific at some time t_0, we have to first identify the relevant reference science, i.e., a science with roughly the same subject matter existing at the same time. Next, we investigate the reference science's development over some longer time span until the time t_0, perhaps something between five and thirty years. We record what sort of systematicity increase the reference science exhibits during this time span. This systematicity increase sets the standard of what is possible in the respective field regarding scientific progress. The sort and strength of the systematicity increase may vary greatly, depending on the specific historical time and the specific reference science. Next, we investigate the development of the putative pseudoscience in the same time span until t_0. Also here, we record the systematicity increase during this time span. We then compare the systematicity increase of the reference science with the systematicity increase of the putative pseudoscience. If the putative pseudoscience scores substantially worse than the reference science, then it is indeed a pseudoscience. If the difference in systematicity increase is only slight, then the putative pseudoscience is a scientific competitor with the reference science.

A few remarks on this demarcation criterion are in order; I shall call it STDC for "systematicity theory's demarcation criterion."

Let us first emphasize the contrast to Popper's criterion (compare section 5.4.1). I will use the acronym PDC for "Popper's demarcation criterion." PDC is global: it is exactly the same criterion for all subject matters and all times. On an abstract level, STDC also is global (in the manner as I described it above). However, in concrete applications, STDC is local because it depends on a specific reference science at a particular time. PDC is static in the sense that a flash picture of a field suffices for its evaluation. By contrast, STDC is dynamic: it evaluates a field's development over a certain time span. PDC evaluates only intrinsic properties of the field in question, whereas STDC is relational, particularly comparative. A field can be evaluated as pseudoscientific only in comparison to a reference science. If STDC is roughly correct, then it is evident why PDC was hopeless from the very beginning. PDC is an ahistorical universal yardstick using flash pictures of the investigated fields only, and as such, it is much too simplistic to do justice to the incredible historically changing variety of sciences and pseudosciences.

Second, it may appear that the vagueness of the length of the time span for the comparison of the putative pseudoscience and the reference science is a weakness of STDC. However, this is not the case. This vagueness is due to the diversity of historical situations in which STDC may be applied. A chosen time span may be unfairly short because of the temporary stagnation of a field; or it may be unfairly long in giving recent drastic advances too little weight. The choice of the appropriate time span may even be a matter of negotiation between the relevant parties, and the result of the comparative evaluation may therefore be unclear. Yes, this may be the case, and the reason may be that the situation is indeed unclear—that it is not possible to decide about the scientific status of some field at a particular historical time.

Third, as it happened several times in this book, things may appear so vague at the abstract level and so complicated and multifarious at the concrete level that the practical value of the given analysis seems to disappear into thin air. I am aware that the given sketch of STDC leaves open many questions that will have to be answered by future research. However, I would like to dispel the possible impression that STDC is virtually vacuous. Although I am not presenting detailed examples of pseudoscience here, my impression is that in many cases, systematicity theory's demarcation criterion can be applied fairly straightforwardly. The reason is that in quite a few pseudoscientific cases there is, over shorter or longer time spans, virtually no systematicity increase at all because there is no real dynamics in any of the dimensions of systematicity at all. In particular, whereas the sciences often try to increase the systematicity of their defense of knowledge claims by new data, risky predictions, new statistical analyses, and so forth, pseudosciences often show very little dynamics and very little initiative in this respect. Furthermore, sciences with vitality often try to expand their scope of application, whereas pseudosciences often stay with their traditional scope, just defending themselves against critical arguments from outside.

Fourth, we should consider the case of fields that were seen as pseudosciences by many contemporaries in spite of a lacking well-defined reference science. It should first be noted that scientists always have the means to evaluate contributions to their own scientific field as good or bad (although any concrete evaluation may be controversial). The basis of these evaluations can be described either in terms of the scientific values that the scientific community is committed to or in terms of the higher value of higher systematicity; I have discussed the intertranslatability of scientific values and degrees of systematicity earlier in section 5.1.3. Under both descriptions, the basis of evaluations is strongly field dependent. To give just one example: what counts as good quantitative accuracy (if applicable) is strongly dependent on the scientific field in question. If a reference science exists for a putative pseudoscience, then it is clear that this field's standard is to be used for the evaluation of the putative pseudoscience. However, if no direct reference science exists, it is not the case that scientists totally lack any evaluative basis for a judgment of a putative pseudoscience. Scientific disciplines in the vicinity of the putative pseudoscience will hint at what is good science, what is bad science, and what is not science at all in some larger area. Furthermore, comparative standards are sometimes not even necessary. This holds already for common sense: more or less completely arbitrary statements that are not backed up by any sort of intelligible argument are simply rejected. The same holds for science: if, for example, the existence of some correlations is claimed for which neither verifiable empirical evidence nor some plausible theoretical explanation is given, scientists will usually react skeptically. In other words, dimension 4 is crucial for any science: scientific belief must somehow be backed up by credible arguments; otherwise it will not be accepted. Bodies or even systems of belief massively deficient in this respect will be judged as nonscientific or as pseudoscientific notwithstanding their own claims to scientificity.

Fifth, for all of those who expect a clear and determinate criterion for the demarcation of the scientific from the nonscientific, all of what I said so far will appear disappointingly vague. One should, however, bear in mind that scientificity is a notion that is extremely dependent on the various disciplines and on time. For instance, in the early eighteenth century, it was still scientifically legitimate (although controversial) to establish a role for God in planetary theory: Newton postulated that God would prevent any seriously accumulating instability of the planetary system by correcting planetary orbits. God's role in planetary theory ended with Laplace declaring that in his theory, he no longer needed the hypothesis of God. Similarly, in nineteenth-century geology and paleontology, God was invoked by some predominantly British authors in the theory of catastrophism. This theory postulated of a number of deluges, analogous to the Flood, in order to explain geological and paleontological data. Again, this was undoubtedly part of science, though controversial. Thus, it is impossible to state in general whether God is a legitimate part of science, especially of scientific explanations; it depends on the particular discipline and on the historic time. This is but one example of the strong historical and disciplinary variability of what is legitimately a part of science and what is not.

In more general terms, the problem of any demarcation criterion is this. Surveying the disciplinary spectrum, on the one end, there is mathematics and within mathematics its fully formalized parts. It is difficult to imagine how any knowledge claim could be defended more transparently and more rigorously than those from this area. On the other end of the disciplinary spectrum there are humanities, whose subject matter is highly elusive but nevertheless extremely interesting because of their relevance for human life. Think of the meaning of literary works, of the meaning of works of art, of the study of foreign cultures, or of the study of historical mentalities. These things are extremely difficult to grasp but certainly worthwhile subjects of systematic investigation, i.e., of science (in the wide sense). Given this extreme disciplinary (and historic) variety, any criterion that demarcates science from nonscience, and from pseudoscience in particular, is bound to be very flexible (in positive terms) or vague (in negative terms) in order to do justice to this variety. However, the criterion should certainly not be so flexible as to be virtually empty. This is the very narrow path predetermined for any proposal of a demarcation criterion. This is a particular problem that any philosophical, general theories of the sciences (in the wide sense) has to confront: to deal in general terms with a subject of immense internal and historical variety without becoming vacuous.

6

Conclusion

THIS BOOK is intended to be a piece of systematic philosophy. It is systematic in two senses. First, "systematic philosophy" is the contrast expression to "history of philosophy." Whereas history of philosophy reconstructs positions, arguments, processes, and such of the past, systematic philosophy poses and discusses philosophical questions of current interest, answers them, supports the answer by arguments, and asks what the answer's consequences for other questions are. The philosophical question treated in this book is "What is science?" Chapter 1 discusses this question; chapter 2 answers this question; chapter 3 argues for this answer; chapter 4 further clarifies and further argues for this answer by comparing it with earlier answers; and chapter 5 discusses some consequences of the answer. So regarding its structure, this is a very simple philosophical book.

Second, this book is systematic in the sense of being systematically organized. It intends to exhibit itself what its subject matter is: it thereby presents itself as a piece of "scientific" philosophy in the sense of professional, academic philosophy. Compare this book's treatment of systematicity with the casual treatment of systematicity as the hallmark of science as it can be found in various sciences; I gave examples in section 2.1.1. Clearly, in the present book, the topic is much more systematically developed than in the cited examples. Without the intention to denigrate them, they represent laypeople's philosophy (laypeople with respect to philosophy), whereas this book represents professional philosophy.

However, as simple as the structure of this book is, there are nevertheless several issues that may hamper understanding this book, or that may appear as problematic aspects of the whole enterprise. Clearly, I cannot anticipate all such stumbling blocks or objections. In concluding this book, I will treat some of those that I am aware of.

First, in this book, the term "science" is used differently from what we are used to. By "we" I mean philosophers and the public alike. "Science," as it is used here, is intended to cover *all* research fields, not just the natural sciences, as I have stressed again and again. "Science" therefore also includes the formal sciences, the social sciences, and the humanities. One has to keep this semantic shift in mind when reading this book because one may easily be misled by the association of "science" with only the natural sciences.

Second, in this book, the question "What is science?" does not presuppose that there is something like the "nature" of science or an "essence" of science, and that the question aims at making explicit this nature or essence. Instead, it turns out that the various sciences and their specialties are so different from one another that it appears as absolutely hopeless to find substantial and universally valid characteristics of them that together might constitute the nature or the essence of science.

Third, in this book, the question "What is science?" is not understood as a question that primarily contrasts science with pseudoscience or metaphysics. This was the main contrast to science dominant in the philosophical tradition of almost a century. Instead, the contrast underlying the question "What is science?" as it is understood here is the contrast to other forms of knowledge, namely, to nonscientific knowledge, and there, primarily to everyday knowledge. This is a major shift in the way the fundamental question of the general philosophy of science is asked. It is important to keep this shift, together with the broadened use of "science," in mind in order to understand the thrust of this book.

Fourth, the central concept of this book, systematicity, is somewhat new, and it has, upon closer inspection, perhaps surprising properties. It cannot be seriously clarified on an abstract level, and it gets more or less diverging meanings when concretized in different contexts. These contexts are nine different dimensions or aspects of science and the huge variety of scientific disciplines, subdisciplines, and research fields. The more concrete, context-dependent concepts of systematicity are connected by family resemblance relations. These properties of the concept of systematicity imprint the same structure upon the network of all of the sciences: all of the different disciplines, subdisciplines, and research fields are connected by multidimensional family resemblance relations. On the one hand, this creates a tenuous sort of unity among all of the sciences. On the other hand, it makes intelligible why any search for an essence of science, common to all sciences and only to them, is bound to fail.

Fifth, the claim that it is the higher degree of systematicity that distinguishes the sciences from (especially) everyday knowledge, immediately elicits many apparent counterexamples, or at least that is my experience in many a discussion period after talks I have given. Some of these counterexamples were serious indeed, and I have tried to take care of them at various places in this book by, for instance, adding a new dimension. However, quite a few putative counterexamples turned out not to be counterexamples at all because they neglected or underestimated the strength of the fourth dimension of systematicity, the systematic defense of knowledge claims. This dimension presupposes that the entity in question is something that indeed embodies knowledge claims, and that these knowledge claims are defended more systematically than other knowledge claims about the same domain. Clearly, both conditions must be met, and this eliminates, for instance, even the most sophisticated forms of stamp collecting and all varieties of glass bead games as putative counterexamples. For the pseudosciences, the elimination procedure is more complicated because they have to be set in the context of an appropriate reference science or of neighboring sciences.

Sixth, a complication in this book arises from the descriptive character of its main thesis, i.e., the descriptive character of the answer to the question "What is science?" Any persuasive argument for a descriptive thesis must ultimately rely on some facts—as opposed to relying on norms. In our case, a descriptive thesis about science must rely on facts about science, as opposed to relying on norms about science, i.e., how science should or must be. As our thesis is very general, encompassing nine subtheses and covering all research fields, the argument for the thesis must take into account many facts about science in the wide sense and is very elaborate and difficult to oversee. This is why chapter 3 turned out so disproportionally long and without a nice, riveting narrative structure.

Finally, in spite of these masses of empirical material apparently supporting the main thesis, one may have doubts about whether the thesis is really empirical. Couldn't it be that the thesis is at heart semantic, i.e., the result of a reflection on the meaning of the term "science" (in the wide sense)? In this case, "systematicity" would be one of the defining features of "science." This hypothesis would presuppose a certain position in the philosophy of language regarding empirical concepts. This position assumes that the features of a given empirical object can be sorted into those that are defining features of the concept under which it falls, and those that the object possesses for accidental reasons. For instance, if John is a bachelor, his being unmarried is a defining feature of the concept of "bachelor," but his being twenty-four years old is not—bachelors come in all ages (above age eighteen). However, I do not subscribe to this position, because for many if not most empirical concepts, such a sorting of features appears hardly possible. Most empirical concepts are neither learned nor used by recourse to features that have been marked as definitional. Philosopher Thomas Kuhn's example is Johnny's learning from his father to distinguish between ducks, geese, and swans. The learning process involves pointing to examples of these waterfowl as well as explicit accentuating particular features of the different kinds. When Johnny has finished the learning process, he will pick out the same birds as ducks, geese, or swans exactly the same way as his father does, and we would describe this process as the learning of the three concepts. However, it is completely open which particular features Johnny uses to identify the members of the different kinds. Moreover, the result of this learning process is also open to Johnny's later replacing his initial criteria for identification with new ones of which he has empirically learned that they are invariably present in the respective kind. Thus, the mastery of an empirical concept may be realized by very different sets of criteria that pick out the concept's referents and nonreferents.

The upshot is the following. Although the *conceptual* difference between definitional features and nondefinitional, i.e., merely empirical, features of objects falling under a concept may be completely transparent, its *application* to a given empirical concept may be impossible. In stark contrast to mathematical concepts, most of which are introduced by explicit definitions, many empirical concepts, both in science and outside of science, are introduced and used in a way such that they do not come with a clear sorting of features into definitional and empirical ones. Of course, this does not exclude the possibility that in a given context, a decision is made which features are treated as definitional. For instance, if an author reviews the work about a certain subject X, she may begin her investigation by declaring her understanding of X, and she may do so by stating which features of X she takes to be definitional. This is, strictly speaking, the introduction of a new technical term that, although it bears strong similarities to the original term in which no features were marked as definitional, is different from the original term. Whether this new technical term will gain wide currency later on and possibly even replace the original term, is still another question.

I believe that in the case of "science," there simply is no clear-cut sorting of its characteristics into definitional (or "analytic") and empirical (or "synthetic") ones in our current language. This does not exclude that this situation may change in the future, i.e., that certain features of science are seen as so intimately connected with the concept of science that they may count as definitional. But at present, the question of whether systematicity belongs to the class of definitional or empirical features of science is not really a well-posed question. The question should thus be dismissed, not answered.

This concludes my review of some possibly problematic points of the systematicity theory. I am certain that there are many more that will come up in discussion. I am looking forward to them, because I see philosophy as an ongoing, open-ended dialogue.

NOTES

NOTES TO PREFACE

p. x, "**a specific attempt that laid out the scientific foundations of geography**": I am referring to the approach developed by the Swiss-Canadian geographer Hans Carol, published in Carol (1956).

p. x, "**Then, I proceeded to develop the answer**": The talk was later published as Hoyningen-Huene (1982).

p. x, "**there is nothing specific about science**": Feyerabend's main work on this topic is Feyerabend (1975). It appeared in two more editions in 1988 and 1993, which are all fairly different from each other. Furthermore, the German editions are all different from one another and different from all English editions.

p. xi, "**I consider this question to be central to the discipline**": I am certainly not alone in this assessment of the question "What is science?" This is, for instance, indicated by book titles of classics in the philosophy of science like *The Structure of Science* (Nagel 1961) or *What is This Thing Called Science?* (Chalmers 1999), or by the title of the introductory chapter of Alexander Bird's book *Philosophy of Science*, "The Nature of Science" (Bird 1998, 1).

p. xi, "**philosophers of science did not really take it up in the past decades**": As this claim about the general neglect of the question "What is science?" is a negative claim, it is impossible to document it conclusively. But here are a few symptomatic examples. In the above-mentioned introductory chapter "Introduction: The Nature of Science" to his book *Philosophy of Science*, Alexander Bird's begins with the sentence: "Our starting point is the question *What is science?*," and he entitles its first section "What is science?" (Bird 1998 1, 2). However, Bird does not attempt to answer this question systematically, that is in this context, in a unified way. Instead, he identifies several of the key issues that are the subject matter of his book (p. 1). Or consider Bas van Fraassen's recent book *The Empirical Stance* (van Fraassen 2002). There is a chapter entitled "What Is Science—and What Is It to Be Secular" that contains section one entitled "What is science?" in which van Fraassen then focuses on the question "What form does scientific inquiry characteristically take?" (p. 156). However, with respect to the original question "What is science?," he dampens down our expectations immediately afterward by continuing: "No pretense of a complete answer to the question, therefore! I am focusing here on *just one aspect* of *one half* of the question" (p. 156, my italics).

p. xi, "**this discipline was so busy in discovering special sciences and their disunity**": See, for example, the seminal article Fodor (1974). I shall come back to the question of the disunity of the sciences at the end of section 2.2.

p. xi, "**The title suggested to me was 'The nature of science'**": The address was later published in the conference proceedings as Hoyningen-Huene (2000a).

p. xi, "**In stark contrast to an earlier book of mine**": I am referring to Hoyningen-Huene (1993) that contains some 1,300 lovely footnotes.

p. xii, "**otorhinolaryngology, whose names I might have mistaken for scholarly names of exotic practices such as anthropophagy**": Otorhinolaryngology is the medical specialty dealing with ear, nose, and throat; "anthropophagy" sounds as scientific as "anthropology" but is the practice of cannibalism, the eating of human flesh by humans.

NOTES TO CHAPTER 1: INTRODUCTION

p. 1, "**The central question to be answered by a general philosophy of science is: What is science?**": Bas van Fraassen seems to agree on this point (van Fraassen 1989, 189).

p. 1, "**However, I should warn historians of science who read this**": If my presentation is really too crude, my overall argument, namely, my present-day answer to the question "What is science?" will still not be seriously impaired (my reputation as a writer of appropriate historical introductions to philosophical theses most probably excepted).

p. 1, "**In the first phase, starting around the times of Plato**": For a first historical overview see Losee (2001, chs. 1–3). For Aristotle, see especially McKirahan (1992).

p. 3, "**Formal logic as the theory of truth-transferring deduction**": A more common characterization of logic uses the term "truth-preserving" instead of "truth-transferring." Why I prefer "truth-transferring" is explained in Hoyningen-Huene (2004, 3).

p. 3, "**this ideal of scientific knowledge has been universally upheld**": It must be noted that this is only true when people wrote about science, but not necessarily when they practiced it. Many of Aristotle's writings contain scientific findings not converted into the form of an axiomatic system—but it is not entirely clear whether due to the preliminary character of his writings or that he did not practice what he preached.

p. 3, "**However, it is discontinuous regarding the means**": There are, of course, other important characteristics of modern science discontinuous with the older tradition, especially the use of mathematics and the new role of the experiment. I am not discussing these features here, because they are not relevant in the present context. Readers interested in these features may turn to Hoyningen-Huene (1989), for example.

p. 3, "**The most famous protagonists of this scientific method**": See, e.g., Losee (2001, chs. 7, 8).

p. 4, "**the conviction of the certainty of scientific knowledge already decays in the late nineteenth century**": This eminently important process is still waiting for in-depth historical research. The only study I am aware of is Schiemann (1997) and mainly concerns one of the key protagonists, Hermann von Helmholtz; an abridged English version has appeared as Schiemann (2009).

p. 4, "**the discovery of non-Euclidean geometries in the course of the nineteenth century is dramatic**": See, e.g., Kline (1980, 81–88).

p. 4, "**it is the so-called historicism of this period that stresses that all knowledge is historically bound and thus fallible**": For a detailed discussion see Iggers (1983) and Iggers (1984, ch. 1).

p. 5, "**that scientific methods with the characteristics posited in the second or third phases simply do not exist**": Paul Feyerabend (1924–1994) has gained worldwide attention with this thesis expressed in highly concentrated form in the title of and historically elaborated in his book *Against Method* (Feyerabend 1975). He also reports that in the early 1950s, Karl Popper started lectures with the following line: "I am a Professor of Scientific Method—but I have a problem: there is no scientific method." (Feyerabend 1995, 88). Thomas Kuhn's (1922–1996) classic *The Structure of Scientific Revolutions* (Kuhn 1962) also contains this thesis, together with an alternative. According to Kuhn, it is not abstract methods or rules that normally govern the course of science but certain substantial results previously achieved, so-called paradigms; for explication and detailed references, see Hoyningen-Huene (1993, chs. 3, 4). Further statements of the absence of a scientific method can be found, for instance, in Schuster (1990, 221); Bauer (1994); Bird (1998, ch. 8); Chalmers (1999): "[T]here is no general account of science and scientific method to be had that applies to all sciences at all historical stages in their development" (p. 247); Weinberg (2001): "I know enough about science to know that there is no such thing as a clear and universal 'scientific method'. All attempts to formulate one since the time of Francis Bacon have failed to capture the way that science and scientists actually work" (p. 43); Haack (2003, 9–10, 24–25, ch. 4, esp. p. 95 where more quotes to the same effect can be found). Jaegwon Kim summarizes the situation as follows: "But what is scientific method? Most contemporary naturalists are likely to wince, if not laugh, at the idea of there being some monolithic 'method' that characterizes all science everywhere" (Kim 2003, 94). This article is available on the Internet at http://www.pdcnet.org/pages/Products/electronic/pdf/7Kim.pdf (accessed Sept. 22, 2011).

p. 6, "**'[T]he strength of the thread does not reside in the fact that some one fiber runs through its whole length....'**": Wittgenstein (1958 [1953], § 67).

p. 7, "**In view of the fact that these discussions involved the best philosophers of science, that they sometimes lasted over several decades, and that the lack of success emerged independently in different areas**": Readers familiar with the work of Thomas Kuhn will easily recognize these features as similar to the ones he uses in order to identify "significant anomalies" in the history of the basic natural sciences: see, e.g., Hoyningen-Huene (1993, 225–26). The effect of significant anomalies is the same in both contexts: they shed doubt upon their presuppositions.

p. 8, "**Other authors pursuing studies of a similar breadth and being confronted with the same difficulty have also resorted to the very broad usage of the term 'science'**": See, for instance, Szostak (2004, 3 note 4).

p. 10, "**Popper saw the quest for a demarcation criterion as one of the most fundamental problem of epistemology**": See Popper (1959 [1934], §§ 4–6).

p. 10, "**many writers followed Popper at least in the sense that they have understood the question, 'What is science?' as a question that aims at demarcating science from pseudoscience**": Two examples must stand for many. At the very beginning of the Introduction to his book *Philosophy of Science*, Alexander Bird announces that his starting point is the question *What is science?* (Bird 1998, 1). He then immediately turns to the question of whether what is called "creation science," i.e., a purported pseudoscience, really is science—thus making the contrast implicit in his understanding of the question "What is science?" apparent. Similarly, in his book *Understanding Philosophy of Science*, James Ladyman claims that an answer to the question "What is science?" is perhaps the most fundamental task of philosophy of science (Ladyman 2002, 4). He then goes on to say that the "problem of saying what is scientific and what is not is called the demarcation problem."

p. 10, "**it will not guide our investigations from the beginning**": For a similar attitude concerning the demarcation criterion, see Haack (2003, 115–16). Interestingly, also Ernest Nagel begins his 1961 classic *The Structure of Science* with an introductory chapter on the contrast between science and common sense (Nagel 1961). In this comprehensive presentation of the philosophy of science of its time, Popper is only cited twice in footnotes referring to a single page of the 1935 German version of his *Logic of Scientific Discovery* (p. 37 note 9, p. 38 note 10), but not with reference to his demarcation criterion.

p. 12, "**The development of a fusion reactor for energy production**": For comprehensive information about the current stage of this development, see the homepage of ITER: http://www.iter.org/ (accessed Sept. 29, 2011). The project's aim is concisely described in the following way: ITER (the International Thermonuclear Experimental Reactor) represents "an experimental step between today's studies of plasma physics and tomorrow's electricity-producing fusion power plants." (http://iter.rma.ac.be/en/iterproject/WhyITER/index.php, accessed Sept. 22, 2011).

p. 12, "**It is concerned with 'planning, designing, constructing and managing earthquake-resistant structures and facilities.'**": This definition is taken from Bertero (1997), available at http://nisee.berkeley.edu/bertero/index.html (accessed Sept. 22, 2011).

p. 12, "**the experimental study of the seismic behavior of certain types of assemblages**": See, for example, Bertero (1997) at http://nisee.berkeley.edu/bertero/html/research_and_development_needs.html (accessed Sept. 22, 2011)

p. 12, "**'[r]esearch alone is not enough; analytical and experimental studies must be augmented by development work.'**": See Bertero (1997) at http://nisee.berkeley.edu/bertero/html/research_and_development_needs.html (accessed Sept. 22, 2011).

p. 12, "**the seismic response of these buildings under severe earthquake ground motions has been an important source of data**": See Bertero (1997) at http://nisee.berkeley.edu/bertero/html/recent_developments_in_seismic_design_and_construction.html (accessed Sept. 22, 2011).

p. 12, "**there is a 'science of chocolate'**": For a book-length treatment see Beckett (2000); for an easily accessible introduction to this topic, see http://en.wikipedia.org/wiki/Chocolate, accessed Sept. 22, 2011); for a sample of recent articles on chocolate research from the Journals of the American Chemical Society, see http://acselementsofchocolate.typepad.com/elements_of_chocolate/ChocolateResearch.html (accessed Sept. 22, 2011).

p. 12, "**typically explored at Departments of Food Sciences at research Universities**": See, e.g., https://www.rdb.ethz.ch/projects/project.php?proj_id=5830&type=search&z_detailed=1&z_popular=1&z_keywords=1 (accessed Sept. 22, 2011).

p. 12, "**research conducted by the research and development departments of the large chocolate manufacturing companies**": See, e.g., the "Mission Statement" of the Hershey Center for Health & Nutrition: it "investigates and promotes the chemistry and health benefits of cocoa, chocolate, nuts and other ingredients. The results of these investigations guide new products and product development for The Hershey Company." This mission statement can be found at http://www.hersheys.com/nutrition-professionals/mission-statement.aspx (accessed Sept. 22, 2011).

NOTES TO CHAPTER 2: THE MAIN THESIS

p. 14, "**In the last century, however, an early articulation of the idea that science is characterized by systematicity**": Even earlier than Dewey's is the quote by George Henry Lewes (1817–1878): "Science is the systematic classification of experience." However, I could not find the source of this quote that is widely cited on the Internet. Only after I had finished the manuscript of this book, I discovered in Schnädelbach (2012, 18, 22), a much earlier source in which science was characterized by its "systematic spirit." This characterization is due to the French mathematician, physicist, and encyclopedist Jean le Rond d'Alembert (1717–1783), who contrasted this scientific spirit with the philosophical spirit described by him as the "spirit of systems." The French originals "*esprit systématique*" and "*esprit de système*" are contained in his 1751 introduction to the famous encyclopedia that he edited together with Denis Diderot. The French original of this "Discours Préliminaire" is available at http://fr.wikisource.org/wiki/Discours_pr%C3%A9liminaire_de_l%E2%80%99Encyclop%C3%A9die (accessed Nov. 26, 2012); an English translation is d'Alambert (1995 [1751]), which is available at http://quod.lib.umich.edu/cgi/t/text/text-idx?c=did;cc=did;rgn=main;view=text;idno=did2222.0001.083 (accessed Nov. 26, 2012). Later, in my discussion of Nicholas Rescher's approach to systematicity in section 4.8.2, I shall demarcate his project from mine by the essentially identical contrast between "systematicity" as derived from "systematic" (corresponding to the "systematic spirit") and "systematicity" as derived from "system" (corresponding to the "spirit of systems").

p. 14, "**he wrote an article entitled 'Logical Conditions of a Scientific Treatment of Morality.'**": See Dewey (1977 [1903]).

p. 15, "**Dewey contends that the notion that science is a body of systematized knowledge....**": I do not know which contexts Dewey refers to with this assertion. However, this phrase may also be due to a stylistic particularity of many philosophical works, namely to start a philosophical work with a fairly general statement that the author believes to be widely accepted, and to implicitly urge the reader's acceptance as well. See, for instance, the first sentence (or sentences) of Aristotle's *Metaphysics*, of Descartes' *Discourse on Method*, of Kant's *Critique of Pure Reason* (Introduction), of Hegel's *Phenomenology of Spirit* (Introduction), of Frege's *Foundations of Arithmetic*, of Wittgenstein's *Philosophical Investigations*, and many more.

p. 15, "**an influential book entitled *An Introduction to Logic and Scientific Method***": Cohen and Nagel (1934). The quotes a little further down in the main text are on p. 394.

p. 15, "**he published a booklet entitled *The Study of the History of Science***": Sarton (1936). The "definition" I am quoting a little further down in the main text is on p. 4. I wish to thank Helmut Heit for bringing this quote to my attention. People interested in Sarton should consult Garfield (1985).

p. 16, "**a short article entitled 'On the History of the International Encyclopedia of Unified Science'**": Morris (1960).

p. 16, "**About the plan for sections 2 and 3, Morris wrote**": Morris (1960, 518).

p. 16, "**section 2 of the *Encyclopedia* is very similarly described**": Morris (1960, 519).

p. 16, "**Volumes III–VIII will especially stress the controversial differences . . . .**": Morris (1960, 519–20).

p. 16, "**Hempel uses the terms 'deductive systematization' and 'inductive systematization'**": See Hempel (1958, sec. 1).

p. 17: "**scientific explanation, prediction, and postdiction all have the same logical character**": The cited passage is on p. 174 of the reprint of Hempel (1958).

p. 17, "'*systematic* **connections among empirical facts**'": reprint of Hempel (1958, 177, my emphasis).

p. 17, "'**all scientific explanation … seeks to provide a *systematic* understanding.…**'": Hempel (1965a, 488, my emphasis).

p. 17, "**Hempel reports a widely held conception of science**": See Hempel (1983, 91, my emphasis).

p. 17, "**Similar to Hempel was Ernest Nagel**": The quotes are on pp. 4 and 5, respectively, of Nagel (1961); the italics are mine.

p. 18, "**because Nagel's book mainly deals with questions of scientific explanation**": In Nagel's view, probably a view widely held in the early 1960s, the philosophy of science deals with three broad areas: "the logical patterns exhibited by explanations in the sciences; the construction of scientific concepts; and the validation of scientific conclusions" (p. 14; see also pp. viii–ix).

p. 18, "**the only philosopher who extensively considered systematicity and its relationship with science in the last one hundred years is Nicholas Rescher**": This conclusion is also supported by Rescher. He notes in 1979 that despite the importance of questions concerning cognitive systematization, "it appears that no work published in the present century affords any substantial treatment of these matters" (Rescher 1979, 2).

p. 18, "**he published a book entitled *Cognitive Systematization: A Systems-Theoretic Approach to a Coherentist Theory of Knowledge***": Rescher (1979) and Rescher (2005). For better or worse, I found these books only when my project was already quite advanced. Neither had I detected them on the Internet, nor did I receive a hint when I asked the HOPOS-L list (History of Philosophy of Science list) in April 2004 the following question: "Does anyone know of other authors, philosophers or scientists, who somehow connect systematicity with being scientific (I know of Hempel, for instance, about his deductive and inductive systematizations)?" This request was distributed to 733 members of the list.

p. 18, "**Alan Sokal and Jean Bricmont 'stress the methodological continuity between scientific knowledge and everyday knowledge.'**": This quote is on p. 56, note 56 of Sokal and Bricmont (1998); the following quote is on the same page (see also Sokal 2008, 178). The philosopher Willard Van Orman Quine expressed something very similar: "The scientist is indistinguishable from the common man in his sense of evidence, except that the scientist is more careful. This increased care is not a revision of evidential standards, but only the more patient and systematic collection and use of what anyone would deem to be evidence" (Quine 1966, 233).

p. 18, "**Amir Alexander, the historian of science—and of mathematics in particular—describes the result of the 'Rebirth of Mathematics' in the early nineteenth century as follows**": Alexander (2006, 721 and 723; my emphasis).

p. 19, "**Mental Models of the Earth: A Study of Conceptual Change in Childhood**": The reference is Vosniadou and Brewer (1992). The following two quotes are from p. 537; the emphasis is mine.

p. 19, "**In his perceptive introduction to the methodology of history entitled *The Pursuit of History***": Tosh (2006); the quotations are from p. 110; the emphasis is mine.

p. 20, "*Introductory Lecture to Pictorial Semiotics* **by Göran Sonesson**": See Sonesson (2004, esp. 34).

p. 20, "***The Encyclopedia of Philosophy* of 1967 characterizes theology in the following way**": See Alston (1967, 287).

p. 20, "**The systematicity of science or of some parts or aspects of it is somehow taken for granted**": Here are a few more examples. In an introduction to science and technology studies, Sergio Sismondo describes the "common view" of science as follows: "Science's method is a set of procedures and approaches that makes research systematic" (Sismondo 2004, 1). The first sentence of the Introduction of Richard Whitley's book *The Intellectual and Social Organization of the Sciences* begins: "Formal knowledge produced through systematic enquiry, and disseminated largely through publication in scientific and technological journals, is increasingly being seen as an economic resource …." (Whitley 2000, ix). In an article that seeks to distinguish historical and experimental science, Carol Cleland sketches the activities of experimental scientist as follows: "They are engaging in systematic, extended experimentation that … is aimed at … minimizing the very real possibility of misleading confirmations and disconfirmations in concrete laboratory settings" (Cleland 2002, 478). In her discussion about what science might be, Sandra Harding first contemplates the possibility that science is a set of sentences. Then she asks: "Or is science more accurately and usefully conceptualized as effective systematic interactions with the world?" (Harding 2003, 57). In his *Understanding Philosophy of Science*, James Ladyman characterizes Francis Bacon as prophetic because of "his vision of science as a systematic and collaborative effort … to produce knowledge" (Ladyman 2002, 18). In her *Introduction to the Philosophy of Science*, Lisa Bortolotti implicitly characterizes a scientist as someone who is "systematically engaged in the empirical investigation of nature" (Bortolotti 2008, 2). And so on.

p. 21, "**I do not mean to imply that there is a *quantitative* measure of systematicity**": In other words, I am using a *comparative* notion of systematicity and not a quantitative one. The former allows for nonquantitative comparisons ("*x* is more systematic than *y*"), whereas the latter would claim to quantify systematicity (e.g., "the systematicity of *x* is 3.9"). Of course, a quantitative notion of systematicity allows for comparisons whereas a comparative notion does not immediately imply the existence of a quantitative one; further conditions must be met. On this topic, see Carnap (1966, chs. 5, 6). In social research, the terminology is different. What is called "comparative" here is called there "ordinal," and what is called "quantitative" here is distinguished there into "interval" and "ratio," respectively. See, e.g., http://www.socialresearchmethods.net/kb/measlevl.htm (accessed Sept. 22, 2011). I shall come back to the comparative character of my thesis in the third remark.

p. 21, "**in the sense of a 'body of … belief that is well-established, ….'**": The quotes are from Bird (2004, 345–46) and concern Kuhn's and other writers' use of the expression "scientific knowledge."

p. 21, "**Knowledge is then understood as a particular kind of belief, namely (roughly), as belief that is true and for whose truth a particular warrant exists**": For an introduction into the philosophical intricacies of the concept of knowledge, see, e.g., Klein (1999).

p. 22, "**the knowledge that is often referred to as 'local' or 'traditional' knowledge**": For an introduction to this topic and further literature, see the proceedings of the 1999 World

Conference on Science, in which a thematic meeting took place on "Science and other systems of knowledge" (Cetto 2000, 432–44) (this publication is also available on the Internet: http://unesdoc.unesco.org/images/0012/001207/120706e.pdf (accessed Sept. 22, 2011). See also the 2002 report by the ICSU Study Group on Science and Traditional Knowledge: "Science and Traditional Knowledge," available on the Internet at http://www.icsu.org/publications/reports-and-reviews/science-traditional-knowledge (accessed Sept. 22, 2011); ICSU stands for "International Council for Science" (its former name was "International Council for Scientific Unions," hence its acronym).

p. 22, "**a smooth transition between prescientific (or nonscientific) knowledge and scientific knowledge**": Whether the transition from prescience to science (wherever this transition is put precisely) is rather smooth or rather discontinuous constitutes a somewhat controversial question. Today's historians of science tend to ascribe more continuity to the transition than earlier historians and philosophers: see, e.g., Kuhn (1970 [1962], 2–3); Popper (1959 [1934], 18, Preface to the first English edition 1959). For philosophers who stress discontinuity, see, e.g., Bachelard and Lecourt (1971, sec. I, B). It seems that one's position regarding this issue is influenced by historical information as well as by one's philosophical position. Hence, if the position developed in this book is plausible, it also provides an argument for continuity.

p. 23, "**personality assessment center**": See, for example, the article "personality assessment" in the *Encyclopædia Britannica Online* http://www.britannica.com/EBchecked/topic/453022/personality-assessment (accessed Sept. 22, 2011).

p. 23, "**the Violent Crime Linkage Analysis System (ViCLAS)**": Information on ViCLAS can be found on its homepage at http://www.rcmp-grc.gc.ca/tops-opst/viclas-salvac-eng.htm (accessed Sept. 22, 2011).

p. 24, "**At first, black holes were theoretical predictions of the general theory of relativity**": For easily accessible information on black holes see, e.g., http://blackholes.stardate.org/ (accessed Sept. 22, 2011).

p. 25, "**Anglo-Saxon philosophical dictionaries and encyclopedias do not feature 'system'**": See, for instance, Edwards (1967), Audi (1995), and Craig (1998). However, Rescher has sketched the history of the term "system" from the ancient Greeks on (Rescher 1979, 5–8). He draws on the German sources Ritschl (1906), Messer (1907), and Stein (1968).

p. 25, "**German philosophical reference works, however, always feature 'system'**": See, for instance, Krings, Baumgartner, and Wild (1973, vol. 5, 1458–75; Mittelstraß (1980, vol. 4, 183–94); Sandkühler (1999, vol. 2, 1576–88).

p. 25, "**its somewhat special use in cognitive science**": Searching for "systematicity" on the Internet resulted in approximately 146,000 hits in August 2011, most of them relating to cognitive science or philosophy of mind. There is even a book that features "systematicity" in its title (Aizawa 2003). For a short introduction to its use and its function in this context, see Aizawa's entry "systematicity" in the *Dictionary of Philosophy of Mind* on the Internet, http://philosophy.uwaterloo.ca/MindDict/systematicity.html (accessed Sept. 22, 2011).

p. 26, "**nor the contrasting terms have enhanced our understanding of the concept of systematicity substantially**": It may be objected that not all items on the list of supposedly contrasting terms are really contrasting terms, because, for example, "not unmethodical" is the same as "methodical," which is a positive characterization. However, I believe that this analysis of double negation in English is not really adequate. To describe some procedure as "not unmethodical" is less than calling it methodical. "Not unmethodical" claims only that the procedure is not *entirely* chaotic, but qualifying it as methodical is a more positive statement, implicitly claiming that there really is *one* method that has been used and that can be stated explicitly. Interestingly enough, other authors also choose this way of explicating the meaning of "systematic" or "system" by way of contrast with contraries: for Kant, see Kant and Smith (2003 [1781/1787], A645/B673 and A832/B860); for Lambert, see Rescher (1979, 10).

p. 28, "**a family resemblance relation between these different concepts in Wittgenstein's sense**": Wittgenstein introduced this notion in his famous *Philosophical Investigations* (Wittgenstein (1958 [1953]), §§ 65–71); for a lucid exposition see Baker and Hacker (1984 [1980]-b, ch. X, esp. sec. 3) and Baker and Hacker (1984 [1980]-a, esp. 130–45).

p. 29, "**the idea of a disunity of science has been much more popular**": Witnesses for this tendency are, for instance, Fodor (1974), Dupré (1983), Macdonald (1986), Carrier (1991), Dupré (1993), and Galison and Stump (1996). However, Dupré (1993) uses an argument similar to mine in order to argue for *disunity*, whereas I argue along the same lines for a *unity* of sorts (see a few lines below for more details about the difference of Dupré's argument and mine). I think that this is not a real contradiction but rather a consequence of different perspectives. Presupposing a strong concept of unity of science (for example, logical reducibility of all sciences to physics), any relation among the sciences that is weaker than logical reducibility results in disunity of science. Without the contrast to a strong unity of science, however, focusing on the unifying aspects of that weaker relation may establish a weaker form of unity of science. In a review of Dupré's book, the biologist Gunther Stent criticizes Dupré's disunity thesis for exactly this reason (Stent 1994, 498); see also Grantham (2004, 136).

p. 29, "**unity of science that was characteristic for the positivist phase**": See, for instance, Oppenheim and Putnam (1958). For a concise introduction to the history and philosophy of the topic, see Cat (1999).

p. 29, "**All of the sciences are united by relations of family resemblance only**": In the literature, I have found several places where other authors have claimed, or at least entertained, the idea of the family resemblance relation among the sciences: Putnam (1994 [1987]), Putnam (1994 [1990]), Dupré (1993), Keil (1996), Kroes (2002, 2nd ed.), and Okasha (2002). The specificity of my approach, in comparison to these authors, derives from the particular argument I am using in order to argue the family resemblance relation. In the form of a rhetorical question, Putnam only very suggestively exclaims: "I have to ask why on earth we should expect the sciences to have more than a family resemblance to one another? ... [T]here is no set of 'essential' properties that all the sciences have in common" (Putnam (1994 [1987], 471–72). At another place, Putnam states again without argument: "Ludwig Wittgenstein taught us that not all concepts have 'necessary and sufficient conditions.' For many concepts, we have only paradigm cases, and more than one paradigm case at that. I believe that 'empirical science' is a concept of this sort" (Putnam (1994 [1990], 481). Dupré argues against any stronger idea of a unity of science and for a family resemblance relation among the diverse sciences on a metaphysical basis. The universe is just too disordered to admit a science that "constitutes, or could ever come up to constitute, a single, unified project" (Dupré 1993, 1). Instead: "My suggestion that science should be seen as a family resemblance concept seems to imply ... that no strong version of scientific unity of the kind advocated by the classical reductionists can be sustained" (Dupré 1993, 242). Keil bases his diagnosis of family resemblance among the sciences (in the wide sense, including the social sciences, arts and humanities) upon the observation that one looks in vain for methods that are universally valid for all of the sciences and for all times (Keil 1996, 35–38, 43–47). However, he does not simply exploit the possibly existing methodological gap between the natural sciences and the rest, but claims that the methodological diversity holds even within the realm of the natural sciences (p. 41). Keil's argument is similar to Feyerabend's (Feyerabend 1975), namely, to present counterexamples to any method supposedly universal and supposedly constitutive of science, but his conclusion is different from Feyerabend's. Although there are no features that are necessary and sufficient for being a science, the sciences are not totally disconnected from each other (pp. 43, 51). Rather, their unity is given by Wittgensteinian family resemblance with respect to methods. By contrast, my argument as developed in the main text proceeds from the observation that the hallmark of scientific knowledge, in comparison to other kinds of knowledge, is an increase of systematicity. Because systematicity is a family resemblance concept and because different aspects of systematicity are exploited by different sciences, the family resemblance relation is transmitted to the sciences. This view is, it seems to me, different from but compatible with the views of both Dupré and Keil. Okasha only entertains the idea of a family resemblance relation among the sciences without committing himself to it (Okasha 2002, 16–17). With respect to the varieties of analytic philosophy, Glock has used the family resemblance concept in a similar way (Glock 2007, 234–35).

p. 30, "**there is no established academic discipline nowadays that informs us about the landscape of all the sciences**": In the past, there were at least several attempts at something that could be called a cartography of the sciences: for Germany, see, e.g., Eschenburg (1792); for the United States, see, e.g., Book II of Vol. I of Peirce et al. (1965) entitled "The Classification of the Sciences."

p. 31, "**The Thomson Reuters Company that, among other things, composes indices**": Information about these indices can be found on the webpage at http://thomsonreuters.com/products_services/science/science_products/a-z/isI_web_of_knowledge? (accessed Sept. 22, 2011).

p. 31, "**Following Thomson Reuters classification of scientific subjects**": The classifications that Thomson Reuters' citation indices use can be found on the following web pages (accessed Sept. 22, 2011):
Science: http://science.thomsonreuters.com/mjl/scope/scope_sci/
Social Science: http://science.thomsonreuters.com/mjl/scope/scope_ssci/
CompuMath: http://science.thomsonreuters.com/mjl/scope/scope_cmci/
Arts & Humanities: http://science.thomsonreuters.com/mjl/scope/scope_ahci/

p. 31, "**this classification is far from unique**": If one looks at indices other than Thomas Reuters, one gets different results. For instance, in December 2003, the Taylor & Francis Group published 961 journals, which are classified in twenty-one main subject categories, ranging from "Agricultural and Biological Sciences" to "Social Science." These categories are resolved into 168 subjects (see their web page, "Journals by Subject"). However, this list probably does not aim at completeness because it is supposed to cover only the products of this publishing group. The *Deutscher Hochschulverband* (German Association of University Professors) has a list of disciplines comprising 411 items. As this list is for internal use only (whom to send which job advertisements), it is not available on the web. Yet another list of "disciplinary domains" can be found in Kürschner (2003) and contains 624 items. The corresponding list of the Alexander von Humboldt Foundation features roughly 1,100 "research areas" (download at the URL http://www.humboldt-foundation.de/pls/web/docs/F22402/application_package_E.zip the set of files, which contains the file, research_area_index.pdf, accessed Nov. 26, 2012).

p. 31, "**all issues subsumed under the rubric of law by the *Social Science Citation Index***": See http://science.thomsonreuters.com/mjl/scope/scope_ssci/ (accessed Sept. 22, 2011).

p. 32, "**one source in which these smaller units are termed 'fields' counts 8,530 such fields**": See Crane and Small (1992, 197). The article also contains a useful discussion of the concept of a scientific discipline and its ambiguity (pp. 198–201).

p. 32, "**disciplines that lack a more or less general consensus about their basics**": Three examples must stand for many: for psychology, see, e.g., Driver-Linn (2003), who describes this discipline as exhibiting "structural fault lines"; for literary theory, see, e.g., Culler (1997, "Appendix: Theoretical Schools and Movements"); for semiotics, see, e.g., Sonesson (2004, esp. 28–40) and Chandler (2004, 5ff). There is a somewhat different version of the latter book, available on the Internet under the title *Semiotics for Beginners*: http://www.aber.ac.uk/media/Documents/S4B/semiotic.html (accessed Sept. 22, 2011); see especially the introduction at http://www.aber.ac.uk/media/Documents/S4B/sem01.html (accessed Sept. 22, 2011).

p. 32, "**I should constrain my thesis about the higher degree of systematicity in controversial research areas to each individual school**": It must be noted that the identification of such schools is far from being unproblematic. What appears from the outside as one school—for instance "psychoanalysis"—may be from the inside a conglomerate of different movements in deep controversy.

p. 33, "**All I can hope to attain is some plausibility for the thesis**": I was made aware of the difficulties to argue for essentially statistical hypotheses in philosophy of science by the work of Faust and Meehl, especially Faust and Meehl (1992), Meehl (1992), and more recently Faust and Meehl (2002).

NOTES TO CHAPTER 3: THE SYSTEMATICITY OF SCIENCE UNFOLDED

p. 35, "**The whole of science is nothing more than a refinement of everyday thinking**": The quote is from Einstein (1982 [1936], 290). Statements similar to Einstein's abound in the literature (if often only as incidental remarks). Ernest Nagel even relates the idea of the refinement of common sense by science to systematicity: "The sciences thus introduce refinements into ordinary conceptions by the very process of exhibiting the systematic connections of propositions about matters of common knowledge" (Nagel 1961, 5–6). Arthur Danto takes this position to be one of the tenets of naturalism: "Science reflects while it refines upon the very methods primitively exemplified in common life and practice" (Danto 1967, 449). When Nancy Nersessian describes her view on "cognitive-historical analysis," she writes: "The underlying presupposition is that the problem-solving strategies scientists have invented and the representational practices they have developed over the course of the history of science are *very sophisticated and refined outgrowths of ordinary reasoning and representational processes*" (Nersessian 1992, 5; in the original text, the whole sentence is in italics). Arthur Fine writes, using our Einstein quote explicitly: "[N]o distinctive mode of thought goes into [science's] making.... Perhaps the first false step in this whole area is the notion that science is special and scientific thinking is unlike any other" (Fine 1998, 19). See also Haack (2003, 9–10).

p. 35, "**that Einstein had a vision in mind that is similar to the one developed here**": In order to avoid possible misunderstandings, the following argument for the higher degree of the systematicity of science is, of course, independent of agreement, or lack of agreement, with Einstein's vision. I am not building my case on Einstein's authority.

p. 35, "**in another of his articles that was published in 1944**": Einstein (1982 [1944]). The quote is on p. 23, my italics. This article is available on the Internet at http://evans-experientialism. freewebspace.com/einstein_russell.htm (accessed Sept. 22, 2011).

p. 36, "**historical natural sciences like cosmology, paleontology, paleoclimatology, or paleoceanography**": Whereas the first two disciplines, dealing with the history of the universe and the history of life on earth, are fairly well known to an educated public, the latter two are less familiar, dealing with the history of Earth's climate and oceans. For information on these disciplines consult, e.g., http://eesc.columbia.edu/disciplines/paleoclimatology-paleoceanography (accessed Nov. 26, 2012).

p. 37, "**Later I realized that science was more systematic than other knowledge-seeking enterprises in more dimensions than just these three**": Although it is unnecessary for the systematic presentation of this project, in the course of this chapter I shall make a few remarks on how I found the dimensions I am using. This may reduce possible misunderstanding by clarifying my intentions connected with each of the dimensions (this is at least what people told me who urged me to include such a genetic sketch). I shall relate these stories in the Notes with reference to the section titles.

p. 39, "**a (ontological) difference between reproducible events or processes as pertaining to the natural sciences, and singular events or processes**": See, e.g., Sarton (1936, 9): "Physical sciences deal with the "laws of nature," with the *repetition of facts* under given circumstances … ; history deals with isolated facts of the past, *facts that cannot be repeated*" (my italics).

p. 39, "**drafting of the Declaration of Independence**": See, for instance, http://www.archives.gov/national_archives_Experience/declaration.html (accessed Sept. 22, 2011).

p. 41, "**for the modern treatment of logic**": See, for instance, Hoyningen-Huene (2004) (it is, of course, a pure accident that I am just picking out this work—there are many more!).

p. 41, "**The axioms must be logically independent of one another, …**": For this characterization of a set of axioms see, e.g., Corry (2004, 95–97).

p. 41, "**extremely systematic, where 'systematic' in this context just means 'in the form of an axiomatic system'**": This usage of "systematic" is quite often found in the literature. See, for example, section 4.3, where Kant's notion of systematicity is discussed.

p. 42, "**Classification organizes an assortment of individual items**": There is an enormous literature on classification, both on a more abstract level and on the level of particular sciences. Because of the obvious and staggering diversity of living beings, questions of classification have been discussed throughout biology's history; for an extensive historical overview see, e.g., Mayr (1982, pt. I). More recently, Mayr and Brock have emphasized "that not all ordering systems are classifications, as is all too frequently assumed by both scientists and philosophers" (Mayr and Brock 2002, 170; consult this article also for a discussion of the other ordering systems relevant for biology). However, the conclusion of this section is not affected by this variety of ordering systems, because it is obvious that these alternative ordering systems are also much more systematic than what we use in our everyday practices.

p. 42, "**Items considered for classification or taxonomy in the sciences are extremely diverse**": Here are a few examples of classifications and nomenclature that are easily accessible on the Internet. For viruses, see the web page of the International Committee on Taxonomy of Viruses, http://www.ictvonline.org/ (accessed Sept. 22, 2011). For the nomenclature of human genes, see http://www.genenames.org/guidelines.html (accessed Sept. 26, 2011). For a classification of proteins by patterns of tertiary structure, see http://kinemage.biochem.duke.edu/teaching/anatax/html/anatax.3a2.html (accessed Sept. 26, 2011). For the staggering number of different systems of nomenclature in different branches of chemistry and their principles, see the respective web page of the International Union of Pure and Applied Chemistry at http://www.chem.qmul.ac.uk/iupac/ (accessed Sept. 22, 2011). Although there are only ninety-two stable chemical elements, the number of known compounds is in the order of ten million! For enzymes, see the web page of the Nomenclature Committee of the International Union of Biochemistry and Molecular Biology at http://www.chem.qmul.ac.uk/iubmb/enzyme/ (accessed Sept. 22, 2011). For the World Health Organization's International Statistical Classification of Diseases and Related Health Problems, see http://www.who.int/classifications/apps/icd/icd10online/ (accessed Sept. 22, 2011). For an example of the classification and nomenclature of a particular group of diseases, see Fardon and Milette (2001) (this paper is available on the Internet at http://www.asnr.org/spine_nomenclature/ (accessed Nov. 26, 2012). For the classification of nursing diagnoses, see Nanda (2004). For the classification of finite simple groups, see, e.g., Wilson (2009) (the preface and the table of contents of this book are available on the Internet at http://www.springerlink.com/content/k61214/front-matter.pdf (accessed Sept. 26, 2011); a more popular presentation for laypeople is Elwes (2006), accessible on the Internet at http://plus.maths.org/content/os/issue41/features/elwes/index (accessed Sept. 27, 2011). Note, however, that the complete classification itself is not contained in either of these publications. It is spread over five hundred or so articles covering more than ten thousand pages written by more than one hundred different authors. Many more examples of classifications, taxonomies, and/or nomenclatures can be retrieved from the Internet by searching for these terms.

p. 42, "**in scientific practice, a number of severe difficulties may arise**": For the problems that arise in biology see, e.g., Ruse (1988, ch. 6); Ereshefsky (1994); Ereshefsky (2002).

p. 42, "**In its earliest full form for plants, published by Linnaeus in 1751**": The book bears the title *Philosophia Botanica*; my reference is Jahn, Löther, and Senglaub (1982, 278).

p. 43, "**above the genus *Bos*, there are no less than twenty-six hierarchical levels**": This information is available, for instance, on the Taxonomy Browser of the National Center for Biotechnology Information, at http://www.ncbi.nlm.nih.gov/Taxonomy/ (accessed Sept. 27, 2011). Another classification counts even thirty-one levels above the genus level of Bos: http://sn2000.taxonomy.nl/. Still another classification project is the "Tree of Life Web Project," available at http://www.tolweb.org/tree/ (accessed Sept. 27, 2011).

p. 43, "**Geographic Names Information System (GNIS)**": See http://gnis.usgs.gov/ (accessed Sept. 27, 2011).

p. 43, "**planetary nomenclature**": See the respective web page of the USGS Astrogeology research program, at http://planetarynames.wr.usgs.gov/ (accessed Sept. 27, 2011).

p. 43, "**In 2009, the authoritative system counted 6,909 living languages**": The information is taken from the "Ethnologue: Languages of the World," available on the web at http://www.ethnologue.com/web.asp (accessed Sept. 27, 2011). The data there are taken from Lewis (2009).

p. 43, "**the title *Systema naturae***": For obvious reasons, this expression is still used today. For instance, the website http://sn2000.taxonomy.nl, mentioned above, bears the title *Systema Naturae 2000*. It aims at a classification of life.

p. 44, "**paleontology, the discipline that deals with the history of life on Earth**": For some information on paleontology, see, e.g., the website of the Museum of Paleontology of the University of California at Berkeley at http://www.ucmp.berkeley.edu/paleo/paleowhat.html (accessed Sept. 27, 2011).

p. 44, "**the division of ancient Egypt history**": See, e.g., Kitchen (1991). This article is available on the Internet at http://hbar.phys.msu.ru/gorm/dating/chroneg.pdf (accessed Nov. 26, 2012).

p. 44, "**the periodization of Earth's history, the so-called geologic time scale**": See, for instance, http://www.geosociety.org/science/timescale/timescl.pdf (accessed Sept. 27, 2011).

p. 44, "**in developmental psychology, a variety of different life span theories have been proposed**": See, for instance, http://www.angelfire.com/life.lab/lifespan.html (accessed Sept. 27, 2011)

p. 44, "**instead of two major transitions, one should have only one, located roughly at the so-called saddle time around the turn to the nineteenth century**": See Koselleck (1972, XV); Zammito (2004, esp. 126ff).

p. 45, "**an increasing tendency toward quantification in many areas of scientific research**": See on this topic, e.g., Carnap (1966, ch. 11). I shall come back in later sections to some of the issues that Carnap treats.

p. 45, "**under which conditions the transition from the qualitative use of a concept to its quantitative use**": On this question and the difference between the qualitative and the quantitative use of concepts, see, e.g., Carnap (1966, ch. 5). On the somewhat different terminology in the social sciences regarding this topic, see http://www.socialresearchmethods.net/kb/measlevl.htm (accessed Sept. 27, 2011).

p. 46, "**in ancient times, especially in astronomy**": See, e.g., Kuhn (1977 [1976], 37); for the bigger picture, see Lindberg (1992, ch. 5): The Mathematical Sciences in Antiquity.

p. 46, "**The quantitative treatment of local motion started in the fourteenth century**": See, e.g., Schuster (1990, 226).

p. 46, "**Chemistry had only begun to become a quantitative science in the late eighteenth century**": See, e.g., Brock (1993, 117).

p. 46, "**Many more disciplines followed this path**": The topic of the increase of quantification, which is a special case of the more general phenomenon of the mathematization of the sciences, is a vast one. For a start, see Booss and Krickeberg (1976), Bradley and Schaefer (1998), and Hoyningen-Huene (1983).

p. 48, "**the nowadays well-established distinction between (physical) solutions and chemical compounds**": Thomas Kuhn has discussed this case in his classic Kuhn (1970 [1962], 130–35); for the wider context, see, e.g., Brock (1993, ch. 4).

p. 48, "**in the 1730s, after more than one hundred years of research throughout Europe, a multitude of different and puzzling electrical effects were known**": I am taking the information about this case from Steinle (2002b) and Steinle (2006). A case with very similar characteristics is Ampère's exploratory experimentation in the new field of electromagnetism: see Steinle (2002a).

p. 49, "**many interesting philosophical questions that can be asked about these narratives**": A particularly influential work dealing with these questions is Danto (1965).

p. 50, "**the history of childhood**": This may be a somewhat unfamiliar item in need of explanation. The ideas of what childhood is differ among cultures, and they change over time. The story of such changes within one culture is told in a history of childhood. The classic example is Ariès (1962).

p. 50, "**principles of historical relevance**": For the following, see Hoyningen-Huene (1993, sec. 1.2, esp. 13–14). As a historian and author of a nine hundred-page volume on the history of Europe since World War I recently put it with regard to his book: "Yet much has had to be omitted or boiled down: as the painter Max Liebermann put it: 'drawing implies leaving out'" (Wasserstein 2007, vii).

p. 51, "**The development of our universe during its early phase when the formation, evolution, and clustering of galaxies and quasars took place**": See Springel et al. (2005). See also Gnedin (2005) in the same issue of *Nature*, which is more accessible for nonspecialists. In the latter paper, a little more context is given for this gigantic computational project, the largest of its kind at the time.

p. 52, "**The largest extinction, however, took place some 251 million years ago, at the so-called Permian-Triassic boundary**": See, e.g., Erwin (1993).

p. 52, "**A comprehensive climate model has recently been devised that couples land, ocean, and sea-ice using realistic paleogeographic and paleotopographic data**": See Kiehl and Shields (2005).

p. 52, "**In a new research field called computer assisted paleontology, researchers reconstruct fragmented and distorted fossil specimens in three-dimensional images**": See Zollikofer and Ponce de León (2005).

p. 53, "**the controversial question whether *Homo neanderthalensis* and *Homo sapiens* represent morphologically discrete, separate species**": See Ponce de León and Zollikofer (2001).

p. 53, "**are introduced as answers to a class of specific why-questions. These questions are identified as 'explanation-seeking why-questions'**": See, e.g., the classic article Hempel (1965a, 334), where the expression "explanation-seeking why-questions" has been used. The contrasting class is, according to Hempel, *reason-seeking* or *epistemic* questions that do not aim at an explanation for a phenomenon but rather reasons in support of some assertion (p. 335). It should be noted that explanation-seeking questions can also come in other forms, like "What is the explanation of x?" or "How come that x is the case?" The introduction of the difference between descriptions and explanations by means of different questions is much older. Aristotle had already introduced the topic of reasons and causes (which is not a fully adequate idiom, but it will have to suffice here) by why-questions: see Aristotle ([1996], book 1, ch. 1).

p. 54, "**has been a linguistic practice in many philosophical and humanist circles**": Perhaps, this stereotyped practice is finally waning—at least, the relation between explanation and understanding is being discussed in analytical circles; see, for example, de Regt (2004).

p. 54, "**two *technical terms* were introduced: 'explaining' (*erklären*) as the characteristic procedure of the natural sciences . . . .**": The contrast of *erklären* versus *verstehen* in the technical sense was introduced by Johann Gustav Droysen (1808–1884) in 1858; see Droysen (1967 [1858], § 14). The most important publication about this contrast in recent decades is probably Wright (1971). For some discussion of the historical background, see Apel (1984).

p. 54, "**In today's dominant worldview of the Western world, there is simply no such thing as an understanding *in the technical sense* of natural phenomena**": This worldview is realism about natural phenomena. For nonrealists, the situation may be different. Under nonrealist assumptions, natural phenomena are also constituted by contributions that are subject-sided as far as their origin is concerned (whatever the epistemic subject is). Hence, an element of human origin is inherent in natural phenomena whose unearthing may ask for understanding in the terminological sense.

p. 55, "**I will therefore employ 'explanation' largely in agreement with the everyday use of this word**": For an analysis of the everyday use of "explanation," see Passmore (1962).

p. 56, "**mechanisms have been a recently much-discussed topic**": See, e.g., the website http://www.philosophy.umd.edu/Faculty/LDarden/Research/bibmech.htm (accessed Nov. 26,2012).

p. 56, "**Each of these types of explanation has been widely discussed in the literature**": One of the classical sources in the twentieth century is Hempel (1965a). For the evolution of the discussion about this type of explanation and its many problems involved, see Salmon (1989). The latter reference includes a chronological bibliography about scientific explanation (pp. 196–219).

p. 56, "**already in the nineteenth century, it was well known that explanations belonging to the generalizing natural sciences are derivations from laws**": See, for instance, Droysen (1967 [1858], § 37).

p. 58, "**The derivation of p_I from the state equation, however, *explains* why the pressure has taken on this value and not another one**": This reveals a subtle difference in the use of state equations or, more generally, of empirical generalizations for descriptive and explanatory purposes, respectively. For an adequate description of the system, the generalization must be empirically correct. For an adequate explanation, however, the empirical correctness is not enough. Rather, the generalization must carry some sort of necessity—simply being accidentally true is not enough. The latter property would not make intelligible why "the pressure has taken on this value *and not another one*," which is just the explanatory burden. This subject has been extensively discussed in the philosophy of science; for a very accessible presentation of this thorny subject, see, e.g., Bird (1998, ch. 1).

p. 59, "**the Standard English philosophical reference works are not very helpful with respect to 'theory' because usually they do not feature that term as an entry**": Edwards (1967) only features "Laws and theories" and is strongly positivist. Audi (1995) deals with theory in the general "Philosophy of science" entry only. Craig (1998) discusses under "Theories, scientific" natural science theories exclusively; in the very short entry "Theory and observation in social sciences," social science theories are only summarily addressed. And in the entry "Structuralism in literary theory," literary theory is only dealt with in its connection to structuralism, but not in more general terms. In June 2004, the web-based *Stanford Encyclopedia of Philosophy* featured "theory" only in combinations with particular theories: see http://plato.stanford.edu/contents.html. German reference works typically feature "*Theorie*"; see, e.g., Braun and Radermacher (1978), Mittelstraß (1980, ff.), Sandkühler (1999), Seiffert and Radnitzky (1989), and Speck (1980). Suppe (1977) and Savage (1990) predominantly deal with natural science type theories.

p. 59, "**the theory of common descent**": In conversation, the biologist Ernst Mayr has repeatedly brought to my attention that the theory of common descent is today treated as a fact, and no longer as a theory. His main interest regarding this point was how this transformation from theory to supposed fact happens. See also Mayr (1988b, 192) and Mayr (1997, 61), in which Mayr goes even further by stating that "a modern evolutionist might say that the theory of evolution is now a fact." For a systematic treatment of this subject, see Hofmann and Weber (2003).

p. 60, "**Theoretical entities are not directly observable and are therefore posits**": Theoretical entities share this feature with what was called in the Middle Ages "occult qualities." Occult qualities fell into disrepute in the scientific revolution of the sixteenth and seventeenth centuries because they appeared to be arbitrary. For several decades, Newton's introduction of gravitational force therefore had difficulties being accepted because gravitation seemed to be an occult quality, with Huygens and Leibniz being his main adversaries. On this intriguing subject see, e.g., Koyré (1965, 115–48), Hutchison (1982), Hutchison (1991), Clarke (1993), and Hutchison (1993).

p. 60, "**these entities always figure in the context of some theory**": Several different possibilities exist. Sometimes, theoretical entities are postulated for the first time by the very same theories (or models) in which they occur. For example, phlogiston in the eighteenth-century chemistry, or the existence of a wave function associated with material particles in Schrödinger's wave mechanics, was postulated in the context of the respective theory. Theoretical entities may or may not survive any subsequent theoretical change. For instance, phlogiston did not survive the rather radical change at the end of the eighteenth century to what is today called modern chemistry; whether wave functions will survive a possible radical change of quantum physics remains to be seen. The survival of theoretical entities through revolutionary events does not imply, however, that they are taken over by the later theory or model in exactly the same form. Electrons, for instance, to which Schrödinger's wave mechanics was also applied, were theoretical entities introduced into atomic physics much earlier, but their character changed in the transition.

p. 60, "**They are hypothetical either in the sense that they are posits believed to be real but not proven to be real, or they are hypothetical in the sense of a consciously counterfactual assumption**": This implies that theoretical entities in general do not possess a uniquely determined ontological status. However, those believed to be real have the same *ontological* status as observable entities, in spite of our insecurity of their existence—*if* they really exist. Such theoretical entities are only different from observable entities in their relation to the sense physiology of human observers.

p. 62, "**a typical explanation of individual human actions does not refer to any general statements, i.e., to empirical generalizations or theories**": The recent discussion about this type of explanation has been strongly influenced by Wright (1971).

p. 62, "**Why did President Truman decide in 1945 to drop atomic bombs on Japan?**": For extensive source material on this question, see on the Internet the section "The Decision to Drop the Atomic Bomb" in the Truman Presidential Museum & Library: http://www.trumanlibrary.org/whistlestop/study_collections/bomb/large/ (accessed Sept. 27, 2011). Recent books discussing this question controversially are Allen and Polmar (2001), Alperovitz (1995), Ferrell (1996), Newman (1995), O'Neal (1990), Sholin (1996), Takaki (1995), Wainstock (1996), Walker (1997), and Walker (2003). See also Rhodes (1986, esp. ch. 19).

p. 63, "**his diary entry of July 25, 1945**": It is available on the Internet: http://www.trumanlibrary.org/whistlestop/study_collections/bomb/large/documents/fulltext.php?fulltextid=15 (accessed Sept. 27, 2011).

p. 64, "**Reductive explanations have been widely and controversially discussed**": There are literally thousands of references. Here are a few from the philosophy of science: Nagel (1961), Wimsatt (1979), Bonevac (1981), Charles and Lennon (1992), Schaffner (1993); from physics: Anderson (1972), Weinberg (1987), Weinberg (1988), Weinberg (1992), Greene (2000), Weinberg (2001 [1995]), Greene (2004), Anderson (2005), Laughlin (2005) (note that apart from Brian Greene, all of these physicists are Nobel prize winners); from chemistry: Pauling (1970) (he won a Nobel price), Primas (1981), Scerri (1994); from biology: Crick (1966), Ayala and Dobzhansky (1974), Mayr (1982), Mayr (1988a) (Crick won a Nobel price; Dobzhansky and Mayr are two of the most important evolutionary theorists of the twentieth century); from the social sciences: Alexander et al. (1987); from formal semantics: Partee (2004).

p. 66, "**levels contain both originally subject-sided (the subject being the epistemic subject) and originally object-sided elements**": I am deliberately using the terms "originally subject-sided" and "originally object-sided" instead of simply "subjective" and "objective." The reason is that the terms "subjective" and "objective" are impregnated by a presupposed realist position, whereas the more complicated terms are neutral in this respect.

p. 66, "**became clear that the postulate of the existence of an unambiguous and universal level structure of the world—whatever its origin may be—is highly contentious**": See, e.g., Brown (1926), Beckner (1974), Wimsatt (1976), Blitz (1990), Blitz (1992), and Emmeche, Køppe, and Stjernfelt (1997).

p. 66, "**there are many different possibilities of distinguishing levels below the level of the organism**": See, for instance, Wimsatt (1974) and Wimsatt (1976).

p. 69, "**Its explanandum, i.e., the event, process, or the state to be explained, is a singular thing, and the explanation has essentially a narrative form**": Note that I am not claiming that *all* explanations offered in the historical disciplines have a narrative form. For instance, the explanation of the single action of some historical actor may follow the pattern of an action explanation, explicated above in subsection 3.2.4, by means of the actor's intentions and beliefs. However, historical explanations typically concern longer trains of events in which many facts must be narrated. There is a large amount of literature on the narrative explanation in history; see, for example, the classic works Danto (1965, ch. XI) and Dray (1971).

p. 69, "**Take as an example the outbreak of World War I**": See, for example, Tuchman (1962).

p. 70, "**Alfred Wegener's continental drift hypothesis, advanced in 1912, was designed to explain a variety of puzzling features on the surface of the Earth**": I take this case as an example for this particular type of historical explanation from Cleland (2002, 481).

p. 71, "**The objects of study in the humanities are cultural products**": See, for instance, the list of humanities and their subject matters at: http://science.thomsonreuters.com/mjl/scope/scope_ahci/ (accessed Nov. 27, 2012). I may add a word of caution at this point concerning this section. Probably more than any other domain of the sciences, the humanities are permeated by deep (and partly bitter) controversies: almost everything is controversial. Correspondingly, it is virtually impossible to describe any activities of the humanities such that the description is acceptable to all parties concerned. On top of that, this domain is so vast that the superficiality of my knowledge of it is not only my personal shortcoming.

p. 72, "**the concept of meaning just employed is neither very clear nor unambiguous**": For some important steps to elucidate the pertinent concept of meaning see, e.g., Taylor (1985 [1971], esp. sec. I.3).

p. 72, "**the popular description of the humanities as being 'united by a commitment to studying aspects of the human condition'**": The quote is from the entry "Humanities" from the Internet encyclopedia Wikipedia: http://en.wikipedia.org/wiki/Humanities (accessed Nov. 12, 2005).

p. 72, "**its meaning is by no means identical with any of its purely physical properties or physical aspects of its use**": The meaning of a cultural product is not even "supervenient" on its physical properties, i.e., it is not determined by its physical properties. One and the same physical object may have different meanings in different cultures, for instance a cow or an airplane. Even physically identical objects may have different meanings in one and the same culture by having different (attributed) histories. A piece of linen, looking completely like another one, may have a very specific cultural meaning by (presumably) being a part of Christ's death linen.

p. 72, "**which human faculty allows exploration of meanings and enables humans to grasp them?**": "Grasping meanings" is a somewhat dangerous metaphor in this context, because it presupposes that meanings are somehow out there to be grasped. This would be a position of meaning realism, which is by no means the only choice. The strictly opposite position would assert that meanings are not at all out there but are constructed by those who supposedly "grasp" them. Intermediate positions are also conceivable. However, as I do not intend to enter the question of the ontology of meaning here, the expression "grasping meaning" should be read here without any ontological implications.

p. 72, "**reflection about observable cultural products … enables us to form hypotheses about their meaning**": In the same way that I tried to avoid ontological questions about meaning, I am now not entering epistemological questions about meaning. In other words, I will now not discuss how this "grasping" of meaning is possible at all or how reliable this process is (although these are fascinating and controversial questions). Rather, I shall content myself with a description of the phenomenon as it exists in the humanities. As throughout this book, my purpose is to explicate the higher degree of systematicity of the sciences, here of the humanities, in comparison to other kinds of knowledge, and for this purpose, I can and must skip those other interesting questions.

p. 72, "**(although their being *cultural* products may not be available to inspection)**": Artifacts can often be observationally identified as cultural products. Natural entities like mountains, animals, and configurations of stars that bear meaning in some culture cannot be identified as such by pure observation of those entities alone.

p. 73, "**When such difficulties present themselves in the humanities, the methodological situation is often described as the 'hermeneutic circle'**": It should be noted that this situation by no means occurs only in the humanities. In general terms, it is found whenever a set of somewhat hypothetical particulars has to be adjusted to a somewhat hypothetical whole. Then one moves back and forth between the parts and the whole in order to achieve optimal adjustment. It is important to note that this adjustment process does not depend on the parts or the whole possessing meaning in the sense as this term is used in the humanities (and as discussed in the beginning of this subsection). For an example from the natural sciences that does certainly not involve meaning in the pertinent sense, see the concept of "reciprocal illumination" that was introduced to describe the methodological situation in phylogenetic reconstructions (see next page).

p. 73, "***Hermeneutics* is both the art of interpretation and the theoretical discipline that studies the process of interpretation**": An introduction to hermeneutics is, e.g., Demeterio III (2001). Unfortunately, this source is no longer available on the Internet.

p. 74: "**One is often confronted with a particular *circle* when one is trying to make sense of a given document**": It may appear that the hermeneutic circle is a specialty of the humanities, i.e., that it never occurs outside of the humanities because it is bound to texts and text-analogs. This, however, is not the case. It occurs whenever one tries to understand a whole out of its parts and starts with incomplete information about whole and parts, whatever the whole and its parts are. For instance, in the natural sciences, it has been discussed in morphology under the rubric of the "principle of reciprocal illumination," introduced by the German morphologist Willi Henning in 1950; see, e.g., Henning (1979, 206, 222), Rieppel (2003, 182), and Rieppel (2004, 16–17). Rieppel even explicitly identifies this principle with the hermeneutic circle: Rieppel (2003, 182 col. 2). I wish to thank Thomas Reydon for bringing these quotes and its biological context to my attention.

p. 73, "**one should be able to increase the understanding of both in a stepwise process. This process comes to an end once one has reached a reflective equilibrium**": The expression "reflective equilibrium" was introduced in 1971 by John Rawls in his influential *A Theory of Justice* in the context of moral philosophy (Rawls 1971, 20, 48–51). As Rawls himself notes, the idea of a "mutual adjustment of principles and considered judgments" had been used earlier by Nelson Goodman in the context of "the justification of the principles of deductive and inductive inference" (Rawls 1971, 20 note 7). For a recent discussion see, e.g., Daniels (2003).

p. 74, "**the character of cultural products as being text-analogs has been stressed in many disciplines of the humanities**": An important source for this procedure has been Taylor (1985 [1971]).

p. 74, "**an understanding of a picture often also involves a hermeneutical circle**": I am relying here on Sonesson (2004). The reference to the hermeneutic circle is on p. 6. On pp. 3–27, this lecture contains a detailed and very instructive exemplary semiotic study of Diego Velázquez's painting "Las Meninas" of 1656, its relation to Pablo Picasso's revised version of "Las Meninas" of 1957, and Richard Hamilton's paraphrase of Picasso's "Las Meninas" of 1973.

p. 75, "**we are constantly interpreting our physical and human environment with respect to their meaning**": This is a central topic in Heidegger's philosophy: see Heidegger (1962 [1927], esp. §§ 31, 32). In fact, Heidegger contends that this permanent interpretative activity is one of the constitutive features of human beings.

p. 76, "**the concept of the 'poetological difference'**": I take this concept and also the example from Gerigk (2002, 17–40) (which is, unfortunately, in German). There is an essential parallel between Gerigk's poetological difference and an important distinction drawn in the philosophy of technology. In a technological artifact, say a machine, one can distinguish between the function of a given part with respect to the overall working of the machine and the function of the machine when used by human beings. For example, the function of the carburetor in a car is to mix air and fuel such that the mixture can be used in the internal combustion engine. The function of the car, however, is, among other things, to transport people and goods on comparatively smooth solid surfaces. Clearly, the parallel to the poetological difference is essential because the distinction between functions of parts for the whole and the function of the whole for human beings can be drawn with respect to any artifact, including fictional texts. There is a somewhat weaker parallel to the important biological distinction between proximate and ultimate explanations introduced by Ernst Mayr; see Mayr (1976 [1961]); for a recent discussion, see Ariew (2003). Proximate explanations concern the operation and interaction of structural elements of organisms, from molecules up to organs and whole individuals. Ultimate explanations concern the evolutionary path that brought some structure or function of an organism into existence.

p. 76, "**the main message of the play was the approximation of the avenger to the one on whom he takes revenge**": See Gerigk (2002, 29).

p. 77, "**Children in their early teens may be aware of the difference between the inner logic of a fictional story and the author's intentions**": I am indebted to Alexander Hoyningen, aged twelve in 2004, and his friends for letting me witness a conversation about the possible content of *Harry Potter*, volumes 6 and 7 (that had not appeared in 2004). They hypothesized that a figure called Snape would play some role in the forthcoming volumes because he had saved Harry's life in volume 1 of the series. They suspected that the author established Harry's thankfulness to Snape in order to let this (fictional) fact play some role in a future volume. Clearly, these adolescents demonstrated awareness of the poetological difference.

p. 77, "**Here, 'theory' denotes an extremely heterogeneous and large variety of works that typically do *not* directly deal with the subject matter of literary studies**": I rely here on Culler (1997, ch. 1).

p. 77, "'**works of anthropology, art history, film studies, gender studies, linguistics, philosophy, political theory, psychoanalysis, science studies, social and intellectual history, and sociology.**'": Culler (1997, 4).

p. 78, "**literary studies are permanently permeated by controversies of all sorts**": Recently, protagonists of literary studies (and of related fields) have been strongly and publicly criticized even by authors not belonging to the humanities, namely by physicists. The physicist Alan Sokal submitted a nonsense article to a well-established journal—where it was published (Sokal 1996c). This article is now known as "Sokal's hoax." In two subsequent articles, Sokal explained what he did and why he did it (Sokal 1996a, 1996b). A little later, he published a book together with his colleague Jean Bricmont expanding his criticism (Sokal and Bricmont 1998). A bitter controversy, commonly called the "Sokal affair" or the "science wars" ensued this diatribe in the course of which, among other things, Sokal and Bricmont were accused of the very same mistakes they accused their targets of. For documentation of the Sokal affair, consult http://physics.nyu.edu/faculty/sokal/ (accessed Sept. 27, 2011), where many of the relevant articles can be downloaded.

p. 78, "**In the sciences, however, a looser usage of 'prediction'**": Of course, I am not the first one who notes this ambiguity in the use of "prediction." See, e.g., Brush (1995, 304).

p. 78, "**there are successful natural sciences whose main epistemic goals do not include producing predictions**": In a discussion at Chicago University in October 2006, evolutionary theorist Leigh Van Valen made me aware of the fact that the historical natural sciences are also able to make predictions (which I had denied in a talk preceding the discussion). An example from paleontology is the prediction that after a catastrophic event in which life in a certain geographical area has been wiped out, it takes quite some time for life to develop there again. This prediction can obviously be gained by inductive generalization from known historical cases of the same sort.

p. 79, "**arguments that make it plausible that reliable longer-term predictions concerning human affairs are impossible**": For a general treatment of obstacles to predictive foreknowledge, see Rescher (1998, ch. 8); for a discussion of prediction in human affairs, see ch. 11 of the same book. Much to my regret, I came across this book only after substantial parts of the present section had already been written.

p. 79, "**One well-known argument stresses the possibilities of 'self-destroying prophecies'**": In the literature, this argument is usually attributed to the sociologist Robert Merton. Merton, however, who indeed discusses the argument, traces it back to the nineteenth-century logician John Venn, who called it "suicidal prophecy" (Merton 1968, 182–84). Merton notes that many people have realized the possibility of that pattern, among them Abraham Lincoln in 1862.

p. 79, "**that longer-term developments of human affairs will depend on future knowledge, i.e. knowledge that we do not possess today**": This argument is due to Karl Popper: Popper (1957), Preface.

p. 79, "**cultural change can be so profound that we may presently lack the concepts that might become necessary to describe our future**": This argument is due to Charles Taylor (Taylor 1985 [1971], sec. IV). It is akin to, but not identical to, Popper's argument.

p. 79, "**Scientific predictions can be sorted into several classes**": This classification, although independently devised, roughly corresponds to Rescher's (Rescher 1998, 88). For a somewhat different classification, see Armstrong (1985, ch. 5)

p. 80: "**accessible and of interest to all cultures of all ages, have been and still are astronomical regularities.**" This is not just speculation, but born from available evidence: see Steele (2000, 5–9).

p. 80, "**premature to classify all systematic records and even predictions of eclipses in some culture as scientific**": Compare Steele (2000, 4).

p. 80, "**in Assyrian and Babylonian letters and reports, some eclipses were foretold**": See Steele (2000, 75–76).

p. 80, "**western heritage**": Following Steele (2000, 8), the Western heritage in astronomy is "founded upon the astronomy of Mesopotamia, that may be traced th[r]ough Greco-Roman, Indian, and Islamic astronomy up to the astronomers of the European Renaissance."

p. 80, "**The earliest of such methods, used at least by 600 BC, was the so-called Saros period**": See for the following Steele (2000, 78–80).

p. 81, "**By the end of the fourth century BC, small eclipses that were not foretold were beginning to be visible**": See Steele (2000, 80).

p. 81, "**The economic explanation for this periodicity appears simple**": The stereotyped explanation runs like this. Start with a situation of low supply, i.e., according to market economics, with high prices. This is an incentive for investment in pig production, which will be felt on the market only somewhat later, namely when the pigs are ready to be sold. Due to increased supply, prices will then fall. This leads to lower profits, which will lead, in turn, to lower investments in pig production, which will, however, only somewhat later be felt on the market. This will then reproduce the initial situation, and the cycle can start again. Abstractly put, the ingredients of this sort of cycle are a negative feedback mechanism that operates with a time lag. However, in reality, the situation is much more complex. For instance, the structure of the pig industry is more complicated than suggested above; feeding costs have to be taken into account because they vary and influence profits, and so on. For details, see the classic papers by Ronald Coase (who won the Nobel prize in economic sciences in 1991) and R. F. Fowler (Coase and Fowler 1935a, 1935b, 1937). Footnote 1 of Coase and Fowler (1935a) refers to papers in which the cycle has indeed been used to forecast pig prices.

p. 81, "**we are using heuristics that are useful but sometimes lead to severe and systematic error**": See, e.g., the classic paper, Tversky and Kahnemann (1974).

p. 82, "**The economist William Stanley Jevons (1835–1882) developed the sunspot theory of the business cycle**": For a very short sketch of the theory, see Mills (1999b, ix); for a somewhat more extended presentation together with some context, see Black (1987). For an extended discussion that shows, as usual, that things are much more complicated than stereotyped history has it, see Gallegati (1994).

p. 83, "'**forecasting with leading indicators**'": For some of the classic papers, see chapters 19 through 23 of volume II of Mills (1999a).

p. 83, "**earliest attempt was the so-called Harvard A-B-C barometer**": I rely here on Samuelson (1987).

p. 84, "**the discovery of the planet Neptune in 1846**": See, for instance, http://www-gap.dcs.st-and.ac.uk/~history/HistTopics/Neptune_and_Pluto.html (accessed Sept. 27, 2011) with further references.

p. 84, "**the bending of light by gravitation**": This story has been told many times in its cliché form. For a semi-popular account, see, for instance, on the Internet http://arxiv.org/pdf/astro-ph/0102462v1.pdf (accessed Nov. 27, 2012); see also Coles (1999). In philosophical circles, there has been a lot of discussion about the historical and epistemological details of that case, which are, again, much more complicated and much less unequivocal than the stereotyped history; see, for instance, Earman and Glymour (1980), Brush (1989), and Mayo (1996, esp. 278–92).

p. 85, "**Models (in the sense relevant here) are used when systems are too complex to be treated by theories or general laws alone**": I am here using information from Suits (1962) and Frigg and Hartmann (2005). It may be objected that at least some of the predictions discussed in this subsection are not really parts of science because they belong to other contexts like weather forecasts or economic forecasts. Thus, their aim is not scientific but practical (compare the discussion of an appropriate aim as a necessary component of being scientific in section 2.1, sixth remark). This is undoubtedly correct. However, such forecasts are indeed called "scientific" because they are produced by means that are the product of scientific research. Correspondingly, the institutions that deliver weather forecasts have their own research and developments departments. For instance, Germany's National Meteorological Service has a section on Meteorological Research: http://www.dwd.de/bvbw/appmanager/bvbw/dwdwwwDesktop?_nfpb=true&_pageLabel=_dwdwww_spezielle_nutzer_forschung&activePage=&_nfls=false (accessed Sept. 29, 2011); the U.S. National Weather Service: http://www.nws.noaa.gov/ (accessed Sept. 27, 2011) has an Environmental Modeling Center: http://www.emc.ncep.noaa.gov/ (accessed Sept. 27, 2011).

p. 85, "**the global climate system involves a large number of variables interacting in various and complex ways**": For an instructive history of the attempts to model the climate system, see Weart (2006).

p. 85, "**there are meteorological models in which even the law of energy conservation is violated**": See, e.g., Weart (2003).

p. 86, "**A global meteorological model consists of a grid, dissolving the whole of Earth's atmosphere into a discrete set of points**": See, e.g., http://www.dwd.de/bvbw/appmanager/bvbw/dwdwwwDesktop?_nfpb=true&_pageLabel=_dwdwww_aufgabenspektrum_vorhersagedienst&T18609318401152164701685gsbDocumentPath=Navigation%2FOeffentlichkeit%2FAufgabenspektrum%2FNumerische__Modellierung%2FAS__NM__intro__node.html%3F__nnn%3Dtrue (accessed Dec. 2, 2012). On the left-hand side of this page, there are links leading to the more specific models discussed below, namely the GME model, the COSMO-EU model, and the COSMO-DE model. Note, however, that these models are constantly being developed; my account is based on information from December 2012.

p. 86, "**a worldwide observation system has been installed**": This system is described at http://www.wmo.int/pages/prog/gcos/index.php?name=AboutGCOS (accessed Sept. 27, 2011).

p. 87, "**there are many more areas where models are used for predictive (and other) purposes**": Economics is one of these areas. For the large variety of different models used there, see Mills (1999a) and Clements and Hendry (2002); for a critical assessment of forecasting in economics and its sociopolitical consequences, see Betz (2006).

p. 87, "**the pioneering and controversial work of Jay Forrester and Dennis Meadows in the 1970s**": See Forrester (1970) and Meadows (1972).

p. 87, "**a model calculation of the consequences of long-term fossil fuel consumption**": See Bala et al. (2005). The results of the model calculation are easily accessible at http://www.llnl.gov/pao/news/news_releases/2005/NR-05-11-01.html (accessed Sept. 27, 2011).

p. 87, "**a supercomputer that was the world's fastest machine in November 2005**": A list of the world's five hundred fastest supercomputers that is updated twice a year is available at http://www.top500.org/ (accessed Sept. 27, 2011).

p. 87, "**many institutions around the world deal with global predictions that are all based on models**": For instance, the Japanese Earth Simulation Center (whose supercomputer was the world's number one between 2002 and 2004) spends a considerable fraction of its capacity on predictive models concerning the atmosphere and the oceans: http://www.jamstec.go.jp/e/ (accessed Sept. 27, 2011); or the Climate Modeling Program of the U.S. Department of Energy: http://www.csm.ornl.gov/chammp/Climate/ (accessed Sept. 27, 2011).

p. 87, "**type of scientific predictions is based on so-called Delphi methods**": For a short description of Delphi methods by one of its inventors, see Rescher (1998, 92–96). For an older book-length study on this subject featuring many examples, see, e.g., Linstone and Turoff (1975). Because of its age, this book has the advantage of being available on the Internet at http://www.is.njit.edu/pubs/delphibook/#toc (accessed Sept. 27, 2011). For a critical review of empirical studies dealing with the effectiveness of the Delphi technique, see Rowe and Wright (1999). Searching the Internet for "Delphi method" returned about 650,000 hits in April 2010, including many detailed studies.

p. 88, "**Forecasting long-term developments, i.e., concerning the next twenty-five years, for example, in technology, science, society, and warfare**": For a nice example from 1960, see Rescher (1998, 94–95).

p. 88, "**The high esteem that science enjoys almost everywhere derives from its reputation to produce a superior form of knowledge**": The expression "*almost* everywhere" has been deliberately chosen and refers to the fact that quite a bit of science criticism, or anti-science, exists, mostly directed against the natural sciences. Two main forms of science criticism can analytically be distinguished that come in many flavors. The first denies science a superior cognitive status by claiming that science has a somehow reduced perspective at the expense of other aspects of the world that also deserve attention. The second line of criticism points at science's negative effects on nature, societies, or individuals that may or may not derive from presumed cognitive deficiencies of science. For a short introduction, see Hoyningen-Huene, Weber, and Oberheim (1999), sections 2.1 and 2.4, available on the Internet at http://www.eolss.net/21st_c.aspx (accessed Sept. 27, 2011).

p. 89, "**Error may arise as the result of individual or collective mistakes ....**": There is a large empirical literature on mechanisms in everyday thinking that are often useful but may also systematically lead to fallacies. One of the classic papers is Tversky and Kahnemann (1974).

p. 90, "**a tendency for those working in the experimental sciences to depreciate the historical natural sciences**": For a defense of the historical natural sciences against this form of depreciation, see Cleland (2002).

p. 90, "**they will have to learn that the standards that are valid in a particular field are typically not so easily**": This is, of course, a very general empirical claim that I cannot substantiate here. One example must stand for many. Sokal and Bricmont (1998) have rightly, I think, criticized the way in which some scholars from the social sciences and the humanities have presented and used particular scientific results. Given the basically very simple standards common in theoretical physics (from where the two authors come) for representing scientific results, there is indeed something to criticize in that respect in the targeted fields. However, when it comes to getting involved in a philosophical discussion, as it happens in chapter 4 of their book, they do certainly not display a higher standard of rigor in the defense of their philosophical claims than is common in average philosophy. Indeed, when discussing Kuhn and Feyerabend, for example, they partly rehearse the same stereotypes that abound in the philosophical literature (pp. 71–85).

p. 90, "**to what in analytical philosophy of science has traditionally been called 'the context of justification'**": For a discussion of the intricacies of the distinction between the "context of discovery" and the "context of justification" see Hoyningen-Huene (1987), Hoyningen-Huene (2006), and the other articles in Schickore and Steinle (2006).

p. 91, "**one should not approach this section with a sharp (traditional) distinction between the context of discovery and the context of justification in mind**": Note, however, that I am not categorically denying that there is a sharp distinction between the two "contexts." Yet, it is not the traditional distinction but a lean version of it, namely, a distinction between two types of questions. For details, see Hoyningen-Huene (1987, 511–12) and Hoyningen-Huene (2006, 128–30).

p. 92, "**the same sort of interpretation of data is totally absent in most of the natural empirical sciences**": However, there are other forms of interpretation in the natural sciences. I shall briefly discuss them in section 3.4.3.

p. 92, "**their justificatory procedures depend on any sort of data that are empirical in character or, in other words, on empirical evidence**": The concept of evidence that I am using here seems to be quite straightforward and uncomplicated, but it is not. Both in the sciences and in philosophy, there have been deep controversies about what can count as evidence and what its precise function for scientific knowledge is; for an overview, see, e.g., Achinstein (2005, esp. the introduction) and Kelly (2006).

p. 93, "**In fact, both components of proofs have undergone historical change: the axioms and the admissible rules of derivation**": See, e.g., Kline (1980).

p. 93, "**Nonevidential considerations also play a role in the empirical sciences, and they have been extensively discussed in recent decades in the philosophy of science**": I have tried to clarify and develop the seminal contributions of Thomas Kuhn to this discussion in Hoyningen-Huene (1993, sections 4.2.c and 7.4.b).

p. 95, "**the values of the variables to be measured must not only be in principle empirically testable, but they must be in the technical reach of real measuring instruments**": This is a major problem in contemporary theoretical particle physics, especially in string theory. Some theoreticians have asserted that it will take at least twenty years until a confrontation of string theory with empirical data will be feasible. In fact, this has even led to an extended controversy about the status of string theory as a scientific theory; see, e.g., Smolin (2006) and Woit (2006).

p. 96, "**This is the essence of the so-called Duhem-Quine thesis**": The thesis goes back to Duhem (1954 [1906], ch. 10, esp. § 2) and Quine (1953, 41). However, contrary to what one would expect from the common label "Duhem-Quine thesis," Duhem and Quine were not advancing exactly the same thesis; see, e.g., Ariew (1984) and Gillies (1993, 98–116). Duhem's chapter, Quine's paper, and Gillies' chapter are conveniently reprinted in Curd and Cover (1998).

p. 96, "**not the same meaning of 'interpretation' that we have used when discussing the specifics of the humanities**": For a discussion of the similarities and differences of the two notions of interpretation, see Faye (2010, esp. section 9.3).

p. 97, "**the 1976 Viking Lander missions to Mars**": Basic information about this mission can be found on NASA's Viking web page http://www.nasa.gov/mission_pages/viking/ (accessed Sept. 27, 2011).

p. 97, "**the experimental results were far from unequivocal**": For a quick overview that emphasizes some aspects that are interesting from a philosophical point of view, see Cleland (2002, 479–80).

p. 97, "**even thirty years after the Viking mission, the discussion about the interpretation of its results has not ended**": See, for instance, the article by Navarro-González et al. (2006).

p. 97, "**This does not only concern the historical natural sciences, like paleontology or cosmology**": For an interesting analysis of the difference between the historical and the experimental natural sciences with respect to typical patterns in their evidential reasoning, see Cleland (2002).

p. 97, "**since the 1990s, theoretical research on tsunamis has developed computer models for tsunami propagation through the open ocean**": I am drawing in this section on Geist, Titov, and Synolakis (2006).

p. 98, "**John Stuart Mill (1806–1873) had clearly described this arrangement in his *System of Logic***": See Mill (1886, esp. Book III, ch. VIII, §2, 255f.; Book VI, ch. VII, §3, 575f).

p. 98, "**particular difficulties are raised if the theories in question are 'incommensurable'**": There is a vast literature on this subject; see, for instance, Hoyningen-Huene and Sankey (2001) and Hoyningen-Huene (2005).

p. 99, "**This sort of experimental setup is of utmost importance in many branches of chemistry, pharmacology, medicine, biology, education, criminology, and other areas**": For biology, this fact has recently been stressed by Weber (2005, 118–26), an authority in the philosophy of experimental biology. He argues that control experiments "are the main strategy for eliminating errors in experimental biology" (p. 146).

p. 99, "**the question of whether a particular gene G is causally relevant for some disease D can be investigated in this way**": A concrete example is the research into the genetic basis of cystic fibrosis by means of the *Cftr* knockout mouse; see, for example, http://www.genome.gov/10005834 (accessed Sept. 27, 2011).

p. 100, "**the question often arises whether a combination of two pharmaceuticals is more effective than treatment with one of them alone**": For a concrete example, see, e.g., http://www.herceptin.com/breast/metastatic/ (accessed Sept. 27, 2011).

p. 100, "**does community service rehabilitate better than short-term imprisonment**": This is the title question of Killias, Aebi, and Ribeaud (2000). I wish to thank Professor Martin Killias for his January 2007 inaugural lecture at the University of Zurich, which brought this study, related studies, and the collaborations to be discussed below to my attention.

p. 100, "**the famous Cambridge-Somerville study, a pioneering longitudinal study of delinquency prevention**": See, e.g., McCord (1992).

p. 101, "**The philosopher J. T., who had a brilliant career both in the United States and Europe, was known to join in and contribute to discussions on almost any subject**": The anecdote is related by Hans Jonas in Jonas (2003, 272–73). Unfortunately, no English translation of this book is available. Sokal's hoax, which I mentioned earlier (note to p. 78), works along the same lines as the case of the philosopher J. T. For extensive documentation on the so-called Sokal affair (or the "*Social Text* affair" as Sokal himself prefers to call the case after the journal in which he published his hoax), see http://physics.nyu.edu/faculty/sokal/ (accessed Sept. 27, 2011).

p. 101, "**This is the task of a huge collaboration in the medical sciences called the Cochrane Collaboration**": Its homepage is http://www.cochrane.org/index.htm (accessed Sept. 27, 2011). On this page, there is a variety of documents describing the activities of the collaboration; for a concise description see Clarke (2004). Regarding its history, see http://www.cochrane.org/docs/cchronol.htm (accessed Sept. 27, 2011).

p. 102, "**the *Cochrane Handbook for Systematic Reviews of Interventions*"**: See Higgins and Green (eds.) (2006).

p. 102, "**the *Cochrane Database of Systematic Reviews*** and **the *Cochrane Central Register of Controlled Trials*"**: See http://www.thecochranelibrary.com/view/0/index.html (accessed Dec. 3, 2012).

p. 102, "**there is a sister organization called the Campbell Collaboration**": Their homepage is http://www.campbellcollaboration.org/ (accessed Sept. 27, 2011).

p. 102, "**'people make well-informed decisions by preparing, maintaining and disseminating systematic reviews in education, crime and justice, and social welfare'**": See http://www.campbellcollaboration.org/about_us/index.php (accessed Sept. 27, 2011).

p. 102, "**the principle was introduced and discussed by Giambattista Vico**": See, e.g., Costelloe (2008).

p. 103, "**Here is an interesting example from biology**": See Wehner, Michel, and Antonsen (1996) and Lambrinos et al. (2000).

p. 104, "**various hypotheses were formed in order to explain the existence of mountains**": For this example, see Oreskes (1999, esp. 25–26). The quote is on p. 26.

p. 104, "**This can be expressed in terms introduced by Karl Popper into the philosophy of science**": See Popper (1959 [1934], ch. 6).

p. 105, "**French mathematician and astronomer Urbain le Verrier very accurately calculated the orbit of planet Mercury**": The standard source is Le Verrier (1859a); it is available on the Internet at http://www.archive.org/stream/comptesrendusheb49acad#page/378/mode/2up (accessed Sept. 27, 2011). However, in this publication, Le Verrier only reports his results. The original publication containing the actual calculations and the results is a heavy-duty 195-page book (Le Verrier 1859b). It is also available on the Internet at http://articles.adsabs.harvard.edu/cgi-bin/nph-iarticle_query?bibcode=1859AnPar...5....1L&page_ind=0&epage_ind=194&type=PRINTER&data_type=PDF_HIGH&email=&emailsize=500&emailsplit=YES&send=GET&verified=YES (accessed Dec. 3, 2012).

p. 105, "**The effect is rather small, seen from the Earth only 5599.7 arcseconds per century**": See, e.g., Will (1993, 4).

p. 105, "**today's value is 42.7 arcseconds per century**": As said above, the observed value of Mercury's perihelion advance is 5599.7 arcseconds per century. The largest part of the effect is due to the peculiar movement of the Earth, namely her axial precession, which contributes to Mercury's perihelion advance 5025.6 arcseconds per century. The remaining 574.1 arcseconds per century are caused by gravitation. The gravitational contributions by the other planets add up to 531.4 arcseconds per century—thus, the remaining 42.7 arcseconds per century. All data are from Will (1993, 4).

p. 105, "**Verrier considered this to be a 'serious difficulty'**": Le Verrier (1859a, 382).

p. 106, "**relativity theory would not have come out so superior**": I am simplifying at this point a little. There were also attempts at alternative explanations, also later in the twentieth century.

p. 107, "**What is the historian's basic task? It is "to choose *reliable* sources...."**": See Howell and Prevenier (2001, 2). The book contains a large research bibliography covering all sorts of methodological issues of historiography. Another useful book about the study of history is Tosh (2006).

p. 107, "**a broad spectrum of so-called historical auxiliary sciences**": See, e.g., Howell and Prevenier (2001, ch. 2).

p. 108, "**The dimension of systematicity that is the subject of the present section is somewhat different from the other ones**": The inclusion of this dimension was suggested to me, independently of each other, by Ken Waters and Noretta Koertge in October 2006 when I gave talks on systematicity theory at their Universities in Minnesota and Indiana, respectively. They were right, and I am grateful to them.

p. 109, "**How far the self-criticism of the sciences goes or should go is, or at least was, a controversial issue in the philosophy of science**": Theo Kuipers has recently reviewed this fact; see Kuipers (2010, 154–55). The book that contains Kuipers' article is not only interesting in showing on what subjects European philosophers of science work but also provides surprising insights into European geography, which may be especially interesting for non-Europeans: see p. 211, bottom.

p. 110, "**starting with Robert Merton's classic paper 'Science and the Social Order,' first published in 1938**": Merton (1973 [1938]). Interestingly, this paper was not published in a sociology journal but in the then comparatively young philosophical journal, *Philosophy of Science*, founded in 1934.

p. 110, "**and especially in Merton's 'The Normative Structure of Science,' first published in 1942**": Merton (1973 [1942]). In fact, the original title was "Science and Technology in a Democratic Order."

p. 110, "**The institutional goal of science is, in Merton's words, 'the extension of certified knowledge'**": (Merton 1973 [1942], 270).

p. 110, "**Merton's 'organized skepticism' is usually seen as the imperative to emphasize 'primarily an institutionally enjoined critical attitude toward the work of fellow scientists'**": I say "is usually seen" because this common view is not quite correct (the quote is from Norman Storer's Prefatory Note to the reprint of Merton's papers on the scientific ethos (Storer 1973, 225). In his papers from 1938 and 1942, organized skepticism does not mean something that is directed inward, toward the scientific community, like the critical attitude toward knowledge claims by fellow scientists. It rather means something directed outward, to other institutions of society. For instance, in 1938, Merton writes: "Another feature of the scientific attitude is organized skepticism, which becomes, often enough, iconoclasm. Science may seem to challenge the 'comfortable power assumptions' of other institutions, simply by subjecting them to detached scrutiny. Organized skepticism involves a latent questioning of certain basis of established routine, authority vested procedures and the realm of the 'sacred' generally" (Merton 1973 [1938], 264). Similar passages are in his 1942 paper (pp. 277–78). Mutual criticism of scientists is indeed briefly discussed in the 1942 paper, but under the rubric of Disinterestedness (p. 276). However, later Merton himself used "organized skepticism" in the sense of "self-engendered criticism and external criticism": see Merton's Foreword to Garfield (1983, vii) (this piece is available on the Internet: http://www.garfield.library.upenn.edu/ci/foreword.pdf, accessed Sept. 27, 2011). In much of the secondary literature, this use has also been projected back into the original 1942 paper; see, e.g., Kuipers (2010, 154–55).

p. 110, "**philosopher Helen Longino has fleshed out what 'critical discursive interactions' are**": Longino (2002, 128–35; the quotes are on pp. 129–31). It is surprising, at least to me, that Longino does not refer back to Merton.

p. 111, "**imperative for scientists to publish their results**": This is Merton's norm of "communism" (in quotes): see Merton (1973 [1942], 273–75). Clearly, there is also another motive in play, namely, that publication helps to avoid unnecessary duplication of scientific work. This, however, is of no concern in this section.

p. 111, "**In the most prestigious journals, the rejection rate is rather high**": I have not checked this systematically for all fields of research. Here are just two examples of high-profile journals. In 2010, the rejection rate of *Nature* was 92.1 percent, and it has been above 90 percent since 2002 (see http://www.nature.com/nature/authors/get_published/index.html, accessed Sept. 27, 2011). The rejection rate of the highly respected *Journal of Philosophy* is 95 percent (see http://el-prod.baylor.edu/certain_doubts/?page_id=823, accessed Sept. 27, 2011, where also the rates of other philosophy journals are listed). Many journals provide rejection rates on their websites.

p. 111, "**also the reviewers do not know who the author is**": This is the theory. In my own reviewing practice of double-blind reviewing in philosophy, in roughly half of the cases, I could not avoid finding out who the author was. There may be many clues in a piece of work that was officially made anonymous pointing rather precisely to the author.

p. 112, "**specific review journals**": A randomly picked example is the journal *Reviews of Modern Physics*; its website is http://rmp.aps.org/ (accessed Sept. 27, 2011). Other examples are provided by the Cochrane Collaboration and the Campbell Collaboration, which I discussed in Section 3.4.4 (p. 145): they produce so-called "systematic reviews" in the medical sciences and the social sciences, respectively.

p. 112, "**As already emphasized by Robert Merton, there are no taboos regarding subjects to be discussed in science**": See Merton (1973 [1938]).

p. 112, "**Big Science is characterized by large-scale instruments and facilities ....**": See the article "Big Science" in *Encyclopædia Britannica* (2011), available on the Internet at http://www.britannica.com/EBchecked/topic/64995/Big-Science (accessed Sept. 27, 2011).

p. 112, "**This paper lists no less than 3,172 authors affiliated with some two hundred institutions**": ATLAS (2011); the paper is available on the Internet at http://arxiv.org/abs/1009.5069v1 (accessed Sept. 27, 2011). The website of ATLAS is http://www.atlas.ch/ (accessed Sept. 27, 2011). I shall briefly come back to the topic of particle accelerators in section 3.8.

p. 112, "**Sociologist of science Karin Knorr-Cetina has investigated the specifics of different scientific communities**": See Knorr-Cetina (1999); for the patterns of discourse in experimental high-energy physics, see especially pp. 174–79. The list of meetings is on p. 174; the following quote is on p. 175.

p. 113, "**the most obvious ways to demarcate this kind of knowledge from scientific knowledge**": Although I am using the word "demarcate" here, I do of course not mean what has traditionally been called the "demarcation criterion" in the philosophy of science. The latter applies to the demarcation between proper science and pseudoscience. I shall come back to this sort of demarcation in section 5.4.

p. 114, "**In automobile development, nowadays one of the most important goals is the decrease of fuel consumption**": See, e.g., the respective pages of a large automobile manufacturer: http://www.daimler.com/dccom/0-5-1200798-1-1200840-1-0-0-0-0-0-36-7165-0-0-0-0-0-0-0.html (accessed Sept. 27, 2011).

p. 114, "**Chess theory usually comes in three branches**": Hooper and Whyld (1992), entry "theory," however, do not use the expression "middle theory": see p. 418. Readers not afraid of Wikipedia may also consult its rich article on chess theory: http://en.wikipedia.org/wiki/Chess_theory (accessed Sept. 27, 2011). Additional information is available, for instance, at http://www.chess-theory.com/ (accessed Sept. 27, 2011).

p. 115, "*Encyclopedia of Chess Openings*": See http://www.chess.com/download/view/encyclopedia-of-chess-openings (accessed Dec. 3, 2012).

p. 115, "*Encyclopedia of Chess Endings*": See, e.g., http://en.wikipedia.org/wiki/Chess_endgame_literature (accessed Dec. 3, 2012).

p. 115, "**Shouldn't chess theory qualify as science according to systematicity theory?**": This was the challenging question Martin Kusch posed to me in May 2004 in the discussion period after a talk on systematicity theory delivered at the University of Cambridge. At that time, the category "epistemic connectedness" was still completely absent from the then nascent systematicity theory. I am grateful to Martin, because the inclusion of epistemic connectedness as one of the systematicity dimensions is partly a result of his highly legitimate question.

p. 117, "**A well-known example from the past is the putatively pure research in number theory**": Further examples from mathematics can be found in Rowlett (2011).

p. 118, "**that I shall not pursue the project of distinguishing nonscientific areas as exemplified above from genuine scientific knowledge by reference to any aims of science because I believe that it does not work**": For quite some time, I have followed this strategy. Strangely enough, the question of what the aims of science are is seldom systematically treated in the current philosophy of science. In most textbooks or anthologies on philosophy of science, one looks in vein in the index for entries like "goal(s) of science" or "aims of science" and similar ones; see, e.g., Bird (1998); Boyd, Gasper, and Trout (1991); Cohen and Nagel (1934); Fetzer (1993); Harré (1985); Klee (1997); O'Hear (1989); Losee (2001); Nagel (1961); Salmon et al. (1992); Suppe (1977); and Toulmin (1953). Typically, the discussion of the aims of science in the philosophy of science, if it takes place at all, is severely restricted in three ways. First, it is usually only science in the sense of the natural (and perhaps the social) sciences that is discussed. The focus of the humanities, though included in the wider sense of "science," may be markedly different from the other ones. Second, even within natural science, it is only theories whose aims are typically discussed. There are, however, many more things to science than theories whose aims are in need of analysis. Third, even the aims of theories are usually not discussed as a subject in its own right, but only as a particular theoretical move in the context of the so-called realism debate. There, the topic of the aims of theories is but a means to articulate the contrast between anti-realist (or instrumentalist or empiricist) and realist views of science. This latter tendency seems to go back to Duhem's *Aim and Structure of Physical Theory* of 1906 (Duhem (1954 [1906], chs. 1, 2). More recently, van Fraassen has revived this attitude in his highly influential *The Scientific Image* (van Fraassen 1980, see esp. 8, 12), and many others have followed him (see, e.g., Ellis (1996 [1985]); Sober (1990, 394)). However, the general questions about the aims of theories contain more than just the alternative between a realist understanding of theories (associated with their explanatory potential by means of theoretical entities) and an anti-realist one (associated with their increasing empirical adequacy only). Also, the last attempt that I have seen does not work: "The goal [of science] is the systematic generation of knowledge that makes a difference"

(Börner 2010, 53). Although I enjoy this statement because of the systematicity involved, it is not good enough to distinguish science from other knowledge-seeking enterprises that also try to systematically generate knowledge that makes a difference like, for instance, criminal investigations. In September 2005, I received the ultimate push away from the idea that "goals of science" could do some work for systematicity theory. That happened in the discussion following a talk on systematicity theory I gave at Cornell University. Professor Harold Hodes cornered me. I am grateful to him.

p. 120, "**This reflects the general tendency of systematicity theory**": We shall see this tendency later in greater detail when discussing the relation between our concept of systematicity and older concept of methodicity. Both concepts aim to characterize science. "Methodicity" will turn out to be a special case of the more general "systematicity"; see section 4.2.2. This sort of relationship is not new in the history of scientific procedures. The transition between what I described in section 1.1 as phase 1 and phase 2 of the answers to the question "What is science?" has the same character. "Deductive proof" is a special case of "scientific method"; the latter weakens and thereby generalizes the former.

p. 120, "**It is a system in the sense of a rather loose assembly**": This resonates with the weak kind of unity of science generated by family resemblances between different concretized concepts of systematicity about which I have already spoken in section 2.2. The kind of weak unity of science discussed there concerns something like a formal aspect, namely systematicity in general, whereas the kind of weak unity of science discussed here concerns an aspect of content, namely, epistemic connections.

p. 121, "**the 'Shell Eco-marathon,' where the goal is to build a car that drives as far as possible with the least amount of energy**": See http://www.shell.com/home/content/ecomarathon/ (accessed Dec. 3, 2012).

p. 121, "**one of the participant groups set a new world record for fuel efficiency at an amazing 5,385 kilometers with hydrogen equivalent to one liter of gasoline**": See http://www.paccar.ethz.ch/news/index (accessed Dec. 3, 2012).

p. 121, "**According to the project director, the motivation of the project was . . . .**": The following quotes are from the homepage of the project: http://www.paccar.ethz.ch/ (accessed Sept. 27, 2011).

p. 122, "**any application of scientific knowledge for nonscientific purposes is epistemically less connected than the scientific knowledge itself**": It is important to note that this really only holds for nonscientific applications. For inner-scientific applications, i.e., the generation of data to be used for theory testing, we have again a high degree of epistemic connectedness, as it is to be expected.

p. 123, "**mathematical game theory**": For a very first orientation about game theory and its many applications, see, e.g., Binmore (2007).

p. 123, "**In an article published in the scientific journal *Third World Quarterly***": Reiche (2011b).

p. 123, "**an article in a German quality newspaper, *Frankfurter Allgemeine Zeitung*, covering the same topic**": Reiche (2011a).

p. 125, "**From the seventeenth century through the twentieth century, science grew roughly exponentially with a doubling time of roughly fifteen to twenty years**": The classic study of scientometrics, the quantitative study of science dealing with the growth of science, is Solla-Price (1963). He gave somewhat higher growth rates. I am relying here on more recent data reported, for instance, by Fahrbach (2011), Schummer (1997), and Vickery (2000, xxii (quoted in Fahrbach (2009, 103)).

p. 125, "**including very general social conditions—like the existence of cities**": On this and other social conditions, see, e.g., the classic work Zilsel (2003 [1942]).

p. 125, "**I claim that science has an ideal of completeness**": As there are groups who discuss the ideal of completeness of scientific knowledge as if they had discovered this subject, I may note that already in my very first publication on systematicity theory, I discussed the ideal of completeness (Hoyningen-Huene 1982, 25).

p. 125, "**Intentions of collective agents are a difficult and recently much-discussed subject**": namely in action theory and in ethics. Particularly in the latter, the important question of shared responsibility arises. Searching the Internet for "shared intentions" produces many hits, and how many depends on your particular search engine. Mine, whose name I don't convey, produced roughly 16,200 hits on May 17, 2011. At any rate, those who are interested in the literature on shared intentions will find plenty.

p. 125, "**talk of cognitive values that hold in scientific communities has gotten wide currency**": This began with the work of Thomas Kuhn, especially Kuhn (1962) and Kuhn (1977). For a recent interesting analysis of Kuhn's epistemic values "using the machinery of social choice theory," see Okasha (2011); he thinks that Kuhn's values are widely accepted in the philosophy of science (p. 84).

p. 127, "**The axiomatization 'should allow for a derivation of *all* the known theorems of the discipline in question'**": The quote is from Corry (2004) and reports the view of the great mathematician David Hilbert at the end of the nineteenth century. This is commonplace today. There exists also a narrow, technical sense of "completeness" in mathematical logic. "Completeness theorems" try to establish it, and it very roughly means that one axiom system can completely reproduce another one.

p. 127, "**stricter than in other sciences because here, successful classifications always have to come with a *proof* of completeness**": See for the following, e.g., Elwes (2006, 3).

p. 127, "**all objects fulfilling a given definition, can be exhaustively listed in detail**": The "in detail" clause is very important: it excludes trivial forms of completeness that would be achieved by introducing residual categories that contain "all the rest."

p. 128, "'**This is exactly the sort of theorem that researchers in many areas of mathematics would absolutely love to prove'**": See Elwes (2006, 3).

p. 128, "**A truly extreme example is the classification of finite simple groups**": For the following presentation, I am using two sources. One is Elwes (2006), an intuitive, nontechnical introduction to the classification of finite simple groups; it is available on the Internet at http://plus.maths.org/content/os/issue41/features/elwes/index (accessed Dec. 3, 2012). The other source is Wilson (2009) which, in its own words, is the first attempt "to bring within a single cover an introductory overview of all the finite simple groups" (p. V). The preface and the table of contents are available on the Internet for free: http://www.springerlink.com/content/k61214/front-matter.pdf (accessed Sept. 27, 2011). From its table of contents, one may get a first impression of the complexity of the set of all finite simple groups. Wikipedia features a list of finite simple groups at http://en.wikipedia.org/wiki/List_of_finite_simple_groups (accessed Sept. 27, 2011) whose correctness I cannot guarantee. Honestly speaking, I cannot guarantee the correctness of Wilson (2009) either, but at least I know that the author is an accomplished mathematician and that the book underwent some anonymous refereeing (p. VIII). For the history of the classification theorem, see Solomon (2001). This article is also available on the Internet at http://www.ams.org/journals/bull/2001-38-03/S0273-0979-01-00909-0/S0273-0979-01-00909-0.pdf (accessed Sept. 27, 2011).

p. 128, "'that no-one in the world today completely understands the whole proof'": See Elwes (2006, 1).

p. 128, "This looks a little ugly to a mathematical mind": See, e.g., a statement of the important mathematician Thompson, which I quote from Solomon (2001, 345): "The great sticking point, though there are several, concerns the sporadic groups. I find it aesthetically repugnant to accept that these groups are mere anomalies."

p. 129, "'undoubtedly one of the most extraordinary theorems that pure mathematics has ever seen'": Elwes (2006, 1).

p. 129, "Physical theories make statements about the state of the universe as close as 10^{-35} seconds after the big bang": I am alluding to the theory of inflation in cosmology. For a popular exposition see, e.g., Greene (2004, chs. 10 and 11, esp. p. 285).

p. 129, "the ultimate theory that they search for: the 'theory of everything' (T.O.E.)": See, e.g., Greene (2000, 16).

p. 129, "their four elements they postulated were 'the fundamental basis of theoretical chemistry until the eighteenth century'": Brock (1993, 13).

p. 129, "the periodic system of elements": See, e.g., Brock (1993, ch. 9). For an in-depth analysis, see, e.g., Scerri (2007).

p. 129, "the human genome project": See its web page at http://www.ornl.gov/sci/techresources/Human_Genome/home.shtml (accessed Sept. 27, 2011); the quotes are from this web page.

p. 130, "the complete proteome set for *Homo sapiens* and *Mus musculus*": See the project's web page at http://www.uniprot.org/news/2011/05/03/release (accessed Sept. 27, 2011).

p. 130, "'a complete proteome is the set of protein sequences ....'": See http://www.uniprot.org/keywords/KW-0181 (accessed Sept. 27, 2011). I have only quoted the first part of the definition, which is enough for our purposes.

p. 130, "to study the evolution of a particular population, namely, the population of finches on one of the Galápagos Islands": For a popular description of the project (with references to the original research literature) see Weiner (1995).

p. 130, "the theory of plate tectonics": See, e.g., Kious and Tilling (1996). This book can be downloaded from the Internet at http://pubs.usgs.gov/gip/dynamic/dynamic.html#anchor19309449 (accessed Sept. 27, 2011), and I am using it in the following. There are countless other publications on plate tectonics.

p. 131, "the lithosphere can be completely described in terms of tectonic plates": See, e.g., http://pubs.usgs.gov/gip/dynamic/slabs.html (accessed Sept. 27, 2011).

p. 131, "there are different types of plate boundaries": See http://pubs.usgs.gov/gip/dynamic/understanding.html#anchor15039288 (accessed Sept. 27, 2011).

p. 131, "'The fact that the tectonic plates have moved in the past and are still moving today ....'": The quote is from http://pubs.usgs.gov/gip/dynamic/unanswered.html (accessed Sept. 27, 2011).

p. 132, "'all known accounts of timed eclipse observations and predictions made by early astronomers....'": See Steele (2000, 3).

p. 132, "**stocktaking of all knowledge in encyclopedias**": For the following, see, e.g., Börner (2010, 18–19).

p. 132, "**two important English language encyclopedias covering all of philosophy**": See Edwards (1967) and Craig (1998).

p. 132, "**the *Stanford Encyclopedia of Philosophy***": See http://plato.stanford.edu/ (accessed Sept. 27, 2011).

p. 132, "**Without attempting to critically assess these sometimes worshipping stories**": The historian of science, Thomas S. Kuhn, has stressed that many popular astonishing stories about revolutionary discoveries turn out to not stand up to historical scrutiny. When analyzed in their appropriate historical context and in sufficient detail, the respective processes often turn out to consist of more and smaller steps whose order is quite comprehensible and which were plausibly in the intellectual reach of the acting scientist. See, e.g., Hoyningen-Huene (1993, 22), which mainly refers back to Kuhn (1984).

p. 135, "**gigantic optical telescopes exceeding the power of today's largest telescopes by a factor of up to one hundred**": See, e.g., an article in *Nature* at the beginning of the "year of astronomy": Kanipe (2009).

p. 135, "**Wilkinson Microwave Anisotropy Probe (WMAP)**": See http://map.gsfc.nasa.gov/ and http://map.gsfc.nasa.gov/mission/ (accessed Sept. 27, 2011).

p. 135, "**the age of the universe was determined to be 13.73 billion years within 1 percent**": NASA's WMAP website even boasts that this figure is "definitive": "WMAP definitively determined the age of the universe to be 13.73 billion years old to within 1 percent" (http://map.gsfc.nasa.gov/, WMAP's Top Ten, # 2, accessed Sept. 27, 2011). Whether this age determination is really definitive remains to be seen; at least, the wording of the quoted sentence should change.

p. 135, "**Sloan Digital Sky Survey**": I take the information from its web page http://www.sdss.org/ (accessed Sept. 27, 2011).

p. 135, "**The data produced by the Sloan Digital Sky Survey have been used in a wide range of astronomical investigations**": See http://www.sdss.org/signature.html (accessed Sept. 27, 2011).

p. 136, "**the first accelerator in which particles traveled a spiral path, called a cyclotron, was put to use in 1932 by Ernest Lawrence**": For this and the following, see, e.g., http://www.lbl.gov/Science-Articles/Archive/early-years.html (accessed Sept. 27, 2011).

p. 137, "**'LHC' stands for Large Hadron Collider**": For information about the LHC, go to its web page at http://lhc.web.cern.ch/lhc/ (accessed at Sept. 27, 2011). Apart from its sheer physical size, there are many more superlatives the LHC experiments boast. Unfortunately, this is not the place to describe them. There are many links on the above web page leading to additional information.

p. 137, "**a paper in which a certain aspect of proton collisions was investigated**": See ATLAS (2011); the paper is available on the Internet at http://arxiv.org/abs/1009.5069v1 (accessed Sept. 27, 2011).

p. 137, "**paleoclimatology, the study of climate prior to the widespread availability of records**": See, e.g., http://www.oar.noaa.gov/climate/t_paleo.html (accessed Sept. 27, 2011).

p. 137, "**Ice cores are taken from continental glaciers by hollow drills. Several ice cores of more than 3,000 meters in depth have been taken**": See, among others, the web pages of the "Greenland Ice Sheet Project 2," which recovered an ice core of 3,053 meters in depth and finished drilling in July 1993 (http://www.gisp2.sr.unh.edu/, accessed Sept. 27, 2011), and of the

"West Antarctic Ice Sheet Divide Ice Core Project," where drilling was finished in January 2011 and resulted in an ice core of 3,331 meters in depth (http://waisdivide.unh.edu/, accessed Sept. 27, 2011). Further information is taken from http://www.ncdc.noaa.gov/paleo/icecore/greenland/summit/document/, accessed Sept. 27, 2011.

p. 137, "**On the study's web page, it is described as follows**": See http://adultdev.bwh.harvard.edu/research-SAD.html (accessed Sept. 27, 2011). For some results of the study, see, e.g., Vaillant (2002) (the first chapter of the book in which the study's design is outlined is available at http://www.mamashealth.com/book/, accessed Sept. 27, 2011). In this book, the results of the longest prospective study of women's development in the world from the "Terman study of gifted children" have been incorporated (see pp. 17, 21–23).

p. 138, "**Take the Leibniz Academy Edition as an example**": The main page of the edition is in German: http://www.leibniz-edition.de/. However, most of the relevant information is available in English at http://www.bbaw.de/bbaw/Forschung/Forschungsprojekte/leibniz_potsdam/en/Startseite; see also UNESCO's "Memory of the World" web page on Leibniz' letters: http://portal.unesco.org/ci/en/ev.php-URL_ID=22464&URL_DO=DO_TOPIC&URL_SECTION=201.html (all accessed Sept. 27, 2011).

p. 138, "**The discovery of penicillin by Alexander Fleming**": See, e.g., Flemings own account of his discovery in Fleming (1946).

p. 138, "**as Fleming writes**": Fleming (1946, 2–3).

p. 139, "**important strategy of contemporary drug development**": See, e.g., http://watcut.uwaterloo.ca/webnotes/Pharmacology/page-1.6.html (accessed Sept. 27, 2011).

p. 139, "**ever since the seventeenth century mathematics' development has been strongly influenced by the challenges posed by natural science**": One of the most impressive examples is, of course, the development of calculus by Newton and Leibniz due to the needs of mechanics.

p. 139, "**Computers and software technology have been invented and developed largely in the context of scientific applications**": There are, of course, countless books describing this development. One example is Ceruzzi (2003). Those who are not afraid of or forbidden to use Wikipedia may consult http://en.wikipedia.org/wiki/History_of_computing_hardware (accessed Sept. 27, 2011).

p. 139, "**bioinformatics**": See, e.g., http://www.ncbi.nlm.nih.gov/About/primer/bioinformatics.html (accessed Sept. 27, 2011).

p. 140, "**business informatics**": See, e.g., http://www.wim.uni-mannheim.de/en/future-students/about-business-informatics/what-is-business-informatics/ (accessed Sept. 27, 2011).

p. 140, "**For instance, radiocarbon dating (also called the C14 method) is**": For introductory information, see http://www.c14dating.com/int.html; further information can be gleaned from the parent page, http://www.c14dating.com/. For exemplary applications of the technique in archaeology and other fields, see http://www.c14dating.com/applic.html (all accessed Sept. 27, 2011).

p. 140, "**Whenever scientific knowledge from other domains can be productively used, it will be systematically imported and utilized**": I don't want to paint an idealized picture of the sciences here. Scientists may also ward off the intrusion of knowledge from other domains when they see it as threatening the integrity of their own field. Especially in the social sciences and the humanities, but not only there, relationships between somehow neighboring disciplines can be quite tense, mainly due to vastly divergent points of view and incompatible basic assumptions (the same holds, of course, for competing schools within a certain field). I very vividly remember a case in the 1980s at the University of Zurich when I unsuccessfully invited a social historian and a sociologist who both worked on migration to an interdisciplinary panel discussion; they simply refused to talk to each other (not for personal reasons). In cases like this, imports from integrity-threatening rival fields are far from welcome. In the above sentence from the main text, I have taken care of this possibility by speaking of "productive use."

p. 141, "**An alternative positive view was developed by the historian and philosopher of science, Thomas S. Kuhn**": For more details, see Hoyningen-Huene (1993, esp. chs. IV and V).

p. 141, "**an autocatalytic process**": Other authors have also used this expression to characterize science's growth mode: see, e.g., Akeroyd (1990, esp. 409).

p. 141, "**Over several centuries, this has indeed been observed for modern natural science**": See section 3.7.1.

p. 142, "**Mathematics is characterized by, among other things, its tendency toward abstraction**": See, among a host of much graver publications on this matter, e.g., Hoyningen-Huene (2004, 182–83).

p. 142, "**when turning to logic, the character of mathematical objects as something special**": Historically speaking, what I am saying here is not quite correct. Logic, at the time of Aristotle, was not seen as a part of mathematics. I am projecting back a modern viewpoint, which is, among historians, usually an unforgivable sin. Reviews of this book in historical journals may duly note this.

p. 142, "**Aristotle invented the concept of logical form as a key element of his codification of syllogistics**": His syllogistic is contained in the *Analytica priora*; an easily accessible edition is, e.g., Aristotle ([1973]).

p. 143, "**In 1879, he published his *Begriffsschrift***": Frege (1879).

p. 143, "**Frege's formula language is as systematic as it is cumbersome because it is two-dimensional**": Readers who are brave enough not be afraid of Wikipedia may get an impression of the notation at http://en.wikipedia.org/wiki/Begriffschrift (accessed Sept. 27, 2011), but only if they promise to return promptly. How dangerous Wikipedia is can already be seen from the missing second "s" in the URL's address.

p. 143, "**representation of the systematic order of the chemical elements in the periodic system**": See, for instance, the "official" table of the International Union of Pure and Applied Chemistry, IUPAC, in Connelly et al. (2005, 2). This book is also available on the Internet at http://old.iupac.org/publications/books/rbook/Red_Book_2005.pdf (accessed Sept. 27, 2011). For an in-depth presentation of the history and the meaning of the periodic table, see Scerri (2007).

p. 144, "**there are quite a few variants of the standard representation of the periodic system**": For "a compilation of more than 700 representations of the periodic system" (Scerri 2007, xiv), see Mazurs (1974); Scerri devotes a short section on the variety of periodic tables in his book on pp. 277–86. For a quick and potentially unreliable look at some of the variants, see http://en.wikipedia.org/wiki/Alternative_periodic_tables (accessed Sept. 27, 2011).

p. 144, "**more than sixty-nine million unique organic and inorganic chemical substances**": For this and the following enormous numbers, see CAS, the Chemical Abstract Service, a division of the American Chemical Society, at http://www.cas.org/content/chemical-substances (accessed Dec. 3, 2012).

p. 144, "**CAS Registry Number**": See http://www.cas.org/content/chemical-substances/faqs (accessed Dec. 3, 2012).

p. 145, "**The latest version of the rules for naming inorganic compounds**": See Connelly et al. (2005). This book is available on the Internet at http://old.iupac.org/publications/books/rbook/Red_Book_2005.pdf (accessed Sept. 27, 2011). The quotes a little further down in the text are on p. 4. A book that presents chemical nomenclature in a less technical way that is accessible to students and teachers (according to its self-description on p. vii), is Leigh, Favre, and Metanomski (1998). This book is available on the Internet at http://old.iupac.org/publications/books/principles/principles_of_nomenclature.pdf (accessed Sept. 27, 2011).

p. 145, "**Organic chemistry has its own nomenclature**": See IUPAC (1993). The online version is at http://www.acdlabs.com/iupac/nomenclature/ (accessed Sept. 27, 2011). Readers who can't get enough of chemical nomenclature may also consult IUPAC's nomenclature homepage at http://www.chem.qmul.ac.uk/iupac/index.html (accessed Sept. 27, 2011).

p. 145, "**Charles Darwin's groundbreaking book *On the Origin of Species***": A convenient facsimile edition of the first edition is Darwin (1964 [1859]). The diagram is available on the Internet at, e.g., http://darwin-online.org.uk/graphics/Origin_Illustrations.html or http://commons.wikimedia.org/wiki/File:Origin_of_Species.svg (both accessed Sept. 27, 2011). For a contemporary, colorful representation of the tree of life, see, e.g., http://tellapallet.com/tree_of_life.htm (accessed Sept. 27, 2011).

p. 145, "**the physical model of the structure of DNA that James Watson and Francis Crick devised**": The story of the discovery of the structure of DNA was written up by James Watson in a 1968 book that is now a classic. A convenient facsimile reprint is contained in Stent (1980).

p. 146, "**'a possible copying mechanism for the genetic material'**": This quote is from the last sentence of Watson and Crick's original 1953 paper announcing "A structure for Deoxyribose Nucleic Acid" (Watson and Crick 1953). The paper is conveniently reprinted in Stent (1980).

p. 146, "**representation of knowledge by maps**": Avoiding the overwhelmingly abundant literature, for the purpose of this paragraph, I used mainly Börner (2010). Scientific mapmaking is dealt with on p. 11; a reconstruction from 1482 of Ptolemy's map is on p. 79. Hundreds of examples of geographic and nongeographic maps can be found throughout the whole book. Those who are still hungry for more material about maps should consult the references section of this book (pp. 212–46).

p. 146, "**Mapmaking has produced hundreds of different sorts of maps**": See, e.g., the David Rumsey Map Collection at http://www.davidrumsey.com/ (accessed Sept. 27, 2011). The web page makes accessible more than 27,000 maps out of more than 150,000 items contained in the collection.

p. 146, "**contained in the *Atlas of Science***": Börner (2010).

NOTES TO CHAPTER 4: COMPARISON WITH OTHER POSITIONS

p. 149, "**so-called infinitesimals, that is, infinitely small quantities that have later been seen as illegitimate**": However, I should add here that in the 1960s, a branch of mathematics called nonstandard analysis was developed that reconstructs and vindicates talk and use of infinitely small quantities. A somewhat illegitimate source for pedestrians is the Wikipedia article on nonstandard analysis at http://en.wikipedia.org/wiki/Non-standard_analysis (accessed Sept. 27, 2011). By contrast, a fully legitimate source is Robinson (1996), a monograph by the founder of nonstandard analysis, Abraham Robinson.

p. 150, "**the classical authors involved in reflections about systems and systematicity, Leibniz, Wolff, Lambert, and Hegel**": Some pertinent information about these authors is available in chapter I of Rescher (1979), entitled "Historical Stagesetting." See also Sturm (2009, ch. III, §§ 6–7).

p. 150, "**Aristotle (384–324 BC) is a good starting point for our discussion**": For a beginner's introduction to Aristotle's philosophy of science, see Losee (2001, 4–13); for an excellent book-length discussion of the issues that are relevant for us, see McKirahan (1992).

p. 150, "**the very first treatise fully devoted to the philosophy of science, *Posterior Analytics***": The standard English edition is Aristotle ([1976]); a good bilingual edition is Aristotle ([1960]).

p. 151, "**humans have the capacity to grasp the necessary truth of these principles without proof**": A discussion of this topic can be found in McKirahan (1992, ch. 18).

p. 151, "**explanation, where Aristotle posits deduction from principles**": There is a superficial similarity to the deductive-nomological account of explanation that does not, however, go very far; see McKirahan (1992, 230–31).

p. 152, "**The work most relevant in the context is his first published book, *Discours de la Méthode***": The English standard edition is Descartes (1984 [1637]). Also relevant is his unpublished work (Descartes 1984 [1620-c.28]), but for our purposes, it is not necessary to draw on this early, unfinished, and only posthumously published work.

p. 152, "**that it has been my singular good fortune to have very early in life fallen in with certain tracks**": Descartes (1984 [1637], pt. 1, § 3).

p. 153, "**I believed that the four following [precepts] would prove perfectly sufficient for me**": Descartes (1984 [1637], pt. 2, § 6).

p. 153, "**Here are the four precepts that Descartes commits himself to**": Descartes (1984 [1637], pt. 2, § 6).

p. 155, "**Kant (1724–1804) is a well-known champion of systematicity**": This section owes very much to extensive e-mail exchanges with Thomas Sturm in December 2007, December 2008, and June 2011. In addition, I would like to thank Thomas for sending me chapter 3 of his dissertation on "Kant's Concept of Science" before it went to print (it appeared as Sturm (2009)); furthermore, for sending me the manuscripts of his entries "Wissenschaft" and "Naturwissenschaft," appearing in the new Kant-Lexikon (Sturm (in press) and Sturm and De Bianci (in press)), which also helped a lot. Doing full justice to the sophistication of this literature would require much more space than is available. At any rate, Thomas's interventions set me straight. I had first thought that in Kant systematicity was just axiomatization.

p. 155, "**'[S]ystematic unity is what first raises ordinary knowledge to the rank of science....'**": Kant and Smith (2003 [1781/1787], A832/B860 ("A" and "B" refer to the original pagination of the first and second edition, respectively)). The *Critique of Pure Reason* is also available on the Internet: http://www.hkbu.edu.hk/~ppp/cpr/toc.html (accessed Sept. 27, 2011). In this section, I am dealing with the "critical" Kant only, i.e., with his publications after 1781 (the year in which the first edition of his *Critique of Pure Reason* appeared).

p. 155, "**'Every discipline if it be a system—that is, a cognitive whole ordered according to principles—is called a science'**": Kant and Friedman (2004 [1786], Preface, p. 3). As Rescher rightly notes, this is put a bit too strongly: "Systematicity is no doubt a *necessary* condition for a science, but scarcely a *sufficient* one, since the rules of an art (sonnet writing, chess playing) can also be systematized" (Rescher 1979, 21–22).

p. 155, "**the 'concept of *system* is perhaps the most central idea of Kant's theory of knowledge'**": Rescher (1983, 83). This paper deals in great detail with the role of systematicity in Kant. However, I do not agree with Rescher about the character of Kant's systematicity concept. According to Rescher, for Kant, "the systemic unity of knowledge is fundamentally akin to the functional integrity of an organism," and "every part of … science-as-a-whole must serve in the role of a contributory subsystem: and organ of the overall organism" (p. 86); see also Rescher (1979, 12–13). This characterization does not seem to be really born out by the quotes of the *Critique of Pure Reason* that Rescher adduces, and it does not at all sit well with the example of geometry that should be such an organism-like system, but is in fact an axiomatic system that is organized thoroughly top-down (Rescher himself mentions geometry as a "paradigm of system that lay before Kant's eyes": p. 84). I think that Kant uses the analogy of an epistemic system to an organism only with respect to the forming and growth of the system, but not with respect to its structure: see, e.g., A833/B861 and A835/B863. An organism is much more like the network model that Rescher propounds explicitly as an alternative to the Euclidean model of systematization in Rescher (1979, 43–50).

p. 155, "'**In accordance with reason's legislative prescriptions, …**'": Kant and Smith (2003 [1781/1787], A832/B860).

p. 156, "**As can be seen from the German original, Kant means the multitude of different pieces of knowledge**": "Modes of knowledge" is the translation of the German "*Erkenntnisse*" that is the plural of "*Erkenntnis*." In the given context, "*Erkenntnis*" could be translated as knowledge, but unfortunately, there is no plural form of "knowledge." Therefore, the translation uses the workaround "modes of knowledge" for the plural of "knowledge."

p. 156, "'**as the result of a haphazard search,' as he puts it elsewhere**": Kant and Smith (2003 [1781/1787], A81/B106).

p. 156, "**his classification of what we call the natural sciences in three main groups: historical doctrine of nature, improperly so-called natural science and properly so-called natural science**": Kant and Friedman (2004 [1786], Preface, p. 4). The quotes following in the main text are from the same location. An excellent commentary on Kant's *Metaphysical Foundations of Natural Science* is Pollok (2001), unfortunately available only in German.

p. 156, "**As an example of such an improperly called science, Kant uses chemistry**": Kant and Friedman (2004 [1786], 4).

p. 156, "**induction can never establish apodictic certainty of some general statement**": In Kant's own terms: "This strict universality of the rule is never a characteristic of empirical rules; they can acquire through induction only comparative universality, that is, extensive applicability" (Kant and Smith 2003 [1781/1787], A91f./B124).

p. 157, "**does not reach the level of science (in his sense) because it lacks laws of any kind**": See, e.g., Pollok (2001, 61).

p. 157, "**'ground' and 'consequence' are important here because they are intended to cover two relations that are thought of as one piece**": See, e.g., Pollok (2001, 63 note 90).

p. 157, "'**The unity of the end to which all the parts relate …**'": Kant and Smith (2003 [1781/1787], A832–833/B860–861).

p. 158, "'**the systematic connection which reason can give to the empirical employment of the understanding …**'": Kant and Smith (2003 [1781/1787], A680/B708).

p. 159, "**it is a family of positions that are in constant historical change**": See, e.g., the contributions to Richardson and Uebel (2007) with many references to further relevant work.

p. 159, "**the pertinent topics that logical empiricism dealt with regarding philosophy of science**": There is, of course, a vast amount of literature on these topics. The pertinent more comprehensive books by logical empiricists include Carnap (1966), Cohen and Nagel (1934), Hempel (1965b), Hempel (1966), and Nagel (1961).

p. 160, "**Carl G. Hempel begins a paper, one of the central publications on this topic**": Hempel (1965c, 137).

p. 160, "**logical empiricists used systematicity terminology in order to describe the role of scientific predictions and explanations**": The first Hempel quote is from Hempel (1965b, 488); the Nagel quote is from Nagel (1961, 4); the second Hempel reference refers to the reprint of Hempel (1958, 173–77).

p. 161, "**Nagel contrasts common sense with science**": The quote is on p. 5 of Nagel (1961).

p. 161, "**among the many dissents between Popper and the logical empiricists, two rather deep**": See Popper's own account of these disagreements in chapter I of his *The Logic of Scientific Discovery*, originally published in 1934 (Popper 1959 [1934]).

p. 162, "**Popper was the first one who seriously contemplated these questions**": at the end of the last chapter of Popper (1959 [1934]).

p. 163, "**in the process of scientific development, different phases can be distinguished that follow some pattern**": See Kuhn's bestselling *The Structure of Scientific Revolutions* (Kuhn 1970 [1962]). There is a very large amount of secondary literature commenting on Kuhn; it belongs to the most frequently cited scholarly publications of all times (at the time of this writing in August 2011, Google scholar featured more than forty-eight thousand quotes for just the English edition of *Structure*). A very accessible reader's guide to Kuhn's *Structure* is Preston (2008); an in-depth analysis of Kuhn's theory and its philosophical underpinnings is Hoyningen-Huene (1993); Kuhn's philosophical significance is discussed in Hoyningen-Huene (1998).

p. 164, "**Kuhn was cautious enough not to claim that this phase model is always strictly followed**": See Kuhn (1970 [1962], 11–12). For more details of the phase model, see, e.g., Hoyningen-Huene (1993, sec. 1.3, 24–27).

p. 164, "**normal science that is especially suited for a comparison with systematicity theory**": On normal science, see Kuhn (1970 [1962], chs. III–V); for commentary, see Hoyningen-Huene (1993, ch. 5).

p. 164, "**Kuhn classifies the typical problems tackled in the phase of normal science in three groups**": See, e.g., Hoyningen-Huene (1993, 81–82).

p. 165, "**the potential of paradigmatic problems and solutions for further research is systematically exploited**": For further details, see, e.g., Hoyningen-Huene (1993, 159–62).

p. 166, "**his famous slogan 'anything goes,' best known from his book *Against Method*, originally published in 1975**": Feyerabend (1975, 28). The later editions of this book, Feyerabend (1988) and Feyerabend (1993), although in some parts substantially different from the first edition, still contain the slogan. Already in 1970, Feyerabend had used the slogan in the preliminary book chapter form of his later book (Feyerabend 1970, 26). For the connection of this slogan with Feyerabend's earlier philosophical development, see Oberheim (2007, esp. 281–83).

p. 166, "**The target of Feyerabend's attack in *Against Method***": Parts of what follows are taken, slightly varied, from Hoyningen-Huene (2000b, 11–13).

p. 166, "**as one of his papers is entitled, 'the limited validity of methodological rules'**": Originally, the paper appeared in 1972 in German as Feyerabend (1972); a translation appeared posthumously as Feyerabend (1999b).

p. 167, "**it came with an ironic footnote about his surprise that people had not noticed that he was joking**": Feyerabend (1970), p. 105 n. 38: "Some of my friends have chided me for elevating a statement such as 'anything goes' into a fundamental principle of epistemology. They did not notice that I was joking. Theories of knowledge as I conceive them *develop*, like everything else. We find new principles, we abandon old ones. Now there are some people who will accept an epistemology only if it has some stability, or 'rationality' as they are pleased to express themselves. Well, they can have such an epistemology, and 'anything goes' will be its only principle." I wish to thank Eric Oberheim who made me aware of this footnote. See also Feyerabend (1978, 39–40, 188).

p. 167, "**Rather casually, one finds an abstract justification**": See, for example, Feyerabend (1975, 295–96).

p. 167, "**Feyerabend is completely aware of the limited scope of his argumentative strategy**": at least in 1993; the quote is from Feyerabend (1993, 1).

p. 167, "**Introduction to the Chinese Edition**": This introduction is reprinted, fortunately enough in English, in Feyerabend (1993, 1–4).

p. 168, "**'[Science] is a collage, not a system'**": Feyerabend (1995, 43). The next quote is from the same page.

p. 168, "**'indicates how future statements about 'the nature of science' may be undermined….'**": The quote is again on p. 1 of Feyerabend (1993).

p. 168, "**historical processes in the sciences are, on the whole, so diverse and multifaceted that there simply are no substantial generalizations under which they can be subsumed**": This conviction of Feyerabend is also at the core of his criticism of what he calls the rationalist (Western) tradition, with its associated abstractness; see, e.g., Hoyningen-Huene (2000b, 13). It should be noted that many historians and philosophers of history share Feyerabend's conviction of—more generally put—the endless diversity of historical processes such that there are no substantial generalizations about them.

p. 169, "**specific historical examples of science in which the main thesis of systematicity theory is violated**": As we shall see in section 5.1, that systematicity theory can also be applied to the genesis and dynamics of science. Systematicity theory states for these processes that they are characterized by an increase of (overall) systematicity. In principle, counterexamples to systematicity theory could also be found in this dynamic area. However, the strategy to make this look rather unlikely would be the same as in the case of the main thesis of systematicity theory discussed in the main text.

p. 170, "**Nicholas Rescher is, as far as I know, the only philosopher in the twentieth century who has more than casually dealt with systematicity**": Rescher (1979); the following three quotes are from this book on p. 1 and twice on p. 2. I wish to thank Nick Rescher for critically reading this section.

p. 170, "**Rescher explains coherentism and its connection to systematization as follows**": See Rescher (1979, 1–2); the rest of the paragraph refers to pp. 3–4.

p. 171, "**This idea of systematization is even more important when it comes to science**": The quotes in this paragraph are from Rescher (1979, 22, 21).

p. 171, "'**the prospect of organizing a body of claims systematically is crucial to its claims to be a science**'": Rescher (1979, 21). Lambert is not mentioned in this section, but it is evident from pp. 8–10 that he is part of this tradition. The reference to Hegel is particularly clear on p. 1 of Rescher (1979): "For Hegel and his school—especially the English Neo-Hegelians—system was not just an important aspect but *the characterizing feature* of our knowledge. The present study of systematization proceeds within the frame of reference set by these large historical claims on its behalf."

p. 171, "**we have to understand the context in which methodological pragmatism is introduced and developed**": A concise characterization of methodological pragmatism and its context can be found in chapter 1 of Rescher (2001). A book-length treatment of this position is his older book Rescher (1977).

p. 172, "**the route taken by Rescher; he calls it methodological pragmatism**": In the following two paragraphs, I am mainly using chapter 1 of Rescher (2001), chapter II of Rescher (1979), and some parts of Rescher (2005).

p. 173, "**systematization is 'a testing-process for acceptability'**": Rescher (1979, 30).

p. 174, "**Take, for example, his statements**": See Rescher (1979, 19, 22).

p. 174, "**sentences like these are extremely close to sentences that one can find in the context of systematicity theory**": Note first that I speak of sentences, not of statements: the sentences may look similar, but the statements made by them differ in the different contexts. Note second that I only say "extremely close." Rescher's first sentence is normative, whereas systematicity theory is, first of all, descriptive (see section 2.1.2, second remark, p. 29); it would not, at least not without further ado, license a normative statement like Rescher's. Rescher's second sentence, transported in the context of systematicity theory, would be a somewhat imprecise statement in omitting the fundamentally comparative nature of systematicity theory's main thesis (see section 2.1.2, third remark, p. 29).

p. 174, "**I use 'systematicity' as derived from the adjective 'systematic'**": Interestingly enough, in the mid-eighteenth century the French mathematician, physicist, and encyclopedist Jean le Rond d'Alembert noted essentially the same difference as the one between Rescher and me as the difference between a "spirit of systems" and a "systematic spirit," the latter being characteristic of the sciences. More details can be found in a footnote on p. 217 which refers back to the main text on p. 14.

NOTES TO CHAPTER 5: CONSEQUENCES FOR
SCIENTIFIC KNOWLEDGE

p. 180, "**The first step of reflection may be an attempt to bring some order into the variety of pertinent phenomena**": According to Reineke (1982b, 106), it is "typical of an early phase of development of a science that a collection, inspection and ordering of empirical data is pursued" (my translation).

p. 180, "**take the case of mathematics**": My sources are Waerden (1963, chs. I–III) (in fact, I used the German version Waerden (1966) of the book; the original was published in Dutch in 1950); Oelsner (1982b); Reineke (1982b); and Lindberg (1992, 13–18).

p. 182, "**Proofs have been introduced by Greek mathematics**": My main source is Waerden (1963, esp. ch. IV) (again, I used the German version Waerden (1966)).

p. 183, "the development of biology in Soviet Russia from the 1940s to the 1950s, connected with the name Lysenko": There are many publications on this topic; see, e.g., Roll-Hansen (2005).

p. 183, "Euclid's *Elements*, written roughly at that end of the fourth century BC": There are numerous editions of this book on the market; the standard translation and commentary is Euclid (2000 [1925]). Unfortunately, the original edition of *Elements* is no longer available. For further commentaries, see., e.g., Lindberg (1992, 86–89) and Waerden (1963, esp. ch. VI).

p. 184, "may, for the sake of argument, introduce a distinction similar to one that has been made famous by historian and philosopher of science Thomas S. Kuhn": The distinction is introduced in his classic Kuhn (1970 [1962]); the first edition was published in 1962. For extended commentary in the distinction, see, e.g., Hoyningen-Huene (1993, chs. 5–7). I am aware of the fact that this distinction is controversial. Here, I am using it temporarily for illustrative purposes and will drop it later.

p. 184, "research in atomic physics between 1915 and 1922": See, e.g., Eckert (1993). Unfortunately, this book has not been translated into English. See also chapter 3, entitled "The Older Quantum Theory" of Jammer (1966), and more recently Seth (2010).

p. 187, "'simple heuristics were more accurate than standard statistical methods that have the same or more information.'": Gigerenzer and Gaissmaier (2011, 453).

p. 188, "even the subtlest experimental setup designed to disclose mind-boggling aspects of the quantum world is treated as an assembly of ordinary physical objects": As is well known to the specialists, this was most important for Niels Bohr's view of quantum theory; see, e.g., Folse (1985), Honner (1987), or Murdoch (1987).

p. 188, "philosopher Arthur Fine has called this custom the 'natural ontological attitude'": See Fine (1984, esp. 95–99).

p. 189, "Don Eigler and Erhard Schweizer shifted thirty-five individual xenon atoms on a nickel surface": See Eigler and Schweizer (1990).

p. 189, "As it was later described, this event 'changed the nanoworld'": See Toumey (2010).

p. 189, "the notion of objectivity in history": There is an extended discussion about this topic in the philosophy of historiography; see, e.g., Bevir (1994).

p. 189, "all of those social sciences and humanities that understand themselves as 'constructivist'": For a critical discussion, see Hacking (1999).

p. 190, "Martin Heidegger's infamous dictum: 'Science does not think'": See Heidegger (1968); for interpretation, see, e.g., Salanskis (1995) (this article is available on the Internet at http://tekhnema.free.fr/2Salanskis.htm, accessed Sept. 6, 2011).

p. 191, "The beginning of the science of astronomy": See, e.g., Oelsner (1982a), Reineke (1982a), and Waerden (1974).

p. 192, "there are vociferous groups claiming the falsehood of physics": See, e.g., http://sciliterature.50webs.com/RelativityDebates.htm (accessed Sept. 7, 2011).

p. 192, "It was only in 1905 when Einstein challenged this assumption in his special theory of relativity": For an elementary introduction to the topic, see http://en.wikipedia.org/wiki/Relativity_of_simultaneity (accessed Sept. 1, 2011).

p. 193, "the common sense concept of biological species has turned out to be much more complicated": See, e.g., Ereshefsky (2010), available at http://plato.stanford.edu/archives/spr2010/entries/species/ (accessed Sept. 7, 2011).

p. 193, "**The area in which these concepts play a decisive role is often described as 'reader-response criticism'**": See, e.g., Tompkins (1980), especially the introduction by Jane Tompkins. The following quote is from this introduction, p. ix.

p. 194, "**how many different kinds of bacteria live symbiotically in our organs**": See, e.g., http://biology.kenyon.edu/slonc/bio3/symbiosis.html (accessed Sept. 7, 2011).

p. 195, "**'There's no accounting for taste'**": This proverb exists in various forms also in other languages, including Latin. I conclude from this fact that the associated problem is widely known, including in (Western) antiquity. I guess that the insight expressed in the proverb was forced upon common sense against its objectivist stance because of too much quarreling arising from its neglect.

p. 195, "**to qualities that at first sight appear to be rooted in the objects themselves, like aesthetic qualities, though they are apparently not**": It may even be that common sense uses "taste" exactly for this purpose: everything is declared a matter of taste in common sense for which the objectivist stance manifestly does not work.

p. 196, "**Descriptive statements have, by themselves, no normative content**": Looking a little more closely, however, reveals that one has to be careful with this assertion. For instance, for something to qualify as a description, certain norms must be fulfilled. Thus, somehow, these norms are present in a descriptive statement. However, these subtler considerations do not come into play in our present context.

p. 197, "**the March 2009 version of the University of *Oxford Centre for Evidence Based Medicine***": See http://www.cebm.net/index.aspx?o=1025 (accessed Aug. 20, 2011); for a definition of terms used in the table, see the glossary at http://www.cebm.net/?o=1116 (accessed Aug. 20, 2011).

p. 197, "**every new submission of a clinical study has to specify its level-of-evidence rating**": See, e.g., the guidelines for the journal *Foot & Ankle Specialist* in DeVries and Berlet (2010).

p. 199, "**'turning theoretical physics into recreational mathematics'**": See Lindley (1993, 19); Lindley is referring to the very critical article Ginsparg and Glashow (1986) (available on the Internet at http://arxiv.org/PS_cache/physics/pdf/9403/9403001v1.pdf, accessed Sept. 28, 2011) and to interviews with physicists Sheldon Glashow and Richard Feynman published in Davies and Brown (1988, 180–91, 192–210). The most noted critical book publications by physicists on string theory are Smolin (2006) and Woit (2006).

p. 199, "**none of the relevant philosophies reaches a unanimous conclusion**": See the excellent (apart from the part on Kuhn) article, Johansson and Matsubara (2011).

p. 200, "**It was Karl Popper (1902–1994) who made the demarcation problem prominent in the twentieth century**": For a concise presentation of Popper's demarcation criterion and its motives, see Popper (1989 [1957], esp. 33–39).

p. 200, "**the two fundamental problems in epistemology**": See, e.g., Popper (2007).

p. 201, "**In the literature, this criterion has not always been properly understood**": Many misconceptions arise from the seduction to understand Popper's "falsifiability of a sentence" as meaning "it is actually possible to show that the sentence is false," which presupposes, of course, the falsehood of the sentence. On this reading, true sentences could not be falsifiable. See, e.g., Theocharis and Psimopoulos (1987). Another misreading identifies the demarcation criterion with Popper's thesis that scientific activity consists in relentlessly and continuously testing scientific hypotheses and giving them up after empirical refutation. However, the criterion does not imply this thesis (of course, the thesis presupposes the criterion, thus implying it). Thus, (successfully) criticizing the thesis does not imply that the criterion is false. For this sort of fallacy, see, e.g., Lakatos (1978, 3–4) or Curd and Cover (1998, 66–69).

p. 202, "**Popper's criterion of demarcation was severely criticized**": For a summary, see, e.g., Laudan (1983, 121–22).

p. 202, "**it is far from clear that, for example, astrology or Freud's psychoanalysis are indeed pseudoscientific when judged according to Popper's standards**": For astrology, see Thagard (1978, 226); for psychoanalysis see Grünbaum (1979).

p. 202, "**there have been very few systematic attempts to articulate a demarcation criterion**": See, e.g., Laudan (1983, 122–24).

p. 203, "**As philosopher of science Larry Laudan put it in 1983**": See Laudan (1983, 124), (italics in original).

p. 203, "**the case of so-called indigenous or traditional knowledge and its relationship to science**": See, e.g., ICSU (2002, 12) (available on the Internet at http://portal.unesco.org/science/en/ev.php-URL_ID=3521&URL_DO=DO_TOPIC&URL_SECTION=201.html, accessed Sept. 16, 2011).

p. 203, "**situations in which a vast majority of scientists reject a field as pseudoscientific, and there were no competitors**": Curd and Cover (1998, 73).

p. 203, "**a criterion that philosopher Paul Thagard proposed in 1978**": See Thagard (1978, 227–28). This article is reprinted in Curd and Cover (1998); the quote is on p. 32. The following quote is on the same pages. Thagard has criticized his 1978 attempt in his later book (Thagard 1988, esp. 168).

p. 206, "**Newton postulated that God would prevent any seriously accumulating instability of the planetary system**": Newton (1952 [1730]), Query 31; the relevant part of the query is available on the Internet: http://web.lemoyne.edu/~giunta/newton.html (accessed Sept. 30, 2011).

p. 206, "**Laplace declaring that in his theory, he no longer needed the hypothesis of God**": See, e.g., "Pierre Simon Laplace (1749–1827)," in Ball (1960); this article is available on the Internet at http://www.maths.tcd.ie/pub/HistMath/People/Laplace/RouseBall/RB_Laplace.html (accessed Sept. 30, 2011).

p. 206, "**God was invoked by some predominantly British authors in the theory of catastrophism**": See, e.g., Laudan (2011) (accessed Sept. 30, 2011).

p. 207, "**to deal in general terms with a subject of immense internal and historical variety without becoming vacuous**": This is the problem that turned the late Paul Feyerabend away from the philosophy of science toward the history of science. He believed that philosophy (of science), by being abstract, could not meaningfully come to terms with the abundance of reality; see Feyerabend (1999a).

NOTES TO CHAPTER 6: CONCLUSION

p.210, "**one may have doubts about whether the thesis is really empirical**": This concern was expressed by one of this book's anonymous referees.

p. 210, "**Philosopher Thomas Kuhn's example is Johnny's learning from his father to distinguish between ducks, geese, and swans**": See Kuhn (1974), reprint pp. 309–18. For a systematic exposition of Kuhn's view regarding these issues, see Hoyningen-Huene (1993, sec. 3.6, esp. 105–11).

p. 211, "**she may do so by stating which features of X she takes to be definitional**": For example, in the *Annual Review of Psychology*, Gigerenzer and Gaissmaier wrote a long review article on "Heuristic Decision Making" (Gigerenzer and Gaissmaier 2011). After the introduction, they pose the question: "What is a heuristic?" (p. 454). Briefly afterward, they state that "[m]any definitions of heuristics exist" (p. 454), i.e., that there is no consensus about the definitional features of heuristics in the relevant scientific community. After a glance at two of them, they continue: "For the purpose of this review, we adopt the following definition: … " (p. 454). Clearly, the two authors do not indicate the expectation that in the future, all researches in the area will adopt their definition as the only valid one. Instead, they mark it as a working definition for a limited purpose.

LITERATURE CITED

Achinstein, Peter (2005): *Scientific Evidence: Philosophical Theories and Applications*. Baltimore: Johns Hopkins University Press.
Aizawa, Kenneth (2003): *The Systematicity Arguments*. Dordrecht: Kluwer.
Akeroyd, F. Michael (1990): "An Oscillatory Model of the Growth of Scientific Knowledge." *British Journal for the Philosophy of Science* 41:407–14.
Alexander, Amir R. (2006): "Tragic Mathematics: Romantic Narratives and the Refounding of Mathematics in the Early Nineteenth Century." *Isis* 97 (4):714–26.
Alexander, J., J. Giesen, B. Münch, and N. Smelser, eds. (1987): *The Micro-Macro Link*. Berkeley: University of California Press.
Allen, Thomas B., and Norman Polmar (2001): *Why Truman Dropped the Atomic Bomb on Japan: Code Name Downfall: The Secret Plan to Invade Japan*. Washington, DC: Ross and Perry.
Alperovitz, Gar (1995): *The Decision to Use the Atomic Bomb and the Architecture of an American Myth*. 1st ed. New York: Knopf.
Alston, William P. (1967): "Philosophy of Religion, Problems of." In *The Encyclopedia of Philosophy*, vol. 6. Edited by P. Edwards. New York: Macmillan, pp. 285–89.
Anderson, P. W. (1972): "More is Different." *Science* 177:393–96.
Anderson, Philip (2005): "Emerging Physics: A Fresh Approach to Viewing the Complexity of the Universe." *Nature* 434 (Apr. 7, 2005):701–02.
Apel, Karl-Otto (1984): *Understanding and Explanation: A Transcendental-Pragmatic Perspective, Studies in Contemporary German Social Thought*. Cambridge, Mass.: MIT Press.
Ariès, Philippe (1962): *Centuries of Childhood: A Social History of Family Life*. New York: Vintage Books.
Ariew, André (2003): "Ernst Mayr's 'Ultimate/Proximate' Distinction Reconsidered and Reconstructed." *Biology and Philosophy* 18 (4):553–65.
Ariew, Roger (1984): "The Duhem Thesis." *British Journal for the Philosophy of Science* 35 (4):313–25.

Aristotle ([1960]): *Posterior Analytics*. Translated by Hugh Tredennick. Edited by G. P. Goold, *Loeb Classical Library*. Cambridge: Harvard University Press.

Aristotle ([1973]): *Categories. On Interpretation. Prior Analytics*. Translated by H. P. Cooke and H. Tredennick, *Loeb Classical Library*. Cambridge, Mass.: Harvard University Press.

Aristotle ([1976]): *Aristotle's Posterior Analytics*. Translated with Notes by Jonathan Barnes. Edited by J. L. Ackrill, *Clarendon Aristotle Series*. Oxford: Clarendon Press.

Aristotle ([1996]): *Metaphysics, Books I–IX*. Translated by Hugh Tredennik, Loeb Classical Library. Cambridge, Mass.: Harvard University Press.

Armstrong, Jon Scott (1985): *Long-Range Forecasting: From Crystal Ball to Computer*. 2nd ed. New York: Wiley.

ATLAS Collaboration (2011): "Search for Quark Contact Interactions in Dijet Angular Distributions in pp Collisions at sqrt(s) = 7 TeV Measured with the ATLAS Detector." *Phys. Lett.* B694:327–45.

Audi, Robert (1995): *The Cambridge Dictionary of Philosophy*. Cambridge: Cambridge University Press.

Ayala, Francisco J., and Theodosius Dobzhansky, eds. (1974): *Studies in the Philosophy of Biology: Reduction and Related Problems*. Berkeley: University of California Press.

Bachelard, Gaston, and Dominique Lecourt (1971): *Épistémologie*. Paris: Presses universitaires de France.

Baker, G. P., and P. M. S. Hacker (1984 [1980]-a): *An Analytical Commentary on Wittgenstein's Philosophical Investigations*. Oxford: Basil Blackwell.

Baker, G. P., and P. M. S. Hacker (1984 [1980]-b): *Wittgenstein: Meaning and Understanding*. Oxford: Basil Blackwell.

Bala, G., K. Caldeira, A. Mirin, M. Wickett, and C. Delire (2005): "Multicentury Changes to the Global Climate and Carbon Cycle: Results from a Coupled Climate and Carbon Cycle Model." *Journal of Climate* 18 (2):4531–44.

Ball, W. W. Rouse (1960): *A Short Account of the History of Mathematics*. New York: Dover Publications.

Bauer, Henry H. (1994): *Scientific Literacy and the Myth of the Scientific Method*. Urbana: University of Illinois Press.

Beckett, Stephen T. (2000): *The Science of Chocolate*. London: Royal Society of Chemistry.

Beckner, Morton (1974): "Reduction, Hierarchies, and Organicism." In *Studies in the Philosophy of Biology: Reduction and Related Problems*. Edited by F. J. Ayala and T. Dobzhansky. Berkeley: University of California Press, pp. 163–77.

Bertero, Vitelmo V. (1997): "Set J: Earthquake Engineering: An Illustrated Introduction to Earthquake Engineering Principles." In *Structural Engineering Slide Library*. Edited by W. G. Godden. Berkeley: University of California.

Betz, Gregor (2006): *Prediction or Prophecy? The Boundaries of Economic Foreknowledge and Their Socio-Political Consequences*. Wiesbaden: Deutscher Universitäts-Verlag.

Bevir, Mark (1994): "Objectivity in History." *History and Theory* 33 (3):328–44.

Binmore, Ken G. (2007): *Game Theory: A Very Short Introduction*. New York: Oxford University Press.

Bird, Alexander (1998): *Philosophy of Science*. London: UCL Press.

Bird, Alexander (2004): "Kuhn, Naturalism, and the Positivist Legacy." *Studies in History and Philosophy of Science* 35:337–56.

Black, R. D. Collison (1987): "Jevons, William Stanley (1835–1882)." In *The New Palgrave: A Dictionary of Economics*. Edited by J. Eatwell, M. Milgate, P. K. Newman, and R. H. I. Palgrave. London: Macmillan.

Blitz, David (1990): "Emergent Evolution and the Level Structure of Reality." In *Studies on Mario Bunge's Treatise*. Edited by P. Weingartner and G. J. W. Dorn. Amsterdam: Rodopi, pp. 153–69.

Blitz, David (1992): *Emergent Evolution: Qualitative Novelty and the Levels of Reality*. Dordrecht: Kluwer.

Bonevac, D. (1981): *Reduction in the Abstract Sciences*. Indianapolis: Hackett.

Booss, Bernhelm, and Klaus Krickeberg (1976): *Mathematisierung der Einzelwissenschaften: Biologie, Chemie, Erdwissenschaften*. Basel: Birkhäuser.

Börner, Katy (2010): *Atlas of Science: Visualizing What We Know*. Cambridge: MIT Press.

Bortolotti, Lisa (2008): *An Introduction to the Philosophy of Science*. Cambridge; Malden, Mass.: Polity.

Boyd, R., P. Gasper, and J.D. Trout, eds. (1991): *The Philosophy of Science*. Cambridge, Mass.: MIT Press.

Bradley, James, and Kurt C. Schaefer (1998): *The Uses and Misuses of Data and Models: The Mathematization of the Human Sciences*. Thousand Oaks, Calif.: Sage Publications.

Braun, Edmund, and Hans Radermacher, eds. (1978): *Wissenschaftstheoretisches Lexikon*. Graz: Styria.

Brock, William H. (1993): *The Norton History of Chemistry*. New York: Norton.

Brown, H. C. (1926): "A Materialist's View of the Concept of Levels." *Journal of Philosophy* 23:113–20.

Brush, S. G. (1989): "Prediction and Theory Evaluation: The Case of Light Bending." *Science* 246:1124–29.

Brush, Stephen G. (1995): "Prediction and Theory Evaluation in Physics and Astronomy." In *No Truth Except in the Details: Essays in Honor of Martin J. Klein*. Edited by A. J. Kox and D. M. Siegel. Dordrecht: Kluwer, pp. 299–318.

Carnap, Rudolf (1966): *Philosophical Foundations of Physics: An Introduction to the Philosophy of Science*. New York: Basic Books.

Carol, Hans (1956): "Zur Diskussion um Landschaft und Geographie." *Geographica Helvetica* 11:111–32.

Carrier, Martin (1991): "On the Disunity of Science or Why Psychology is not a Branch of Physics." In *Einheit der Wissenschaften*. Edited by Akademie der Wissenschaften zu Berlin. Berlin: de Gruyter, pp. 39–59.

Cat, Jordi (1999): "Unity of Science." In *Routledge Encyclopedia of Philosophy*. Edited by E. Craig. London: Routledge.

Ceruzzi, Paul E. (2003): *A History of Modern Computing*. 2nd ed. Cambridge, Mass.: MIT Press.

Cetto, Ana María, ed. (2000): *World Conference on Science. Science for the Twenty-First Century: A New Commitment*. Paris: UNESCO.

Chalmers, Alan F. (1999): *What is This Thing Called Science?* 3rd ed. St. Lucia: University of Queensland Press.

Chandler, Daniel (2004): *Semiotics: The Basics*. London: Routledge.

Charles, David, and Kathleen Lennon, eds. (1992): *Reduction, Explanation, and Realism*. Oxford: Clarendon.

Clarke, Desmond M. (1993): "Dormitive Powers and Scholastic Qualities: A Reply to Hutchison." *History of Science* 31:317–25.

Clarke, Mike (2004): "Cochrane Collaboration—Systematic Reviews and the Cochrane Collaboration." http://www.cochrane.org/docs/whycc.htm (accessed Sept. 29, 2011).

Cleland, Carol E. (2002): "Methodological and Epistemic Differences between Historical Science and Experimental Science." *Philosophy of Science* 69 (3):474–96.

Clements, Michael P., and David F. Hendry, eds. (2002): *A Companion to Economic Forecasting, Blackwell Companions to Contemporary Economics*. Malden, Mass.: Blackwell Publishers.

Coase, R. H., and R. F. Fowler (1935a): "Bacon Production and Pig-Cycle in Great Britain." *Economica* New Series 2 (6):142–67.

Coase, R. H., and R. F. Fowler (1935b): "The Pig-Cycle: A Rejoinder." *Economica* New Series 2 (8):423–28.

Coase, R. H., and R. F. Fowler (1937): "The Pig-Cycle in Great Britain: An Explanation." *Economica* New Series 4 (13):55–82.

Cohen, Morris R., and Ernest Nagel (1934): *An Introduction to Logic and Scientific Method*. New York: Harcourt.

Coles, Peter (1999): *Einstein and the Total Eclipse*. Trumpington: Icon.

Connelly, Neil G., Ture Damhus, Richard M. Hartshorn, and Alan T. Hutton (2005): *Nomenclature of Inorganic Chemistry: IUPAC Recommendations 2005*. Cambridge: RCS Publishing.

Corry, Leo (2004): *David Hilbert and the Axiomatization of Physics (1898–1918): From Grundlagen der Geometrie to Grundlagen der Physik*. Edited by J. Z. Buchwald, vol. 10. *Archimedes*. Dordrecht: Kluwer.

Costelloe, Timothy (2008): "Giambattista Vico." In *Stanford Encyclopedia of Philosophy*: http://plato.stanford.edu/entries/vico/ (accessed Sept. 29, 2011).

Craig, Edward, ed. (1998): *Routledge Encyclopedia of Philosophy*. 9 vols. London: Routledge.

Crane, Diana, and Henry Small (1992): "American Sociology Since the Seventies: The Emerging Identity Crisis in the Discipline." In *Sociology and Its Publics: The Forms and Fates of Disciplinary Organization*. Edited by T. Halliday and M. Janowitz. Chicago: University of Chicago Press, pp. 197–234.

Crick, F. (1966): *Of Molecules and Men*. Seattle: University of Washington Press.

Culler, Jonathan (1997): *Literary Theory: A Very Short Introduction*. Oxford: Oxford University Press.

Curd, Martin, and J. A. Cover, eds. (1998): *Philosophy of Science: The Central Issues*. New York: Norton.

D'Alembert, Jean le Rond (1995 [1751]): *Preliminary Discourse to the Encyclopedia of Diderot*. Translated by R. N. Schwab & W. E. Rex. Chicago: University of Chicago Press.

Daniels, Norman (2003): "Reflective Equilibrium." In *The Standford Encyclopedia of Philosophy*. Edited by E. N. Zalta. http://plato.stanford.edu/archives/sum2003/entries/reflective-equilibrium/ (accessed Sept. 27, 2011).

Danto, Arthur C. (1965): *Analytical Philosophy of History*. Cambridge: Cambridge University Press.

Danto, Arthur C. (1967): "Naturalism." In *Encyclopedia of Philosophy*, vol. 5. Edited by P. Edwards. New York: Macmillan, pp. 448–50.

Darwin, Charles (1964 [1859]): *On the Origin of Species by Means of Natural Selection or the Preservation of Favored Races in the Struggle for Life: A Facsimile of the First Edition*. Cambridge: Harvard University Press.

Davies, P. C. W., and J. R. Brown (1988): *Superstrings: A Theory of Everything?* Cambridge: Cambridge University Press.

de Regt, Henk W. (2004): "Discussion Note: Making sense of Understanding." *Philosophy of Science* 71 (1):98–109.

Demeterio III, Feorillo P.A. (2001): "Introduction to Hermeneutics." *Diwatao* 1 (1).

Descartes, René (1984 [1620-c.28]): "Regulae ad directionem ingenii (Rules for the Direction of the Mind)." In *The Philosophical Writings of Descartes*. Edited by J. Cottingham, R. Stoothoff, and D. Murdoch. Cambridge: Cambridge University Press.

Descartes, René (1984 [1637]): "Discours de la méthode pour bien conduir sa raison et chercher la vérité dans les sciences plus la dioptrique, les meteores, et la geometrie, qui sont des essais de cete methode (Discourse on the Method for Properly Conducting Reason and Searching for Truth in the Sciences, as well as the Dioptrics, the Meteors, and the Geometry, which are essays in this method)." In *The Philosophical Writings of Descartes*. Edited by J. Cottingham, R. Stoothoff and D. Murdoch. Cambridge: Cambridge University Press.

DeVries, J. George, and Gregory C. Berlet (2010): "Understanding Levels of Evidence for Scientific Communication." *Foot & Ankle Specialist* 3 (4):205–09.

Dewey, John (1977 [1903]): "Logical Conditions of a Scientific Treatment of Morality." In *John Dewey: The Middle Works, 1899–1924. Volume 3: 1903–1906*. Edited by J. A. Boydston. Carbondale: Southern Illinois University Press, pp. 3–39.

Dray, William H. (1971): "On the Nature and Role of Narrative in Historiography." *History and Theory* 10:153–71.

Driver-Linn, Erin (2003): "Where Is Psychology Going? Structural Fault Lines Revealed by Psychologists' Use of Kuhn." *American Psychologist* 58 (4):269–78.

Droysen, Johann Gustav (1967 [1858]): *Outline of the Principles of History (Grundriss der Historik)*. New York: H. Fertig.

Duhem, Pierre (1954 [1906]): *The Aim and Structure of Physical Theory*. Princeton: Princeton University Press.

Dupré, John (1983): "The Disunity of Science." *Mind* 92:321–46.

Dupré, John (1993): *The Disorder of Things: Metaphysical Foundations of the Disunity of Science*. Cambridge: Harvard University Press.

Earman, John, and Clark Glymour (1980): "Relativity and Eclipses: The British Eclipse Expedition of 1919 and Their Predecessors." *Historical Studies in the Physical Sciences* 11 (1):49–85.

Eckert, Michael (1993): *Die Atomphysiker. Eine Geschichte der theoretischen Physik am Beispiel der Sommerfeldschule*. Braunschweig: Vieweg.

Edwards, Paul, ed. (1967): *The Encyclopedia of Philosophy*. New York: Macmillan.

Eigler, D. M., and E. K. Schweizer (1990): "Positioning Single Atoms with a Scanning Tunnelling Microscope." *Nature* 344 (Apr. 5, 1990):524–26.

Einstein, Albert (1982 [1936]): "Physics and Reality." In *Albert Einstein: Ideas and Opinions*. Edited by C. Selig. New York: Three Rivers Press, pp. 290–323.

Einstein, Albert (1982 [1944]): "Remarks on Bertrand Russell's Theory of Knowledge." In *Albert Einstein: Ideas and Opinions*. Edited by C. Seelig. New York: Three Rivers Press (originally in *The Philosophy of Bertrand Russell*. Edited by P. A. Schilpp. 1944), pp. 18–24.

Ellis, Brian (1996 [1985]): "What Science Aims to Do." In *The Philosophy of Science*. Edited by D. Papineau. Oxford: Oxford University Press, pp. 166–93 (originally in *Images of Science*. Edited by P. Churchland and C. Hooker. Chicago: University of Chicago Press, 1985, pp. 48–74).

Elwes, Richard (2006): "An Enormous Theorem: The Classification of Finite Simple Groups." *Plus Magazine* 41 (Dec. 2006).

Emmeche, Claus, Simo Køppe, and Frederik Stjernfelt (1997): "Explaining Emergence: Towards an Ontology of Levels." *Journal for General Philosophy of Science* 28:83–119.

Ereshefsky, Marc (1994): "Some Problems with the Linnean Hierarchy." *Philosophy of Science* 61:186–205.

Ereshefsky, Marc (2002): "Linnean Ranks: Vestiges of a Bygone Era." *Philosophy of Science* 69 (3, Supplement):S305-S315.

Ereshefsky, Marc (2010): "Species." In *The Stanford Encyclopedia of Philosophy (Spring 2010 ed.)*. Edited by E. N. Zalta, http://plato.stanford.edu/archives/spr2010/entries/species/.

Erwin, Douglas H. (1993): *The Great Paleozoic Crisis: Life and Death in the Permian, The Critical Moments in Paleobiology and Earth History Series*. New York: Columbia University Press.

Eschenburg, Johann Joachim (1792): *Lehrbuch der Wissenschaftskunde, ein Grundriß enzyklopädischer Vorlesungen*. Berlin: Nicolai (reprint from the collection of the University of Michigan Library).

Euclid (2000 [1925]): *The Thirteen Books of the Elements. Translated with introduction and commentary by Sir Thomas L. Heath*. 2nd ed. 3 vols. New York: Dover.

Fahrbach, Ludwig (2009): "Pessimistic Meta-Induction and the Exponential Growth of Science." In *Reduction—Abstraction—Analysis. Proceedings of the 31th International Ludwig Wittgenstein-Symposium in Kirchberg, 2008*. Edited by A. Hieke and H. Leitgeb. Heusenstamm: ontos, pp. 95–111.

Fahrbach, Ludwig (2011): "How the Growth of Science Ends Theory Change." *Synthese* 180 (2):139–55.

Fardon, David F., and Pierre C. Milette (2001): "Nomenclature and Classification of Lumbar Disc Pathology." *SPINE* 26 (5):E93-E113.

Faust, David, and Paul E. Meehl (1992): "Using Scientific Methods to Resolve Questions in the History and Philosophy of Science: Some Illustrations." *Behavior Therapy* 23:195–211.

Faust, David, and Paul E. Meehl (2002): "Using Meta-Scientific Studies to Clarify or Resolve Questions in the Philosophy and History of Science." *Philosophy of Science* 69 (3, Supplement):S185-S196.

Faye, Jan (2010): "Interpretation in the Natural Sciences." In *EPSA Epistemology and Methodology of Science: Launch of the European Philosophy of Science Association*. Edited by M. Suárez, M. Dorato and M. Rédei. Dordrecht: Springer, pp. 107–17.

Ferrell, Robert H. (1996): *Harry S. Truman and the Bomb: A Documentary History*. Worland: High Plains.

Fetzer, James H. (1993): *Foundations of Philosophy of Science: Recent Developments*. New York: Paragon House.

Feyerabend, Paul K. (1970): "Against Method: Outline of an Anarchistic Theory of Knowledge." In *Analyses of Theories and Methods of Physics and Psychology. Minnesota Studies in the Philosophy of Science 4*. Edited by M. Radner and S. Winokur. Minneapolis: University of Minnesota Press, pp. 17–130.

Feyerabend, Paul K. (1972): "Von der beschränkten Gültigkeit methodologischer Regeln." In *Dialog als Methode (Neue Hefte für Philosophie, Heft 2/3)*. Edited by R. Bubner, K. Cramer and R. Wiehl. Göttingen: Vandenhoek und Ruprecht, pp. 124–71.

Feyerabend, Paul K. (1975): *Against Method: Outline of an Anarchistic Theory of Knowledge*. London: New Left Books.

Feyerabend, Paul K. (1978): *Science in a Free Society*. London: NLB.

Feyerabend, Paul K. (1988): *Against Method*. Rev. ed. London: Verso.

Feyerabend, Paul K. (1993): *Against Method*. 3rd ed. London: Verso.

Feyerabend, Paul K. (1995): *Killing Time: The Autobiography of Paul Feyerabend*. Chicago: University of Chicago Press.

Feyerabend, Paul K. (1999a): *Conquest of Abundance: A Tale of Abstraction Versus the Richness of Being*. Chicago: University of Chicago Press.

Feyerabend, Paul K. (1999b): "On the Limited Validity of Methodological Rules." In *P. K. Feyerabend: Knowledge, Science and Relativism: Philosophical Papers*, vol. 3. Edited by J. Preston. Cambridge: Cambridge University Press, pp. 138–80.

Fine, Arthur (1984): "The Natural Ontological Attitude." In *Scientific Realism*. Edited by J. Leplin. Berkeley: University of California Press, pp. 83–107.

Fine, Arthur (1998): "The Viewpoint of No-One in Particular." *Proceedings and Addresses of the American Philosophical Association* 72 (2):9–20.

Fleming, Alexander (1946): "History and Development of Penicillin." In *Penicillin: It's Practical Application*. Edited by A. Fleming. London: Butterworth, pp. 1–23.

Fodor, Jerry A. (1974): "Special Sciences, or The Disunity of Science as a Working Hypothesis." In *Readings in Philosophy of Psychology*, vol. I. Edited by N. Block. Cambridge: Harvard University Press, pp. 120–33 (originally in *Synthese* 28: 97–115 (1974)).

Folse, H. J. (1985): *The Philosophy of Niels Bohr. The Framework of Complementarity*. Amsterdam: North-Holland.

Forrester, Jay W. (1970): *World Dynamics*. Portland: Productivity Press.

Frege, G. (1879): *Begriffsschrift, eine der arithmetischen nachgebildete Formelsprache des reinen Denkens*. Halle: Louis Nebert. English transl. in *Frege and Gödel: Two Fundamental Texts in Mathematical Logic*. Edited by J. van Heijenoort, Cambridge: Harvard University Press, 1970.

Frigg, Roman, and Stephan Hartmann (2005): "Scientific Models." In *The Philosophy of Science: An Encyclopedia*. Edited by S. Sarkar and J. Pfeifer. New York: Routledge.

Galison, Peter, and David Stump, eds. (1996): *The Disunity of Science: Boundaries, Contexts, and Power*. Stanford: Stanford University Press.

Gallegati, Mauro (1994): "Jevons, Sunspot Theory and Economic Fluctuations." *History of Economic Ideas* 2 (2):23–40.

Garfield, Eugene (1983): *Citation Indexing—Its Theory and Application in Science, Technology, and Humanities*. Philadelphia: ISI Press.

Garfield, Eugene (1985): "The Life and Career of George Sarton: The Father of the History of Science." *The Journal of the History of the Behavioral Sciences* 21 (2):107–17.

Geist, Eric L., Vasily V. Titov, and Costas E. Synolakis (2006): "Tsunami: Wave of Change." *Scientific American* 294 (Jan. 2006):42–49.

Gerigk, Horst-Jürgen (2002): *Lesen und Interpretieren*, vol. 2323, *UTB für Wissenschaft*. Göttingen: Vandenhoeck & Ruprecht.

Gigerenzer, Gerd, and Wolfgang Gaissmaier (2011): "Heuristic Decision Making." *Annual Review of Psychology* 62:451–82.

Gillies, Donald (1993): *Philosophy of Science in the Twentieth Century: Four Central Themes.* Oxford: Blackwell.

Ginsparg, Paul, and Sheldon L. Glashow (1986): "Desperately Seeking Superstrings." *Physics Today* 39 (5):7–9.

Glock, Hans-Johann (2007): "Could Anything be Wrong with Analytic Philosophy?" *Grazer Philosophische Studien* 74:215–37.

Gnedin, Nickolay Y. (2005): "Digitizing the Universe." *Nature* 435 (June 2, 2005):572–73.

Grantham, Todd A. (2004): "Conceptualizing the (Dis)unity of Science." *Philosophy of Science* 71 (2):133–55.

Greene, Brian R. (2000): *The Elegant Universe: Superstrings, Hidden Dimensions, and the Quest for the Ultimate Theory.* 1st ed. London: Vintage.

Greene, Brian R. (2004): *The Fabric of the Cosmos: Space, Time, and the Texture of Reality.* New York: Knopf.

Grünbaum, Adolf (1979): "Is Freudian Psychoanalytic Theory Pseudo-Scientific by Karl Popper's Criterion of Demarcation?" *American Philosophical Quarterly* 16 (2):131–41.

Haack, Susan (2003): *Defending Science—Within Reason: Between Scientism and Cynicism.* Amherst, N.Y.: Prometheus Books.

Hacking, Ian (1999): *The Social Construction of What?* Cambridge: Harvard University Press.

Harding, Sandra (2003): "A World of Sciences." In *Science and Other Cultures: Issues in Philosophies of Science and Technology.* Edited by R. Figueroa and S. Harding. London: Routledge.

Harré, Rom (1985): *The Philosophies of Science: An Introductory Survey.* Oxford: Oxford University Press.

Heidegger, Martin (1962 [1927]): *Being and Time.* New York,: Harper.

Heidegger, Martin (1968): *What Is Called Thinking?* New York: Harper & Row.

Hempel, Carl G. (1958): "The Theoretician's Dilemma: A Study in the Logic of Theory Construction." In *Concepts, Theories and the Mind-Body Problem. Minnesota Studies in Philosophy of Science*, vol. II. Edited by H. Feigl, M. Scriven and G. Maxwell. Minneapolis: University of Minnesota Press, pp. 37–98. Reprinted in Carl G. Hempel: *Aspects of Scientific Explanation.* New York: Free Press, 1965, pp. 173–226.

Hempel, Carl G. (1965a): "Aspects of Scientific Explanation." In *Aspects of Scientific Explanation and other Essays in the Philosophy of Science.* Edited by C. G. Hempel. New York: Free Press, pp. 331–496.

Hempel, Carl G. (1965b): *Aspects of Scientific Explanation and Other Essays in the Philosophy of Science.* New York: The Free Press.

Hempel, Carl G. (1965c): "Fundamentals of Taxonomy." In *Aspects of Scientific Explanation and Other Essays in the Philosophy of Science.* Edited by C. G. Hempel. New York: The Free Press, pp. 139–54.

Hempel, Carl G. (1966): *Philosophy of Natural Science.* Englewood Cliffs, New Jersey: Prentice-Hall.

Hempel, Carl G. (1983): "Valuation and Objectivity in Science." In *Physics, Philosophy and Psychoanalysis: Essays in Honor of Adolf Grünbaum.* Edited by R. S. Cohen and L. Laudan. Dordrecht: Reidel, pp. 73–100 (reprinted in Hempel, Carl G., and James H. Fetzer (editor) (2001): *The Philosophy of Carl G. Hempel.* Oxford: Oxford University Press, pp. 372–95).

Hennig, Willi (1979): *Phylogenetic Systematics*. Urbana: University of Illinois Press.
Higgins, Julian P.T., and Sally Green, eds. (2006): "Cochrane Handbook for Systematic Reviews of Interventions 4.2.6 [updated September 2006]." http://www.cochrane.org/resources/handbook/hbook.htm (accessed Sept. 29, 2011).
Hofmann, James R., and Bruce H. Weber (2003): "The Fact of Evolution: Implications for Science Education." *Science & Education* 12:729–60.
Honner, J. (1987): *The Description of Nature. Niels Bohr and the Philosophy of Quantum Physics*. Oxford: Clarendon.
Hooper, David, and Ken Whyld (1992): *The Oxford Companion to Chess*. 2nd ed. Oxford: Oxford University Press.
Howell, Martha C., and Walter Prevenier (2001): *From Reliable Sources: An Introduction to Historical Methods*. Ithaca, N.Y.: Cornell University Press.
Hoyningen-Huene, Paul (1982): "Zur Konstitution des Gegenstandsbereichs der Geographie bei Hans Carol." *Geographica Helvetica* 37 (1):23–34.
Hoyningen-Huene, Paul, ed. (1983): *Die Mathematisierung der Wissenschaften*. Zürich: Artemis.
Hoyningen-Huene, Paul (1987): "Context of Discovery and Context of Justification." *Studies in History and Philosophy of Science* 18:501–15.
Hoyningen-Huene, Paul (1989): "Naturbegriff—Wissensideal—Experiment. Warum ist die neuzeitliche Wissenschaft technisch verwertbar?" *Zeitschrift für Wissenschaftsforschung* 5:43–55.
Hoyningen-Huene, Paul (1993): *Reconstructing Scientific Revolutions: Thomas S. Kuhn's Philosophy of Science*. Chicago: University of Chicago Press.
Hoyningen-Huene, Paul (1998): "On Thomas Kuhn's Philosophical Significance." *Configurations* 6:1–14.
Hoyningen-Huene, Paul (2000a): "The Nature of Science." In *World Conference on Science. Science for the Twenty-First Century: A New Commitment*. Edited by A. M. Cetto. Paris: UNESCO, pp. 52–56.
Hoyningen-Huene, Paul (2000b): "Paul K. Feyerabend: An Obituary." In *The Worst Enemy of Science? Essays in memory of Paul Feyerabend*. Edited by J. Preston, G. Munévar and D. Lamb. Oxford: Oxford University Press, pp. 1–15.
Hoyningen-Huene, Paul (2004): *Formal Logic: A Philosophical Approach*. Translated by A. Levine. Pittsburgh: Pittsburgh University Press.
Hoyningen-Huene, Paul (2005): "Three Biographies: Kuhn, Feyerabend, and Incommensurability." In *Rhetoric and Incommensurability*. Edited by R. Harris. West Lafayette: Parlor Press, pp. 150–75.
Hoyningen-Huene, Paul (2006): "Context of Discovery Versus Context of Justification and Thomas Kuhn." In *Revisiting Discovery and Justification: Historical and Philosophical Perspectives on the Context Distinction*. Edited by J. Schickore and F. Steinle. Dordrecht: Springer, pp. 119–31.
Hoyningen-Huene, Paul, and Howard Sankey (2001): *Incommensurability and Related Matters*, vol. 216. *Boston Studies in the Philosophy of Science*. Dordrecht: Kluwer Academic Publishers.
Hoyningen-Huene, Paul, Marcel Weber, and Eric Oberheim (1999): *Background Document for the World Conference on Science: Science for the Twenty-First Century: A New Commitment*. Paris: International Council for Science.

Hutchison, Keith (1982): "What Happened to Occult Qualities in the Scientific Revolution?" *Isis* 73:233–53.

Hutchison, Keith (1991): "Dormitive Virtues, Scholastic Qualities, and the New Philosophies." *History of Science* 29:245–78.

Hutchison, Keith (1993): "Keith Hutchison Responds." *History of Science* 31:325–27.

ICSU, International Council for Science (2002): *Science, Traditional Knowledge and Sustainable Development*, vol. 4. *ICSU Series on Science for Sustainable Development*.

Iggers, Georg G. (1983): *The German Conception of History: The National Tradition of Historical Thought from Herder to the Present*. Rev. ed. Middletown, Conn.: Wesleyan University Press.

Iggers, Georg G. (1984): *New Directions in European Historiography*. Rev. ed. Middletown, Conn.: Wesleyan University Press.

IUPAC, Commission on Nomenclature of Organic Chemistry (1993): *A Guide to IUPAC Nomenclature of Organic Compounds (Recommendations 1993)*. Oxford: Blackwell

Jahn, Ilse, Rolf Löther, and Konrad Senglaub (1982): *Geschichte der Biologie: Theorien, Methoden, Institutionen, Kurzbiographien*. Jena: VEB Gustav Fischer Verlag.

Jammer, Max (1966): *The Conceptual Development of Quantum Mechanics*. New York: McGraw-Hill.

Johansson, Lars-Göran, and Keizo Matsubara (2011): "String Theory and General Methodology: A Mutual Evaluation." *Studies In History and Philosophy of Science Part B: Studies In History and Philosophy of Modern Physics* 42 (3):199–210.

Jonas, Hans (2003): *Erinnerungen. Nach Gesprächen mit Rachel Salamander*. Frankfurt: Insel.

Kanipe, Jeff (2009): "New Eyes, New Skies." *Nature* 457 (Jan. 1, 2009):18–25.

Kant, Immanuel, and Michael Friedman (2004 [1786]): *Metaphysical Foundations of Natural Science, Cambridge Texts in the History of Philosophy*. Cambridge: Cambridge University Press.

Kant, Immanuel, and Norman Kemp Smith (2003 [1781/1787]): *Critique of pure reason*. Rev. 2nd ed. New York: Palgrave Macmillan.

Keil, Geert (1996): "Ist die Philosophie eine Wissenschaft?" In *Sich im Denken orientieren. Für Herbert Schnädelbach*. Edited by S. Dietz, H. Hasted, G. Keil and A. Thyen. Frankfurt: Suhrkamp, pp. 32–51.

Kelly, Thomas (2006): "Evidence." In *Stanford Encyclopedia of Philosophy*: http://plato.stanford.edu/entries/evidence/ (accessed Sept. 29, 2011).

Kiehl, Jeffrey T., and Christine A. Shields (2005): "Climate Simulation of the Latest Permian: Implications for Mass Extinctions." *Geology* 33 (9):757–60.

Killias, Martin, Marcello Aebi, and Denis Ribeaud (2000): "Does Community Service Better Rehabilitate than Short-Term Imprisonment? Results of a Controlled Experiment." *The Howard Journal* 39 (1):40–57.

Kim, Jaegwon (2003): "The American Origins of Philosophical Naturalism." *The Journal of Philosophical Research, Philosophy in America at the Turn of the Century*, Special APA Centennial Supplement: 83–98.

Kious, W. Jacqueline, and Robert I. Tilling (1996): *This Dynamic Earth: The Story of Plate Tectonics*. New York: New York University Press.

Kitchen, K. A. (1991): "The Chronology of Ancient Egypt." *World Archeology* 23 (2):201–08.

Klee, Robert (1997): *Introduction to the Philosophy of Science: Cutting Nature at its Seams*. Oxford: Oxford University Press.

Klein, Peter D. (1999): "Knowledge, concept of." In *Routledge Encyclopedia of Philosophy*. Edited by E. Craig. London: Routledge.

Kline, Morris (1980): *Mathematics: The Loss of Certainty*. New York: Oxford University Press.

Knorr-Cetina, K. (1999): *Epistemic Cultures: How the Sciences Make Knowledge*. Cambridge, Mass.: Harvard University Press.

Koselleck, Reinhart (1972): "Einleitung." In *Geschichtliche Grundbegriffe. Historisches Lexikon zur politischen Sprache in Deutschland*, vol. I. Edited by O. Brunner, W. Conze and R. Koselleck, pp. XIII–XXVII.

Koyré, Alexandre (1965): *Newtonian Studies*. London: Chapman & Hall.

Krings, Hermann, Hans Michael Baumgartner, and Christoph Wild, eds. (1973): *Handbuch philosophischer Grundbegriffe*. 6 vols. München: Kösel.

Kroes, Peter (2002): *Ideaalbeelden van wetenschap: Een inleiding tot de wetenschapsfilosofie*, 2nd ed. Amsterdam: Boom.

Kuhn, Thomas S. (1962): *The Structure of Scientific Revolutions*. Chicago: University of Chicago Press.

Kuhn, Thomas S. (1970 [1962]): *The Structure of Scientific Revolutions*. 2nd ed. Chicago: University of Chicago Press.

Kuhn, Thomas S. (1974): "Second Thoughts on Paradigms." In *The Structure of Scientific Theories*. Edited by F. Suppe Urbana: University of Illinois Press, pp. 459–82. Reprinted in Thomas S. Kuhn: *The Essential Tension: Selected Studies in Scientific Tradition and Change*. Chicago: University of Chicago Press, pp. 293–319.

Kuhn, Thomas S. (1977): "Objectivity, Value Judgement, and Theory Choice." In *The Essential Tension: Selected Studies in Scientific Tradition and Change*. Chicago: University of Chicago Press, pp. 320–39.

Kuhn, Thomas S. (1977 [1976]): "Mathematical versus Experimental Traditions in the Development of Physical Science." In *The Essential Tension: Selected Studies in Scientific Tradition and Change*. Edited by T. S. Kuhn. Chicago: University of Chicago Press, pp. 31–65.

Kuhn, Thomas S. (1984): "Revisiting Planck." *Historical Studies in the Physical Sciences* 14:231–52.

Kuipers, Theo A.F. (2010): "The Gray Area for Incorruptible Scientific Research: An Exploration by Merton's Norms Conceived as 'Default-Norms.'" In *EPSA Epistemology and Methodology of Science: Launch of the European Philosophy of Science Association*. Edited by M. Suárez, M. Dorato and M. Rédei. Dordrecht: Springer, pp. 149–64.

Kürschner (2003): *Kürschners Deutscher Gelehrten-Kalender. Bio-bibliographisches Verzeichnis deutschsprachiger Wissenschaftler der Gegenwart*. 19th ed. München: K. G. Saur.

Ladyman, James (2002): *Understanding Philosophy of Science*. London: Routledge.

Lakatos, Imre (1978): "Introduction: Science and Pseudoscience." In *Imre Lakatos: The Methodology of Scientific Research Programmes*. Edited by J. Worrall and G. Currie. Cambridge: Cambridge University Press, pp. 1–7.

Lambrinos, Dimitrios, Rolf Möller, Thomas Labhart, Rolf Pfeifer, and Rüdiger Wehner (2000): "A Mobile Robot Employing Insect Strategies for Navigation." *Robotics and Autonomous Systems* 30 (1–2):39–64.

Laudan, Larry (1983): "The Demise of the Demarcation Problem." In *Physics, Philosophy and Psychoanalysis: Essays in Honor of Adolf Grünbaum*. Edited by R. S. Cohen and L. Laudan. Dordrecht: Reidel, pp. 111–27.

Laudan, Rachel (2011): "Uniformitarianism and Catastrophism." In *Science Encylopedia*, http://science.jrank.org/pages/49560/uniformitarianism-catastrophism.html.

Laughlin, Robert B. (2005): *A Different Universe: Reinventing Physics from the Bottom Down*. New York: Basic Books.

Le Verrier, Urbain J. (1859a): "Lettre de M. Le Verrierà M. Faye sur la théorie de Mercure et sur le mouvement du périhélie de cette planète." *Comptes rendus hebdomadaires des séances de l'Académie des sciences (Paris)* 49:379–83.

Le Verrier, Urbain J. (1859b): *Theorie Du mouvement de Mercure, Annales de l'Observatoire imperial de Paris; t. 5*. Paris: Mallet-Bachelier.

Leigh, G. J., H. A. Favre, and W. V. Metanomski (1998): *Principles of Chemical Nomenclature: A Guide to IUPAC Recommendations*. Oxford: Blackwell Science.

Lewis, M. Paul, ed. (2009): *Ethnologue: Languages of the World*. 16th ed. Dallas: SIL International.

Lindberg, David C. (1992): *The Beginnings of Western Science: The European Scientific Tradition in Philosophical, Religious, and Institutional Context, 600 B.C. to A.D. 1450*. Chicago: University of Chicago Press.

Lindley, David (1993): *The End of Physics: The Myth of a Unified Theory*. New York: BasicBooks.

Linstone, Harold A., and Murray Turoff, eds. (1975): *The Delphi Method: Techniques and Applications*. Reading, Mass.: Addison-Wesley Pub. Co., Advanced Book Program.

Longino, Helen E. (2002): *The Fate of Knowledge*. Princeton, N.J.: Princeton University Press.

Losee, John (2001): *A Historical Introduction to the Philosophy of Science*. 4th ed. Oxford: Oxford University Press.

Macdonald, G. (1986): "The Possibility of the Disunity of Science." In *Fact, Science and Morality. Essays on A.J. Ayers Language, Truth and Logic*. Edited by G. Macdonald and C. Wright. Oxford: Basil Blackwell, pp. 219–46.

Mayo, Deborah G. (1996): *Error and the Growth of Experimental Knowledge*. Chicago: University of Chicago Press.

Mayr, Ernst (1976 [1961]): "Cause and Effect in Biology." In *Evolution and the Diversity of Life: Selected Essays*. Edited by E. Mayr. Cambridge, Mass.: Harvard University Press, pp. 359–71.

Mayr, Ernst (1982): *The Growth of Biological Thought: Diversity, Evolution, and Inheritance*. Cambridge: Harvard University Press.

Mayr, Ernst (1988a): "The Limits of Reductionism." *Nature* 331:475.

Mayr, Ernst (1988b): *Toward a New Philosophy of Biology. Observations of an Evolutionist*. Cambridge: Harvard University Press.

Mayr, Ernst (1997): *This is Biology*. Cambridge, Mass.: Harvard University Press.

Mayr, Ernst, and W. J. Brock (2002): "Classifications and Other Ordering Systems." *Journal of Zoological Systematics and Evolutionary Research* 40:169–94.

Mazurs, Edward G. (1974): *Graphic Representations of the Periodic System during One Hundred Years*. Alabama: University of Alabama Press.

McCord, Joan (1992): "The Cambridge-Somerville Study: A Pioneering Longitudinal Experimental Study of Delinquency Prevention." In *Preventing Antisocial Behavior: Interventions from Birth through Adolescence*. Edited by J. McCord and R. E. Tremblay. New York: Guilford, pp. 196–206.

McKirahan, R. D. (1992): *Principles and Proofs: Aristotle's Theory of Demonstrative Science*. Princeton: Princeton University Press.

Meadows, Donella H. et al. (1972): *The Limits to Growth; a Report for the Club of Rome's Project on the Predicament of Mankind.* New York: Universe Books.

Meehl, Paul E. (1992): "Cliometric Metatheory: The Actuarian Approach to Empirical, History-Based Philosophy of Science." *Psychological Reports* 71:339–467.

Merton, Robert K. (1968): *Social Theory and Social Structure.* New York: Free Press.

Merton, Robert K. (1973 [1938]): "Science and Social Order." In *The Sociology of Science: Theoretical and Empirical Investigations.* Edited by N. W. Storer. Chicago: University of Chicago Press, pp. 254–66.

Merton, Robert K. (1973 [1942]): "The Normative Structure of Science." In *The Sociology of Science: Theoretical and Empirical Investigations.* Edited by N. W. Storer. Chicago: University of Chicago Press, pp. 267–78.

Messer, August (1907): "Besprechung von Otto Ritschl: System und systematische Methode in der Geschichte des wissenschaftlichen Sprachgebrauchs und der philosophischen Methodologie." *Göttinger gelehrte Anzeigen* 169 (8):659–66.

Mill, John Stuart (1886): *A System of Logic: Ratiocinative and Inductive. Being a Connected View of the Principles of Evidence and the Methods of Scientific Investigation.* 8th ed. London: Longmans, Green, and Co. (1st ed. 1843).

Mills, Terence C., ed. (1999a): *Economic Forecasting.* 2 vols. *The International Library of Critical Writings in Economics 108.* Cheltenham, UK; Northampton, Mass.: E. Elgar.

Mills, Terence C. (1999b): "Introduction." In *Economic Forecasting*, vol. I. Edited by T. C. Mills. Cheltenham, UK; Northampton, Mass.: E. Elgar, pp. ix–xvi.

Mittelstraß, Jürgen, ed. (1980 ff.): *Enzyklopädie Philosophie und Wissenschaftstheorie.* Mannheim: Bibliographisches Institut.

Morris, Charles (1960): "On the History of the International Encyclopedia of Unified Science." *Synthese* 12:517–21.

Murdoch, Dugald (1987): *Niels Bohr's Philosophy of Physics.* Cambridge: Cambridge University Press.

Nagel, Ernest (1961): *The Structure of Science: Problems in the Logic of Scientific Explanation.* London: Routledge & Kegan Paul.

Nanda (2004): *Nursing Diagnoses: Definitions and Classification 2005–2006*: North American Nursing Diagnosis Association.

Navarro-González, Rafael, Karina F. Navarro, José de la Rosa, Enrique Iñiguez, Paola Molina, Luis D. Miranda, Pedro Morales, Edith Cienfuegos, Patrice Coll, François Raulin, Ricardo Amils, and Christopher P. McKay (2006): "The Limitations on Organic Detection in Mars-like Soils by Thermal Volatilization–Gas Chromatography–MS and Their Implications for the Viking Results." *Proceedings of the National Academy of Sciences* 103:16089–94.

Nersessian, Nancy J. (1992): "How Do Scientists Think? Capturing the Dynamics of Conceptual Change in Science." In *Cognitive Models of Science. Minnesota Studies in the Philosophy of Science*, vol. XV. Edited by R. N. Giere. Minneapolis: University of Minnesota Press, pp. 3–44.

Newman, Robert P. (1995): *Truman and the Hiroshima Cult.* East Lansing: Michigan State University Press.

Newton, Isaac (1952 [1730]): *Opticks; or, A treatise of the reflections, refractions, inflections & colours of light. Based on the 4th ed., London, 1730.* New York: Dover.

O'Hear, Anthony (1989): *Introduction to the Philosophy of Science.* Oxford: Clarendon Press.

O'Neal, Michael (1990): *President Truman and the Atomic Bomb: Opposing Viewpoints.* San Diego, Calif.: Greenhaven Press.

Oberheim, Eric (2007): *Feyerabend's Philosophy.* Berlin: de Gruyter.

Oelsner, Joachim (1982a): "Vorderasien: Astronomie." In *Geschichte des wissenschaftlichen Denkens im Altertum.* Edited by F. Jürß. Berlin: Akademie-Verlag, pp. 60–70.

Oelsner, Joachim (1982b): "Vorderasien: Mathematik." In *Geschichte des wissenschaftlichen Denkens im Altertum.* Edited by F. Jürß. Berlin: Akademie-Verlag, pp. 50–60.

Okasha, Samir (2002): *Philosophy of Science: A Very Short Introduction.* Oxford: Oxford University Press.

Okasha, Samir (2011): "Theory Choice and Social Choice: Kuhn versus Arrow." *Mind* 120 (477):83–115.

Oppenheim, Paul, and Hilary Putnam (1958): "The Unity of Science as a Working Hypothesis." In *Minnesota Studies in the Philosophy of Science II.* Edited by H. Feigl, M. Scriven and G. Maxwell. Minneapolis: University of Minnesota Press, pp. 3–36.

Oreskes, Naomi (1999): *The Rejection of Continental Drift: Theory and Method in American Earth Science.* New York: Oxford University Press.

Partee, Barbara Hall (2004): *Compositionality in Formal Semantics: Selected Papers of Barbara Partee, Explorations in Semantics.* Malden, Mass.: Blackwell Pub.

Passmore, J. (1962): "Explanation in Everyday Life, in Science, and in History." *History & Theory* 2:105–23.

Pauling, L. (1970): "Fifty Years of Progress in Structural Chemistry and Molecular Biology." *Daedalus* 99:988–1014.

Peirce, Charles Sanders, Charles Hartshorne, Paul Weiss, and Arthur Walter Burks (1965): *Collected Papers of Charles Sanders Peirce.* Cambridge: Belknap Press of University Press.

Pollok, Konstantin (2001): *Kants "Metaphysische Anfangsgründe der Naturwissenschaft." Ein kritischer Kommentar, Kant-Forschungen Bd. 13.* Hamburg: Meiner.

Ponce de León, Marcia S., and Christoph P. E. Zollikofer (2001): "Neanderthal Cranial Ontogeny and Its Implications for Late Hominid Diversity." *Nature* 412 (Aug. 2, 2001):534–38.

Popper, Karl R. (1957): *The Poverty of Historicism.* London: Routledge.

Popper, Karl R. (1959 [1934]): *The Logic of Scientific Discovery.* London: Hutchinson.

Popper, Karl R. (1989 [1957]): "Science: Conjectures and Refutations." In *Conjectures and Refutations: The Growth of Scientific Knowledge.* Edited by K. R. Popper. London: Routledge, pp. 33–65.

Popper, Karl R. (2007): *The Two Fundamental Problems of the Theory of Knowledge.* London: Routledge.

Preston, John (2008): *Kuhn's "The Structure of Scientific Revolutions": A Reader's Guide.* London: Continuum.

Primas, H. (1981): *Chemistry, Quantum Mechanics and Reductionism. Perspectives in Theoretical Chemistry.* Berlin: Springer.

Putnam, Hilary (1994 [1987]): "The Diversity of the Science." In *Words and Life.* Edited by J. Conant. Cambridge: Harvard University Press, pp. 463–80 (originally published as "The Diversity of the Sciences: Global versus Local Methodological Approaches in *Metaphysics and Morality: Essays in Honor of J. J. C. Smart.* Edited by P. Pettit, R. Sylvan, and J. Norman. Oxford: Basil Blackwell, 1987).

Putnam, Hilary (1994 [1990]): "The Idea of Science." In *Words and Life*. Edited by J. Conant. Cambridge: Harvard University Press, pp. 481–91 (originally published in Midwest Studies in Philosophy, vol. 15: *The Philosophy of the Human Sciences*. Edited by P. French, T. Uehling, and H. Wertstein. Notre Dame: Notre Dame Press, 1990).

Quine, Willard Van Orman (1953): "Two Dogmas of Empiricism." In *From A Logical Point of View*. Cambridge Mass.: Harvard University Press, pp. 20–46.

Quine, Willard Van Orman (1966): *The Ways of Paradox, and Other Essays*. New York: Random.

Rawls, John (1971): *A Theory of Justice*. Cambridge: Harvard University Press.

Reiche, Danyel (2011a): "Brisantes politisches Spiel: Im Libanon baut Sport nicht Spannungen ab, sondern vertieft sie noch—wie im Fußball-Pokalfinale." *Frankfurter Allgemeine Zeitung Nr.* 115, May 18, 2011:23.

Reiche, Danyel (2011b): "War Minus the Shooting? The Politics of Sport in Lebanon as a Unique Case in Comparative Politics." *Third World Quarterly* 32 (2):261–77.

Reineke, Walter Friedrich (1982a): "Ägypten: Astronomie." In *Geschichte des wissenschaftlichen Wissens im Altertum*. Edited by F. Jürß. Berlin: Akademie-Verlag, pp. 111–16.

Reineke, Walter Friedrich (1982b): "Ägypten: Mathematik." In *Geschichte des wissenschaftlichen Wissens im Altertum*. Edited by F. Jürß. Berlin: Akademie-Verlag, pp. 105–11.

Rescher, Nicholas (1977): *Methodological Pragmatism*. Oxford: Basil Blackwell.

Rescher, Nicholas (1979): *Cognitive Systematization: A Systems-Theoretic Approach to a Coherentist Theory of Knowledge*. Oxford: Basil Blackwell.

Rescher, Nicholas (1983): "Kant on Cognitive Systematization." In *Kant's Theory of Knowledge and Reality: A Group of Essays*. Edited by N. Rescher. Washington: University Press of America, pp. 83–113.

Rescher, Nicholas (1998): *Predicting the Future: An Introduction to the Theory of Forecasting*. Albany: State University of New York Press.

Rescher, Nicholas (2001): *Cognitive Pragmatism: The Theory of Knowledge in Pragmatic Perspective*. Pittsburgh: University of Pittsburgh Press.

Rescher, Nicholas (2005): *Cognitive Harmony: The Role of Systemic Harmony in the Constitution of Knowledge*. Pittsburgh: Pittsburgh University Press.

Rhodes, Richard (1986): *The Making of the Atomic Bomb*. New York: Simon & Schuster.

Richardson, Alan W., and Thomas E. Uebel, eds. (2007): *The Cambridge Companion to Logical Empiricism*. New York: Cambridge University Press.

Rieppel, Oliver (2003): "Semaphoronts, Cladograms and the Roots of Total Evidence." *Biological Journal of the Linnean Society* 80:167–86.

Rieppel, Oliver (2004): "The Language of Systematics, and the Philosophy of 'Total Evidence'." *Systematics and Biodiversity* 2 (1):9–19.

Ritschl, Otto (1906): *System und systematische Methode in der Geschichte des wissenschaftlichen Sprachgebrauchs und der philosophischen Methodologie*. Bonn: C. Georgi.

Robinson, Abraham (1996): *Non-standard Analysis*. Rev. ed., *Princeton Landmarks in Mathematics and Physics*. Princeton, N.J.: Princeton University Press.

Roll-Hansen, Nils (2005): *The Lysenko Effect: The Politics of Science*. Amherst, N.Y.: Humanity Books.

Rowe, Gene, and George Wright (1999): "The Delphi Technique as a Forecasting Tool: Issues and Analysis." *International Journal of Forecasting* 15:353–75.

Rowlett, Peter (2011): "The Unplanned Impact of Mathematics." *Nature* 475 (July 14, 2011):166–69.

Ruse, Michael (1988): *Philosophy of Biology Today*. Albany: State University of New York Press.

Salanskis, Jean-Michel (1995): "Die Wissenschaft denkt nicht." *Tekhnema: Journal of Philosophy and Technology* 2.

Salmon, M. H., J. Earman, C. Glymour, J. G. Lennox, P. Machamer, J. E. McGuire, J. D. Norton, W. C. Salmon, and K. F. Schaffner (1992): *Introduction to the Philosophy of Science*. Englewood Cliffs, N.J.: Prentice-Hall.

Salmon, Wesley C. (1989): "Four Decades of Scientific Explanation." In *Scientific Explanation. Minnesota Studies in the Philosophy of Science*, vol. XIII. Edited by P. Kitcher and W. C. Salmon. Minneapolis: University of Minnesota Press, pp. 3–219.

Samuelson, Paul A. (1987): "Paradise Lost & Refound: The Harvard ABC Barometers." *Journal of Portfolio Management* 4 (Spring):4–9 (reprinted in Mills, Terence C., ed. (1999): *Economic Forecasting*, vol. I. Cheltenham: Elgar, pp. 11–16).

Sandkühler, Hans Jörg, ed. (1999): *Enzyklopädie Philosophie*. 2 vols. Meiner: Hamburg.

Sarton, George (1936): *The Study of the History of Science*. New York: Dover.

Savage, C. Wade, ed. (1990): *Scientific Theories (Minnesota Studies in the Philosophy of Science 14)*. Minneapolis: University of Minnesota Press.

Scerri, Eric R. (1994): "Has Chemistry Been at Least Approximately Reduced to Quantum Mechanics?." In *PSA 1994*. Edited by D. Hull, M. Forbes and R. M. Burian. East Lansing Mich.: Philosophy of Science Association, pp. 160–70.

Scerri, Eric R. (2007): *The Periodic Table: Its Story and Its Significance*. Oxford: Oxford University Press.

Schaffner, K. F. (1993): *Discovery and Explanation in Biology and Medicine*. Chicago: University of Chicago Press.

Schickore, Jutta, and Friedrich Steinle, eds. (2006): *Revisiting Discovery and Justification: Historical and Philosophical Perspectives on the Context Distinction*. Dordrecht: Springer.

Schiemann, Gregor (1997): *Wahrheitsgewissheitsverlust. Hermann von Helmholtz' Mechanismus im Anbruch der Moderne. Eine Studie zum Übergang von klassischer zu moderner Naturphilosophie*. Darmstadt: Wissenschaftliche Buchgesellschaft.

Schiemann, Gregor (2009): *Hermann von Helmholtz's Mechanism: The Loss of Certainty: A Study on the Transition from Classical to Modern Philosophy of Nature*, vol. 17. Archimedes. [Dordrecht]: Springer.

Schnädelbach, Herbert (2012): *Was Philosophen wissen und was man von ihnen lernen kann*. München: Beck.

Schummer, Joachim (1997): "Scientometric Studies on Chemistry I: The Exponential Growth of Chemical Substances, 1800–1995." *Scientometrics* 39 (1):107–23.

Schuster, John A. (1990): "The Scientific Revolution." In *Companion to the History of Modern Science*. Edited by R. C. Olby, G. N. Cantor, J. R. R. Christie and M. J. S. Hodge. London: Routledge, pp. 217–42.

Seiffert, Helmut, and Gerard Radnitzky, eds. (1989): *Handlexikon zur Wissenschaftstheorie*. München: Ehrenwirth.

Seth, Suman (2010): *Crafting the Quantum: Arnold Sommerfeld and the Practice of Theory, 1890–1926*. Cambridge, Mass.: MIT Press.

Sholin, Bill (1996): *Truman's Decision*. Bonney Lake: Mountain View.

Sismondo, Sergio (2004): *An Introduction to Science and Technology Studies*. Malden: Blackwell.

Smolin, Lee (2006): *The Trouble with Physics: The Rise of String Theory, the Fall of a Science, and What Comes Next*. Boston: Houghton Mifflin.

Sober, Elliott (1990): "Contrastive Empiricism." In *Scientific Theories. Minnesota Studies in the Philosophy of Science*, vol. XIV. Edited by C. W. Savage. Minneapolis: University of Minnesota Press, pp. 392–412.

Sokal, Alan (1996a): "A Physicist Experiments With Cultural Studies." *Lingua Franca* May/June:62–64.

Sokal, Alan (1996b): "Transgressing the Boundaries: An Afterword." *Dissent* 43 (4):93–99.

Sokal, Alan (1996c): "Transgressing the Boundaries: Towards a Transformative Hermeneutics of Quantum Gravity." *Social Text* 46/47:217–52.

Sokal, Alan D. (2008): *Beyond the Hoax: Science, Philosophy and Culture*. Oxford; New York: Oxford University Press.

Sokal, Alan D., and Jean Bricmont (1998): *Fashionable Nonsense: Postmodern Intellectuals' Abuse of Science*. New York: Picador.

Solla-Price, Derek J. de (1963): *Little Science, Big Science*. New York: Columbia University Press.

Solomon, Ronald (2001): "A Brief History of the Classification of Finite Simple Groups." *American Mathematical Society. Bulletin. New Series* 38 (3):315–52.

Sonesson, Göran (2004): "Current Issues in Pictorial Semiotics. Lecture 1: The Quadrature of the Hermeneutic Circle: Historical and Systematic Introduction to Pictorial Semiotics." Cyber-semiotic Institite: http://www.chass.utoronto.ca/epc/srb/cyber/Sonesson1.pdf (accessed Sept. 27, 2011).

Speck, Josef, ed. (1980): *Handbuch wissenschaftstheoretischer Begriffe. 3 Bände*. Göttingen: UTB Vandenhoeck.

Springel, Volker et al. (2005): "Simulations of the Formation, Evolution and Clustering of Galaxies and Quasars." *Nature* 435 (June 2, 2005):629–36.

Steele, John M. (2000): *Observations and Predictions of Eclipse Times by Early Astronomers*. Edited by J. Z. Buchwald, *Archimedes: New Studies in the History and Philosophy of Science and Technology*, vol. 4. Dordrecht: Kluwer.

Stein, Aloys von der (1968): "Der Systembegriff in seiner geschichtlichen Entwicklung." In *System und Klassifikation in Wissenschaft und Dokumentation, Studien zur Wissenschaftstheorie, Bd. 2*. Edited by A. Diemer. Meisenheim am Glan: A. Hain.

Steinle, Friedrich (2002a): "Challenging Established Concepts: Ampère and Exploratory Experimentation." *Theoria* 17 (2):291–316.

Steinle, Friedrich (2002b): "Experiments in History and Philosophy of Science." *Perspectives on Science* 10 (4):408–32.

Steinle, Friedrich (2006): "Concept Formation and the Limits of Justification: 'Discovering' the Two Electricities." In *Revisiting Discovery and Justification*. Edited by J. Schickore and F. Steinle. Dordrecht: Kluwer, pp. 183–95.

Stent, Gunther S. (1994): "Promiscuous Realism." *Biology and Philosophy* 9:497–506.

Stent, Gunther S., ed. (1980): *James D. Watson: The Double Helix. A Personal Account of the Discovery of the Structure of DNA. Text, Commentary, Reviews, Original Papers*. New York: Norton.

Storer, Norman W., ed. (1973): *Robert K. Merton: The Sociology of Science: Theoretical and Empirical Investigations*. Chicago: University of Chicago Press.

Sturm, Thomas (2009): *Kant und die Wissenschaften vom Menschen*. Paderborn: Mentis.

Sturm, Thomas (in press): "Wissenschaft." In *Kant-Lexikon*. Edited by G. Mohr, J. Stolzenberg and M. Willaschek. Berlin: de Gruyter.

Sturm, Thomas, and Silvia De Bianci (in press): "Naturwissenschaft." In *Kant-Lexikon*. Edited by G. Mohr, J. Stolzenberg and M. Willaschek. Berlin: de Gruyter.

Suits, Daniel B. (1962): "Forecasting and Analysis with an Econometric Model." *American Economic Review* LII (1):104–32 (reprinted in Mills, Terence C., ed. (1999): *Economic Forecasting*, vol. I. Cheltenham: Elgar, pp. 72–100).

Suppe, Frederick (1977): *The Structure of Scientific Theories*. 2nd ed. Urbana: University of Illinois Press.

Szostak, Rick (2004): *Classifying Science: Phenomena, Data, Theory, Method, Practice*. Edited by J. M. Owen, vol. 7. *Information Science and Knowledge Management*. Dordrecht: Springer.

Takaki, Ronald T. (1995): *Hiroshima: Why America Dropped the Atomic Bomb*. Boston: Little, Brown, and Co.

Taylor, Charles (1985 [1971]): "Interpretation and the Sciences of Man." In *Philosophy and the Human Sciences: Philosophical Papers*, vol. 2. Edited by C. Taylor. Cambridge: Cambridge University Press, pp. 15–57 (originally in *Review of Metaphysics* 25(1): 3–51 (1971)).

Thagard, Paul (1988): *Computational Philosophy of Science*. Cambridge, Mass.: MIT Press.

Thagard, Paul R. (1978): "Why Astrology is a Pseudoscience." In *PSA 1978: Proceedings of the 1978 Biennial Meeting of the Philosophy of Science Association, volume one: Contributed Papers*. Edited by P. D. Asquith and I. Hacking. East Lansing: Philosophy of Science Association, pp. 223–34.

Theocharis, T., and M. Psimopoulos (1987): "Where Science Has Gone Wrong." *Nature* 329:595–98.

Tompkins, Jane P., ed. (1980): *Reader-Response Criticism: From Formalism to Post-Structuralism*. Baltimore: Johns Hopkins University Press.

Tosh, John with Séan Lang (2006): *The Pursuit of History: Aims, Methods and New Directions in the Study of Modern History*. 4th ed. Harlow: Pearson.

Toulmin, Stephen (1953): *The Philosophy of Science: An Introduction*. London, New York: Hutchinson's University Library.

Toumey, Chris (2010): "35 Atoms that Changed the Nanoworld." *Nature Nanotechnology* 5 (Apr. 2010):239–41.

Tuchman, Barbara Wertheim (1962): *The Guns of August*. New York: Macmillan.

Tversky, Amos, and Daniel Kahnemann (1974): "Judgment under Uncertainty: Heuristics and Biases." *Science* 185:1124–31.

Vaillant, George E. (2002): *Aging Well: Surprising Guideposts to a Happier Life from the Landmark Harvard Study of Adult Development*. Boston: Little, Brown.

van Fraassen, Bas C. (1980): *The Scientific Image*. Oxford: Clarendon.

van Fraassen, Bas C. (1989): *Laws and Symmetry*. Oxford: Clarendon Press.

van Fraassen, Bas C. (2002): *The Empirical Stance*. New Haven: Yale University Press.

Vickery, Brian C. (2000): *Scientific Communication in History*. Lanham: Scarecrow.

Vosniadou, Stella, and William F. Brewer (1992): "Mental Models of the Earth: A Study of Conceptual Change in Childhood." *Cognitive Psychology* 24:535–85.

Waerden, B. L. van der (1963): *Science Awakening: Egyptian, Babylonian and Greek Mathematics*. New York: John Wiley.

Waerden, B. L. van der (1966): *Erwachende Wissenschaft: Ägyptische, babylonische und griechische Mathematik*. Basel: Birkhäuser.

Waerden, B. L. van der (1974): *Science Awakening II: The Birth of Astronomy*. Leyden: Noordhoff.

Wainstock, Dennis (1996): *The Decision to Drop the Atomic Bomb*. Westport: Praeger.

Walker, J. Samuel (1997): *Prompt and Utter Destruction: Truman and the Use of Atomic Bombs Against Japan*. Chapel Hill: University of North Carolina Press.

Walker, Paul D. (2003): *Truman's Dilemma: Invasion or The Bomb*. Gretna: Pelican.

Wasserstein, Bernard (2007): *Barbarism and Civilization: A History of Europe in Our Time*. Oxford: Oxford University Press.

Watson, James D., and Francis H. C. Crick (1953): "Molecular Structure of Nucleic Acids: A Structure for Deoxyribose Nucleic Acid." *Nature* 171:737–38.

Weart, Spencer (2003): "Arakawa's Computation Device." http://www.aip.org/history/climate/arakawa.htm (accessed Sept. 27, 2011).

Weart, Spencer (2006): "General Circulation Models of the Atmosphere." http://www.aip.org/history/climate/GCM.htm#L00 (accessed Sept. 27, 2011).

Weber, Marcel (2005): *Philosophy of Experimental Biology*. Cambridge: Cambridge University Press.

Wehner, Rüdiger, Barbara Michel, and Per Antonsen (1996): "Visual Navigation in Insects: Coupling of Egocentric and Geocentric Information." *The Journal of Experimental Biology* 199:129–40.

Weinberg, S. (1987): "Newtonianism, Reductionism and the Art of Congressional Testimony." *Nature* 330:433–37.

Weinberg, S. (1988): "Weinberg Replies." *Nature* 331:475–76.

Weinberg, S. (1992): *Dreams of a Final Theory*. New York: Pantheon.

Weinberg, Steven (2001): *Facing Up: Science and Its Cultural Adversaries*. Cambridge, Mass.: Harvard University Press.

Weinberg, Steven (2001 [1995]): "Reductionism Redux." In *Facing Up: Science and Its Cultural Adversaries*. Edited by S. Weinberg. Cambridge: Harvard University Press, pp. 107–22 (originally in *The New York Review of Books*, Oct. 5, 1995).

Weiner, Jonathan (1995): *The Beak of the Finch: A Story of Evolution in Our Time*. New York: Vintage Books.

Whitley, Richard (2000): *The Intellectual and Social Organization of the Sciences*. 2nd ed. Oxford [England]; New York: Oxford University Press.

Will, Clifford M. (1993): *Theory and Experiment in Gravitational Physics*. Rev. ed. Cambridge: Cambridge University Press.

Wilson, Robert A. (2009): *The Finite Simple Groups*, vol. 251. *Graduate Texts in Mathematics*. London: Springer.

Wimsatt, W. C. (1974): "Complexity and Organization." In *PSA 1972*. Edited by K. F. Schaffner and R. S. Cohen. Dordrecht: Reidel, pp. 67–86.

Wimsatt, W. C. (1976): "Reductionism, Levels of Organization, and the Mind-Body Problem." In *Consciousness and the Brain: A Scientific and Philosophical Inquiry*. Edited by G. G. Globus, G. Maxwell, and I. Savodnik. New York: Plenum Press, pp. 199–267.

Wimsatt, William C. (1979): "Reduction and Reductionism." In *Current Research in Philosophy of Science*. Edited by P. D. Asquith and H. E. Kyburg, Jr. East Lansing: Philosophy of Science Association, pp. 352–77.

Wittgenstein, Ludwig (1958 [1953]): *Philosophical Investigations*. Translated by G. E. M. Anscombe. Oxford: Blackwell.

Woit, Peter (2006): *Not Even Wrong: The Failure of String Theory and the Search for Unity in Physical Law*. New York: Basic Books.

Wright, G. H. von (1971): *Explanation and Understanding*. Ithaca, N.Y.: Cornell University Press.

Zammito, John (2004): "Koselleck's Philosophy of Historical Time(s) and the Practice of History." *History and Theory* 43 (1):124–35.

Zilsel, Edgar (2003 [1942]): *The Social Origins of Modern Science*. Dordrecht: Kluwer.

Zollikofer, Christoph P. E., and Marcia S. Ponce de León (2005): *Virtual Reconstruction: A Primer in Computer-Assisted Paleontology and Biomedicine*. Hoboken, N.J.: Wiley-Liss.

INDEX

Academy Edition, *see* Leibniz
action, human, 62–63, 229
Adams, John Couch, 84
Adler, Alfred, 200
Alexander, Amir, 18
anomalies, significant, 215
anthropophagy, xii, 204
anti-science, 236
ants, 103
anything goes, 166–167, 252, 253
archeology, 140
Aristotle, 1, 2, 3, 41, 129, 134, 142, 143, 150–152, 192, 214, 227, 248
artifacts, 64, 71, 107–108, 231, 232
astrology, 202, 203, 257
astronomy, 46, 47, 80–81, 84, 96, 131–132, 134–135, 188, 190–191, 195, 206, 226, 234, 246, 255
ATLAS collaboration, 112, 137, 241
atomic bomb, 62–63, 229
autocatalytic process, 141
automobile development, 114, 121–122, 241–242
axiomatization, *see* system, axiomatic

Bacon, Francis, 3, 215, 219
bacteria, 194, 255
basic sentences, 201
Beiner, Marcus, xii
Bertero, Vitelmo, 12
Bertrand of Hildesheim, 101

bioinformatics, 139, 247
biology, 64, 66, 103, 140, 188, 193, 224–225, 229–230, 238, 239
  molecular, 99, 129–130, 140, 249
  Soviet, 183
biophysics, 140
Bird, Alexander, xiii, 213, 215–216
black holes, 24, 220
Bock, Walter J., 224
Bohr, Niels, 184–185, 255
Börner, Katy, 242–243
Bortolotti, Lisa, 219
botany, 134
Boyd, Richard, xii
Boyle's gas law, 48, 57–58, 67, 94–95, 136
Bricmont, Jean, 18, 233, 236–237
bridge principles, 201
brute force approaches, 138–139
Bschir, Karim, xii
business cycle, sunspot theory of, 82, 234
business informatics, 140, 247

C14 method, 140
Cambridge-Somerville study, 100–101, 238
Campbell Collaboration, 102, 239
cancer treatment studies, 100
Carnap, Rudolf, 16
Carol, Hans, 213
CAS registry number, 144, 249
catastrophism, 206, 257

causal influence, 98–102, 106
CERN, 112
certainty, 2, 3, 4, 93, 151, 214
Chalmers, Alan F., 215
chemistry, 39, 46, 48, 64, 129, 140, 143–145, 156–157, 226, 229, 245, 248–249, 251
chess theory, 114–115, 117, 122–123, 242
chocolate science, *see* science of chocolate
circle, hermeneutical, *see* hermeneutics
citation index, 31–32, 222–223
classification, 42–43, 126–129, 224–225
    and completeness, 126–127, 129
    in mathematics, 127–129
    Linnaean, 42, 225
Cleland, Carol E., 219
clinical studies, *see* research, clinical
Coase, Ronald H., 234
Cochrane Collaboration, 102, 239
cognitive science, *see* science, cognitive
Cohen, Morris R., 15
coherentism, 170–171
common sense, *see* science and common sense
completeness, ideal of, 124–132, 244
compounds, chemical, 48
computer, 46, 51, 52, 53, 87, 93, 97, 103, 139–140, 146, 227, 236, 238, 247, 247
consensus, scientific, 32
constructivism, 189, 255
context of discovery, *see* context of justification
context of justification, 90–91, 237
continental drift, 70, 230
Copernican theory, 192, 195, 196
correlations, 82–83, 206
cosmology, 51–52, 68, 129, 135, 227, 245, 246
creation science, 203, 216
Crick, Francis, 145
criminology, 100, 102, 238
critical discourse, 108–113, 241
cryptography, 117
cultural change, 79, 234
cultural products vs. natural objects, 71–72, 230, 231
cyclotron, 136
cystic fibrosis, 238

d'Alembert, Jean le Rond, 132, 217, 254
Danto, Arthur C., 223
Darwin, Charles, 134, 145
Darwinian theory, *see* theory, evolutionary

data
    collection, 134–139
    interpretation of, *see* interpretation
    raw, 96
decision making, 186–187
Declaration of Independence, 39, 224
definition, 210–211, 258
Delphi methods, 87–88, 236
demarcation problem, 10, 162, 176, 199–207, 215–216, 256, 257
Descartes, René, 3, 152–155, 166
descriptions, 28, 37–53, 141–142
    and abstractness, 38–39, 40
    historical, 39–40, 49–53
    historical vs. generalized, 40
descriptive vs. prescriptive, 21–22, 33–34, 196, 199, 210, 254, 256
development, human, 137–138, 247
Dewey, John, 14, 17, 217
diagram in *Origin of Species*, 145, 249
Diderot, Denis, 132
disciplines, scientific, *see* scientific disciplines
discovery, chance, 138
diseases, classification of, 225
disunity of science, *see* science, disunity of
DNA, structure of, 145–146, 249
Doppler effect, 96
Droysen, Johann Gustav, 227
Dufay, Charles, 49
Duhem, Pierre, 237, 242
Duhem-Quine thesis, 96, 237
Dupré, John, 221

Earth sciences, *see* geology
earthquake engineering, 12, 122, 216
eclipses, 132, 245
    prediction of, 80–81, 234
economics, 64, 81, 82, 83, 85, 114, 121, 140, 234, 235
Eddington, Arthur Stanley, 84
Eigler, Don, 189
Einstein, Albert, 35, 84, 105–106, 192–193, 223–224
Eisner, Werner, xii
electrical effects, 48–49, 226
electromagnetism, 226
elements, chemical, 129
Empedocles, 129
empirical concepts, definition of, 210–211

encyclopedias, 132, 246
Endres, Kirsten, xii
entities, theoretical, 60, 66, 95, 188–189, 201, 228
epistemic connectedness, 7, 113–124
    and systematicity, 119–120, 242
epistemic culture, 112
epistemic values, 94
essentialistic spirit, 7
Euclidean geometry, 3, 151, 170, 182, 183–184, 255
events, reproducible vs. singular, 39–40
evidence
    empirical, 92, 237
    levels of, 197
evidence based medicine, 197, 256
evolutionary theory, *see* theory, evolutionary
existential statements, 202
experiment
    controlled, 98–102
    role of, 135–136
explanation, 53–78, 227, 228
    deductive-nomological, 56
    historical, 68–71, 230,
    inductive-statistical, 56
    narrative, 69, 230
    neurophysiologic, 67
    of human action, 62–63, 67, 69, 230
    proximate, 232
    reductive, 63–68, 229
    ultimate, 232
    vs. description, 53, 58
    vs. understanding, 54–55, 72, 227
explanations
    and mechanisms, 56, 228
    and models, 56
    using empirical generalizations, 56–58, 68
    using theories, 59–61, 68, 77

fallibility, 6
falsifiability, 104, 201–202, 256–257
family resemblance, 7, 26, 28–29, 30, 119, 178, 209, 221–222
Feyerabend, Paul K., x, 5, 165–168, 213, 215, 222, 252, 253, 257
Feynman, Richard, 256
finches on Galápagos Islands, 130, 245
Fine, Arthur, 188–189, 223
finite simple groups, classification of, 128–129, 244–245
Fleming, Alexander, 138

footnotes, endnotes, 123–124
forecast of weather, 82, 86–87, 235
forecasting with leading indicators, 83
forensics, 23
formulae, chemical, 144–145
Forrester, Jay, 87
Frege, Gottlob, 143
Freud, Sigmund, 200, 202
Freund, Urs, xiii
fusion reactor, 12, 216

Gaissmaier, Wolfgang, 258
Galilei, Galileo, 3, 60, 134, 136, 167, 192
Galle, Johann Gottfried, 84
game theory, 123, 243
genealogy, 115
generalizations
    empirical and classification, 48
    empirical, 47–49
    empirical and explanations, 56–58
genome project, human, 129–130, 245
Geographic Names Information System (GNIS), 43
geography, x, 43, 146, 213, 225, 240
geologic time scale, 44, 226
geology, 44, 70, 104, 130–131, 134, 206, 226, 239, *see also* plate tectonics
Gerigk, Horst-Jürgen, 232
Gigerenzer, Gerd, 258
Glashow, Sheldon, 256
Glock, Hans-Johann, 222
God, 206, 257
Goodman, Nelson, 232
graph, 143
gravitation, 24, 235, 239
    Newton's theory of, 60–61, 84, 229
growth, exponential, 243–244

Habsburg family, 118, 124
Harding, Sandra, 219
harmonics, 46
Hartmann, Stephan, xii
Harvard A-B-C barometer, 83, 234
Hegel, Georg Wilhelm Friedrich, 150, 171, 254
Heidegger, Martin, 190, 232
Heit, Helmut, xiii, 217
Hempel, Carl Gustav, 16–17, 54, 56, 160, 227
Henning, Willi, 232
hermeneutics, 73–74, 231

heuristics, 81, 89, 187, 234, 255, 258
  rational, 154–155
Hilbert, David, 244
historical
  doctrine of nature, 156–157, 159
  judgment, selectivity of, 50
  phases, 1–6
  processes, computational reconstruction, 51–53
  relevance, 50–51, 131, 227
  relevance, factual, 50, 131
  relevance, narrative, 50, 131
  relevance, pragmatic, 50–51
historicism, 4, 215
historiography, *see* sciences, historical
history, 19, 28–29, 43–44, 219, 226–227,
  ancient Egypt, 44, 226
  contemporary, 115
  Earth's, 44
  political, 44
  world, 44
hoax, Sokal's, 233, 238
Hodes, Harold, xiii, 243
HOPOS, 218
Hoyningen, Alexander, 232–233
humanities, 4, 54–55, 62–63, 71–75, 79, 91–92, 117, 132, 138, 140, 194, 207, 227, 230, 231, 247–248
Hume, David, 98, 196
Huygens, Christiaan, 229

IBM, 189
ice cores, 108, 137, 246–247
ICSU, 220
ideal gas, *see* Boyle's gas law
ideal, epistemic, 2, 3
incommensurability, 98, 238
indicators, leading, 82–83, 234
induction, 156, 161–162, 202, 251
infinitesimals, 149, 249–250
information technology, 139–140
intentions, 62–63, 244
international relations, 48
Internet, xiii, 132
interpretation, 96–97, 108, 237, 238, *see also* understanding

Jefferson, Thomas, 39
Jevons, William Stanley, 82

Jonas, Hans, 238
journalism, political, 115
justification, context of, 90–91

Kant, Immanuel, 25, 155–159, 171, 250, 251
Keil, Geert, 221–222
Kellermann, Gero, xiii
Kepler's laws, 105
Killias, Martin, xiii, 238
Kim, Jaegwon, 215
knock-out organism, 99
Knorr-Cetina, Karin, 112–113
knowledge, 21, 219
  claims, defense of, 79, 88–108, 179, 190, 197–198, 206, 210
  everyday, 8
  exploitation from other domains, 139–140
  generality of, 118
  generation, 132–141
  local or traditional, 22, 203, 219–220, 257
  representation of, *see* representation of knowledge
  scientific, 9, 21
  system of, 170–171, 174
  theory of, 171–172, 173
Koertge, Noretta, xiii, 240
Kuhn, Thomas S., 17, 109, 141, 163–165, 184–186, 210–211, 215, 219, 244, 246, 252
Kusch, Martin, xiii, 242

Ladyman, James, 216, 219
Lakatos, Imre, 109
Lambert, Johann Heinrich, 150, 171, 254
languages, classification of, 43, 225
Laplace, Pierre-Simon, 206
Large Hadron Collider, 112, 137, 246
Las Meninas, 232
Laudan, Larry, 203
law (discipline), 31
law
  Boyle's, 48
  of constant proportions, 48
  of free fall, 60–61
  of gravitation, 60–61
  of inertia, 192, 195–196
  phenomenological, 47
Lawrence, Ernest, 136
Le Verrier, Urbain, 84, 105

Leibniz, Gottfried Wilhelm, 138, 150, 247, 229, 247
level (in reductive explanation), 66, 67, 230
Lewes, George Henry, 217
LHC, *see* Large Hadron Collider
Liebermann, Max, 227
life span theories, *see* psychology, developmental
light bending due to gravitation, 84, 235
Lincoln, Abraham, 233
linguistics, 43, 64
Linnaeus, 42, 43, 225
literary theory, 77–78, 193, 195, 223
literature, study of, 75–78, 233
logic, 3, 41, 142–143, 214, 224
logical empiricism, 159–161, 248, 252
logical form, 142, 143, 248
Lohse, Simon, xiii
Longino, Helen, 110–111, 241
longitudinal study, 137–138
Lysenko, Trofim, 183, 255

main thesis, 14, 20, 21–25, 27
    comparative character, 22, 133, 219
    descriptive character, 21–22, 33, *see also* descriptive vs. prescriptive
    normative implications, 33–34
maps, geographical, 146, 249
Mars, 97
mathematics, 4, 18–19, 41, 78, 93, 117, 119, 123, 127–129, 139, 142–143, 149, 152, 153, 180–182, 183–184, 188, 206–207, 215, 218, 242, 244, 247, 248, 249–250, 254
    Egyptian and Mesopotamian, 180–182, 254
    role of in science, 45–47, 103–107, 139, 214, 247
Mayr, Ernst, 224, 228, 232
Meadows, Dennis, 87
meaning (of cultural products), 71–72, 77, 230–231, 232
mechanics, classical, 4
mechanisms, 56, 228
medicine, *see* research, medical
Mendeleev, Dmitri, 129
Mercury, perihelion advance 105–106, 239
Merton, Robert K., 110, 112, 233, 240, 241
metaphysics, 162, 200, 201–202, 209, 221
meteorology, 11, 86–87
method
    of difference, 99

    scientific, *see* scientific method(s)
microexplanations, 64, 65, 66–67
Mill, John Stuart, 98
model, 56, 66, 85–87, 103, 145–146, 235, 236
    climate, 52, 87, 227, 235, 236
    economic, 235
    meteorological, 86–87, 122, 235
    world, 87
Morris, Charles, 16
Muller, Fred A., xii

Nagel, Ernest, 15, 17, 18, 160, 161, 216, 218, 223
nanoworld, 189, 255
narratives, 49–50, 131
natural ontological attitude, 188–189, 255
naturalistic fallacy, 196
Neptune, 84, 235
Nersessian, Nancy J., 223
Neurath, Otto, 16
neuroeconomics, 140
neuroscience, 103
Newton, Isaac, 3, 60, 84, 105, 192, 206, 229, 247
nomenclature, 42–43, 141–142, 144–145, 225
    chemical compounds, 145, 249
    enzymes, 225
    genes, 224–225
    planetary, 43, 225
nonrealism, 227–228
normative consequences of systematicity theory, 196–199
norms in scientific communities, 110–111
number theory, 59
nursing diagnoses, classification of, 225

Oberheim, Eric, xiii, 253
objectivist stance, 194–195, 256
objectivity
    common sense conception of, 189–190
    in history, 189, 255
Okasha, Samir, 222, 244
Oppenheim, Paul, 54, 56
organized skepticism, 110, 240–241
otorhinolaryngology, xii, 214

paleoceanography, 36, 224
paleoclimatology, 79, 137, 224, 246
paleontology, 52–53, 69–70, 206, 226, 233
    computer assisted, 52–53, 227
particle accelerators, 136–137, 246

peer review system, 111
penicillin, 138, 247
periodic system, 129, 143–144, 245, 248
periodization, 43–45
   and completeness, 127
   problems of, 44–45
Permian-Triassic boundary, 52, 227
personality assessment, 23, 220
pharmacology, 139, 247
philosophy, systematic, 208
phlogiston, 229
physical objects, common sense conception of, 188–189
physics, 30, 39, 40, 46, 64, 66, 112–113, 129, 136–137, 139, 184–185, 191–193, 226, 229, 241, 255,
pig-cycle, 81, 234
Piper, Adrian, xiii
Plaisance, Katie, xiii
planetary nomenclature, *see* nomenclature, planetary
plate tectonics, theory of, 70, 130–131, 245
Plato, 1, 2, 3
Platonic solids, *see* polyhedra, regular
Pliny the Elder, 132
poetological difference, 76–77, 232
polyhedra, regular, 128
Popper, Karl, 10, 54, 56, 104, 109, 161–163, 200–202, 204–205, 215, 216, 256–257
Pothast, Ulrich, xiii
Potter, Harry, 232–233
pragmatism, methodological, 171, 254
prediction, 36, 78–88, 233, 234
   based on correlations, 82–83
   based on Delphi methods, 87–88
   based on empirical regularities, 80–83
   based on models, 83, 85–87
   based on theories or laws, 83–84
   concerning human affairs, 79, 233–234
prescientific knowledge, transition to scientific knowledge, 22–23
product development, 114, 118, 120–121, 122
progress, scientific, *see* scientific progress
proof, 2, 3, 22, 93, 151, 181–182, 183, 237, 254
prophecy, self-destroying, 79, 233
proteins,
   classification of, 225
   folding of, 24
proteome, 130, 245

protocol sentences, 159–160
pseudoscience, 7, 8, 10, 162, 176, 199–207, 209, 210, 215–216, 257
   competitor for, *see* reference science
psychiatry, 137–138
psychoanalysis, 200, 202, 223, 257
psychology, 23, 137–138, 223, 247
   cognitive, 19, 186–187
   developmental, 43, 44, 218, 226, *see also* human development
   individual, 200
Ptolemy, 146, 249
Putnam, Hilary, 221

qualities, occult, 228–229
quantification, 45–47, 226
Quine, Willard Van Orman, 218, 237

radiocarbon dating, 140, 247
randomized trials, *see* treatment-control studies
Ransdell, Joseph M., xiii
rationalism, 153, 171, 253
raw data, 96
Rawls, John, 232
reactions, chemical, 144
reader-response criticism, 193, 195, 256
realism, 188, 194–195, 227, 242
reciprocal illumination, 231
reductionism, 64, 65, 161, 221
   sociological, 64
reference science, 203, 204, 205–206
refinement, 26, 29–30, 35, 223
reflection, 72–75, 77–78, 231
reflective equilibrium, 73, 232
regularities, empirical, 47
Reiche, Danyel, 123–124
Reineke, Walter Friedrich, 254
Reiss, Julian, xii
rejection rate, 111, 241
relativity theory, 24, 84, 105–106, 192, 220, 240, 255
representation of knowledge, 141–147
Rescher, Nicholas, xiii, 18, 170–175, 217, 218, 220, 250–251, 253, 254
research
   and development, 11–12
   clinical, 11, 197–198, 256
   medical, 99, 102, 137–138
   pharmacological, 11

Reydon, Thomas, xiii, 231, 232
Richter, Peter, xiii
Rieppel, Oliver, 232

saddle time, 44, 226
Sankey, Howard, xiii
Saros period, 80–81, 234
Sarton, Georg, 15–16
scanning tunneling microscope, 189
Scholz, Markus, xiii
schools, scientific, 32, 76, 184, 223, 248
Schummer, Joachim, xiii
Schweizer, Erhard, 189
science
   aims of, 116–118, 242–243
   and common sense, 187–196
   applied, 12
   as a collage, 168–169
   as autocatalytic process, 141, 248
   as historical, 2, 32–33, 93
   big, 112–113, 241
   breaking with common sense, 191–193
   cartography of, 30–31, 222
   classification of, 30–31, 222–223
   cognitive, 26, 220
   consensus in, 32
   contrast to everyday knowledge, 8, 9–10, 209
   definition of, 167–168
   demonstrative, 151
   disunity of, 29, 214, 221
   dynamics of, 162, 163–164, 177, 183–187, 198
   empirical, 91–92, 134, 157
   engineering, 12, 64, 114, 117, 118, 121–122
   essence of, 10–11, 209, 221
   experimental, 47–48, 79, 90, 97, 98–100, 135–139, 191, 219, 236, 238
   exponential growth, 141
   food, 12, 216
   formal, 92, 93, *see also* mathematics
   generalizing empirical, 47, 54, 228
   genesis of, 177, 180–183
   granite, xii
   growth of, 124–125
   historical auxiliary, 107, 240
   historical natural, 51, 69–70, 78–79, 90, 97, 108, 137, 140, 233, 236, 238
   historical, 107–108, 115, 117, 118, 124, 131, 240, 246
   laboratory, *see* science, experimental

   lipid, xi
   natural, 4, 5, 149, 156–157
   normal, 164–165, 184–185, 252
   of chocolate, 12, 122, 216–217
   political, 48, 115, 123–124
   rational, 157
   revolutionary, 184
   social structure of, 109
   social, 4, 54–55, 62–63, 65, 100–101, 102, 104, 230, 247–248
   studies, 9
   unity of, 29, 119–120, 124, 168, 169, 171, 209, 221, 243
   wide sense of, 8–9, 209
scientific
   disciplines, 30–32, 222–223
   fields, 30–32, 223
   method(s), 3, 4–5, 22–23, 141, 152–155, 166, 214, 215, 221, 253
   progress, 168
Scientific Revolution, 70, 139
scientometrics, 243–244
secondary qualities, 192, 195
semantics, formal, 230
semiotics, pictorial, 20, 74–75, 219, 223
Shaha, Maya, xiii
Shakespeare, William, 76
Shell Eco-marathon, 121–122, 243
simultaneity, 192–193, 194, 195
Sirtes, Daniel, xiii
Sismondo, Sergio, 219
Sloan Digital Sky Survey, 135
sociology, 4, 24, 32, 59, 64, 67, 77, 91, 104, 110, 112, 116, 194, 233, 240, 241, 248
software technology, 139–140
Sokal, Alan, 18, 233, 236–237, 238–239
Sonesson, Göran, 20, 232
sources, historical, 107
species, biological, 28, 42, 52–53, 69–70, 97, 127, 129, 130, 134, 141–142, 145, 193, 227, 249, 256
stamp collecting, 210
Stanford Encyclopedia of Philosophy, 132
state equation, 57–58, 67, 228
state variables, 48, 49
statistics, 106
Steele, John, 234
Stent, Gunther, 221
Stephanus, Robert, xiii

Stöckler, Manfred, xii
string theory, 59, 149, 198–199, 237, 256
Sturm, Thomas, xiii, 250
substances, chemical, 144, 248
sunspot theory, *see* business cycle
syllogistics, 142, 248
system, 25, 150, 171, 174–175, 217, 220, 254
    axiomatic, 41, 127, 119, 120, 152, 214, 224, 244, 250, 251
    climate, 85
    economic, 85
    for Kant, 155–156, 157–158, 250–251
    periodic, *see* periodic system
systematic reviews, 102, 197, 239
systematic spirit, 217
systematicities, aggregation of, 169, 178–180
systematicity
    and methodicity, 149, 154, 234
    and order, 28
    and scientific progress, 186–187
    change in time, 29
    concept of, 25–30, 169, 174, 209, 217, 220–221
    degrees of, 169, 174, 178
    dimensions of, 27–28, 35–37, 224
    increase in, 169, 177–180, 183–187, 194, 253
    Kant's notion, 158–159, 250–251
    taken for granted, 16, 17, 18–20, 219
systematicity theory
    and truth, 173
    arguments for, 30–34, 148–149
    as descriptive, *see* descriptive vs. prescriptive
    as empirical or semantic, 210–211
    as generalization, 120, 148–150, 158, 163

taste, 195, 256
taxonomy, *see* classification
technology, 140
    philosophy of, 232
telescope, 134–135, 246
text, fictional, 75–77
text-analog, 54, 71–72, 74, 92, 231–232
Thagard, Paul, 203–204
Thales, 182
theology, 20, 219
Theophrastus, 134
theory, 59–61, 98, 228
    and completeness, 127
    choice, 93–94
    comparison, 105–106

    dramatic, 188
    evolutionary, 130, 145, 228
    hypothetical character of, 59–60
    literary, *see* literary theory
    of common descent, 59
    of everything, 129, 198, 245
    of history, Marxist, 200
thermodynamics, 96
Thompson, John G., 245
Thomson Reuters Company, 31, 222
Tosh, John, 19
transition area between science and nonscience, 11, 120–121
treatment-control studies, 99–102, 197
    in the social sciences, 100–102
tree of life, 145, 249
Truman, Harry S., 62–63, 229
truth
    as correspondence, 171–172
    common sense conception of, 189–190
    criteria, 172–173
Tsou, Jonathan, xii
tsunami, 97, 237

understanding (*verstehen*), 54–55, 72
UNESCO, 247
unification, 67–68, 70–71
unity of science, *see* science, unity of
Uranus, 84

validity, external, 101
values, cognitive, in science, 125–126, 185–186, 205–206, 244
van Fraassen, Bas C., 213–214, 242
van Valen, Leigh, 233
Venn, John, 233
verifiability criterion of meaning, 162
verum factum principle, 102–103
ViCLAS, 23–24, 220
Vico, Giambattista, 102
victimology, 23
Viking Lander mission, 97, 238
virology, 180
viruses, classification of, 224

Wasserstein, Bernard, 227
Waters, Ken, xiii, 240
Watson, James, 145
Weber, Marcel, xiii, 238

Wegener, Alfred, 70
Weinberg, Steven, 215
Whitley, Richard, 219
Wittgenstein, Ludwig, 6, 7, 28, 29, 221
Wolff, Christian, 150, 171
World Conference on Science, xi, 219–220

World War I, 38, 69, 230
World Wide Web, *see* Internet
written documents, study of, 72–75

zoology, 134, 140

www.ingramcontent.com/pod-product-compliance
Ingram Content Group UK Ltd.
Pitfield, Milton Keynes, MK11 3LW, UK
UKHW041229200426
11947UKWH00035B/735